SAVAG

SAVAGE ECOLOGY

WAR AND GEOPOLITICS AT THE END OF THE WORLD

JAIRUS VICTOR GROVE

Duke University Press—*Durham and London*—2019

Designed by Courtney Leigh Baker
Typeset in Warnock Pro by Westchester Publishing Services

Library of Congress Cataloging-in-Publication Data
Names: Grove, Jairus Victor, [date] author.
Title: Savage ecology : war and geopolitics at the end of the world / Jairus Victor Grove.
Description: Durham : Duke University Press, 2019. | Includes bibliographical references and index.
Identifiers: LCCN 2018055115 (print) | LCCN 2019005433 (ebook) | ISBN 9781478005254 (ebook) | ISBN 9781478004219 (hardcover : alk. paper) | ISBN 9781478004844 (pbk. : alk. paper)
Subjects: LCSH: War—Environmental aspects. | Geopolitics—Environmental aspects. | Political violence—Environmental aspects. | Climatic changes—Effect of human beings on. | War and society.
Classification: LCC QH545.W26 (ebook) | LCC QH545.W26 G76 2019 (print) | DDC 363.7—DC23
LC record available at https://lccn.loc.gov/2018055115

Duke University Press gratefully acknowledges the generous support of the University of Hawai'i at Manoa, which provided funds toward the publication of this book.

Cover art: Kuwait, 1991. © Bruno Barbey/Magnum Photos.

O I see now, flashing, that this America is only you and me,

Its power, weapons, testimony, are you and me,

Its crimes, lies, thefts, defections, slavery, are you and me,

Its Congress is you and me—
the officers, capitols, armies, ships, are you and me,

Its endless gestations of new States are you and me,

The war—that war so bloody and grim—
the war I will henceforth forget—was you and me,

Natural and artificial are you and me,

Freedom, language, poems, employments, are you and me,
Past, present, future, are you and me.

WALT WHITMAN, "As I Sat Alone by Blue Ontario's Shores"

CONTENTS

ACKNOWLEDGMENTS

This book wasn't supposed to be written. That it was written means I ran up an incredible debt with mentors, friends, and family. Only now as a professor and advisor do I understand how much the labor given was beyond anything like professional responsibility and just how scarce the time was that many gave. I was a hard student, I struggled, and any aptitude I had for this kind of work was not obvious. I benefited because early on a few people took an interest when most thought I was an annoyance and a waste of time. Beyond all the citations these folks get throughout the rest of the book, I want to say something about the debt I can never pay back, beginning with my mom.

My mom watched me finish most school days in tears and then start most days physically sick with dread. Homework was impossible. Trying to recopy things from books felt like peeling the skin off my own face. Writing felt like punishment. My mom got me through every assignment, wrote notes to rarely understanding teachers explaining why my homework was in her handwriting, and waited out the most hostile of the teachers who felt inspired to convince me and my mom that I had no business being in school at all. Some insisted I would not finish high school. Others just wanted to make sure I understood how truly lazy they thought I was. Before there was an Americans with Disabilities Act, my mom fought for me to have access to a computer, adequate time to finish assignments, and the basic respect necessary to survive in the classroom. Before most teachers knew what a learning disability was, long before we could say something affirmative like neurodiverse or my favorite, neuroqueer, my mom found someone who could teach me to physically write, almost spell, and at least learn to use some rudimentary grammar. When the prevailing wisdom was to put me on medication, she talked me through the side effects and let me choose not to pursue medication. She was a single mother who by her nature hated confrontation and yet she was a fierce advocate and a limitless emotional support so that I could survive a school system designed for me to fail.

In addition to the hard stuff, my mom also cultivated in me a love for 1940s screwball comedies and a devotion to Alfred Hitchcock, which proved, serendipitously, to be an invaluable background for the pursuit of philosophy. There were a few teachers along the way who made school worth showing up for: Mark Webber, my debate coach; Dorothy Blodgett, who inspired my love of history; and Jim Herman, who encouraged my curiosity for the biological sciences: all helped get me through high school.

In college, three amazing scholars refused to let me quit and pushed me to go to graduate school. All three suffered through my tortured writing and all three found ideas buried in the mess of my thinking that they convinced me I should pursue. Without all of them I would have never bothered applying to graduate school and without them I certainly would not have been admitted. They wrote recommendation letters for me, edited statements of purpose, and gave me stacks of books to challenge and inspire me. I was a lot of work and they just kept showing up. Robin Kilson, Begoña Aretxaga, and Barbara Harlow pushed me hard and never gave up on me. Robin also taught me how to make sangria, talk to academics, and look the horrors of our histories dead on, and she showed me what it looked like to struggle with a debilitating disease on top of the cruelty of antiblackness and never give an inch. She wanted me to go to Harvard and she wanted me to be a historian but she was quite happy the day I told her I was admitted to Hopkins. She was also skeptical of all the "theory" I liked but she would have read every page of this book and argued with me about every paragraph, chuckling at me and half-smiling through the whole thing. Barbara took me on as an advisee, despite the fact that I was a history major outside her department, and she supervised my senior thesis. She taught me how to read closely, tried to teach me how to write, and when I was convinced she hated me and thought I was an idiot, she told me that she wanted to write me a recommendation letter for graduate school. And in her sardonic, more serious than joking way she told me that I would be an idiot not to apply. Begoña encouraged me to take my first graduate class while I was still an undergrad at UT Austin. She introduced me to a wonderland of theorists and new ideas and then supervised my MA thesis at the University of Chicago while she was a visiting scholar there. I only entered that program because she insisted that I not stop going to school after I had been unable to accept an offer to Johns Hopkins because I needed to take a job in Chicago. She was an incredible scholar and a generous advisor and reader. Begonia understood the intimacy and creative horror of violence better than anyone and had a nuance and reflexivity for the ethical terrain of war that always left me in awe.

Despite all the labor these folks put in, I was no less work in my PhD program. Bill Connolly and Jane Bennett suffered through late papers, bad drafts, papers that were longer than novellas, and quite a few other challenges along the way. I owe them my degree, my career, and my love for political theory. They bought us our first stroller, organized a summer dissertation defense, and were invaluable in getting me a job. They taught me how to run a seminar, how to teach and advise graduate students, and showed me a generosity in the nature of things that I still struggle to hold on to but look for nonetheless. When I showed up at my first conferences, there was always an advance team of their intellectual friends from Mike Shapiro to Mort Schoolman to Tom Dumm to Bonnie Honig to Stephen White ready and willing to ask encouraging questions at nearly empty 8:00 a.m. panels. Without fail they were there again at the end of the day to buy me a beer after middling presentations. I am further indebted to Bonnie, who read and commented on the full manuscript.

In the weird world of international relations, Daniel Deudney refused to ever let me hide at a conference. Dan introduced me to every single person we passed in the hall at the International Studies Association (ISA) conference. And every year after my graduation he made sure I was on a panel at the ISA even if it meant getting permission to have an extra chair at panels he was invited to. Dan is one of the last truly big thinkers in international relations (IR). In an age of hyperspecialization and hyperregionalization, he has never apologized for his audacity to take on the political structure of the entire planet and even sometimes the entire cosmos. I will always look up to him for that boldness and the intellect he commands to pursue it.

Siba Grovogui treated me like family from the moment I arrived at Johns Hopkins. I learned as much in his office talking about his journey from Guinea to Baltimore that brought him to Hopkins as I did in his seminars. More than anyone, he cultivated in equal parts a cutting disrespect for international relations and a commitment to its enterprise in hopes of a less cruel world. Siba taught me to read the canon with new eyes and helped drag me across the finish line of my PhD.

After showing up in Hawai'i, I had the privilege of attending Chuck Lawrence's mentoring seminar for junior faculty. Chuck and Mari are real live superheroes still fighting injustice with an endurance that seems entirely unaffected by the darkness of our times. It was on their living room floor, trying to absorb everything that was being said by Chuck and Mari and activist scholars like Malia Akutagawa, Aaron Sala, and Kapua Sproat about what it meant to live radical and often endangered lives, that I finally understood the difference between mere knowledge and wisdom. What I came to understand was how little of the latter I possessed. It was a humbling experience for which I am very grateful.

Thanks to Nandita Sharma and Gaye Chan, who remind me over and over to do something with my political discontent. They are a source of inspiration for how to lead a righteous, artistic, and meaningful life all in equal measure.

Without qualification, my department has supported and mentored my research and teaching every step of the way. Kathy Ferguson has led our department and been a constant source of support since my arrival. Michael Shapiro, Jon Goldberg-Hiller, Sankaran Krishna, and Debbie Halbert have read many drafts, inspired ideas, and counseled me through career decisions, life decisions, political defeats, and everything in-between. Thank you to Noelani Goodyear-Ka'opua and Noenoe Silva, who have also taught me a great deal about how to be a scholar but who have also patiently and graciously helped me understand what it means to be a scholar in Hawai'i. Thanks to them, I will never take for granted those who bear the greatest costs of my occupation of these islands. I look forward to the day when I can leave, apply for a work visa, and return to an independent Hawai'i, if they will have me back. Nevi Soguk, Carolyn Stephenson, Dick Chadwick, Manfred Henningson, Katharina Heyer, and Larry Nitz all generously gave me their time, advice, and support. These senior colleagues have never once pulled rank or made me feel like I was anything other than their colleagues. That is a rare privilege for a junior scholar and I cannot imagine the department without any of them. I am also indebted to Laurie Onizuka, whose virtuosity as a departmental administrator I call on every single day. Laurie is a great friend and colleague.

Thanks to Team War, which helps keep fun alive in the insanity of our weird field. Antoine Bousquet and Nisha Shah are the finest of interlocutors and great friends.

Many of the ideas in this book have been shaped by the incredible graduate students I have the privilege to work with. The yearlong Cybernetics Reading Group with Katie Brennan, Rex Troumbly, and Guanpei Ming was instrumental to this book and the next book. We are more than three years into the Phenomenological Reading Group for Useless Theory thanks to the commitment and brilliance of Aaron Cornelison, Jonathan Hui, Sophie Kim, Ryan Koch, and Amy Sojot. I cannot wait to read all your gestating books.

Thank you to Gitte du Plessis, who not only read the entire manuscript but had feedback on all of it. Thank you to Ken Gofigan Kuper, who inspires me to see possibilities for worlds that can fight back against the Eurocene. I hope Ken will hire me as a national security advisor when Guahan becomes independent.

And thank you to all the graduate students who trusted me enough to put me on their committees. I have benefited so much from your work: Sammy Badran, Katie Brennan, Michelle Brown, Duyen Bui, Anna Butcheret, Jonathan Dial, Heather Frey, Brian Gordon, Julia Guimares, Jon HuiKeonwoo Kim, Sean Hyoun Kim, Christopher Klein, Ryan Koch, Ken Gofigan Kuper, Ashley Lukens, Matt Markman, Micheal Miller, Guanpei Ming, Youngwoo Moon, Ali Musleh, Sharain Naylor, Emily Pesicka, FX Plasse-Coutre, Phil Reynolds, Benton Rodden, Aaron Rosa, Jacob Satoriadis, Ben Schrader, Tani Sebro, Riddih Shah, Owen Sheih, Yujing Shentu, Amy Sojot, Zoe Vorsino, Irmak Yazici, and Aubrey Yee.

In graduate school I benefited from an amazing extended cohort of scholars. Almost every weekend, we found ourselves in a backyard cooking, debating, thinking, and watching an illicit firepit burn down to coals. So many ideas took place in Jake Greear's shed that would later find their place in these pages. I miss being bested repeatedly in open combat by Jake Greear, Jeremy Arnold, Mabel Wong, Simon Glezos, Hadley Leach, Rob Higney, Cristie Ellis, Daniel Levine, Michael McCarthy, Chas Philips, Adam Culver, Daniela Ginsberg, Brighu Sing, Alex Lefbevre, and Alex Barder.

Thanks to James Der Derian, a mentor, a true friend, and the person I would want at my back in any kind of fight. Thanks to Mark Salter, who pulled me out of the deep end at my first academic event as an assistant professor and has been my friend ever since. Thanks to Debbie Lisle, whose limitless energy and intellectual force keep kicking IR in the teeth all the while being hilarious.

Thank you to David Weil for the insane six weeks on the campaign trail in central Florida. While I will never be a measured thinker, David will always remind me of its virtue.

And this will quickly become a book unto itself. The following have been significant inspirations and friends at panels, at home, and over the weird intellectual community of blogs and Facebook: Shain Raily, Iris Marion Young, Rohan Kalyan, Heather Johnson, John Protevi, Lauren Berlant, Jonathan Lear, MacKenzie Wark, Miguel de Larrinaga, Megan Mackenzie, Roland Bleiker, Grace Ho, Anand Pandian, Naveeda Kahn, Peter Skafish, Laura Shephard, Patricia Owens, Cynthia Weber, Sam Opondo, Noah Viernes, Dan Monk, Lester Spence, Sam Chambers, Tim Morton, Lisa Disch, Alan Finlayson, David Howarth, Lori Marso, Michael Hanchard, Rom Coles, Daniel Smith, Aletta Norvall, Jennifer Culbert, Mary Tuti Baker, Lida Maxwell, Colin Wight, Alex

Wendt, Ben Mieches, Claudia Aradau, Dan Nexon, Levi Bryant, David Panagia, Kam Shapiro, Terrell Carver, Lida Maxwell, Lisa Disch, Helen Kinsella, Jeff Vandermeer, Kevin Kenny, Veena Das, Paola Marrati, Can Mutlu, Adam Sandor, Jeff Hussyman, João Nogueira, Tony Burke, Audra Mitchell, Ron Deibert, Victor Coutinho Lage, Carlos Braga, Natália Félix de Souza, and Aaron Goodfellow.

And thanks to the debaters who taught me how to think before I could write: Asher Haig, Tristan Morales, Varant Yegparian, Kevin Kuswa, Sarah Apel, David Michael Mullins, Ralph Paone, Kirk Evans, Dallas Perkins, Lindsay Harrison, Sherry Hall, Matt Zamias, Bill Russell, Roger Solt, Scott "Duck" Deatherage, Ross Smith, Chris MacIntosh, Jessica Clark, Michael Horowitz, Jon Paul Lupo, Edmund Zagorin, Brian McBride, Stap Beaton, Jonathan McCartney, Loe Hornbuckle, Jon Brody, Leslie Wexler, Matt Powers, Andrew Leong, Dayvon Love, Eli Anders, Eli Jacobs, Michael Klinger, Elliot Tarlov, Michael Gottlieb, Claire McKinney, Joel Rollins, Mark Webber, Nick Copeland, Karla Leeper, Jason Russell, Will Repko, Biza Repko, Colin Kahl, Brett Griffin, Loren Dent, Anjali Vats, Dan Luxemburg, John Fritch, Chris Lundberg, and Aimi Hamraie.

Special thanks to Courtney Berger for being an amazing editor and Sandra Korn for all her hard work in getting the manuscript into publishable condition. I also want to thank Susan Albury for supervising and directing all of the complexity of making the manuscript into a book. Also thank you to the anonymous reviewers, who helped make the manuscript a much better book.

Thank you to friends who are truly more like family: Jonathan Scolnick, Ricky Garner, Laura Nathan, Anthony Miller, Adrian and Caitlin Ramirez, Paul Boston, and my last debate coach, Joel Rollins.

Thank you to my uncle Donn, who made sure I had the Huck Finn childhood every kid should have, and to June and Pa for being the shelter and support my mom and I needed. Thank you to Grandpa Jim for long talks on a roof in Manzanillo, Mexico, after dark watching the bats and talking to me about the mysteries of the universe. You introduced me to William James and inspired the curiosity to follow questions wherever they lead, even into the darkness. Thanks to my grandmother Beverly, who made me get out the unabridged dictionary every time I asked how to spell a word.

Thank you to my sister and brothers, Zelda, Jonas, and Sean. Being an only child until I was sixteen was lonely and quiet. You all make the world more fun.

Gratitude doesn't conceptually capture the essential necessity of Nicole Grove, my partner in ideas and my partner in life, in all this. We have traveled to just about every corner of the earth, trying to think, research, write, and survive with our children, Oona Tuesday Grove and Scout Ellison Grove, in tow. I cannot imagine doing any of this without you. Nicole is responsible for the best of me and is the scholar I look up to most. She has barely taken an extra breath while continuously making degrees, dissertations, children, articles, books, and our family. I will always be in awe of your fierce and unstoppable pursuit of life. You are our vampire slayer and the legacy in whose path our daughters will follow. To Oona and Scout, every day, I look forward to seeing the new worlds you will make.

A SECTION OF the introduction appeared in "Ecology as Critical Security Method," *Critical Studies on Security* 2, no. 3 (2014): 366–69. An earlier section of chapter 1 appeared in "The Geopolitics for Extinction," in *Technology and World Politics: An Introduction*, edited by Daniel McCarthy (New York: Routledge, 2017). An earlier version of chapter 4 appeared as "An Insurgency of Things: Foray into the World of Improvised Explosive Devices," *International Political Sociology* 10, no. 4 (2016): 332–51. An earlier version of chapter 5 appeared in "The Geopolitics of Blood," in *Making Things International*, edited by Mark Salter, 184–200 (Minneapolis: University of Minnesota Press, 2015). Part of chapter 6 appeared as "Something Darkly This Way Comes: The Horror of Plasticity in the Age of Control," in *Plastic Materialities: Politics, Legality, and Metamorphosis in the Work of Catherine Malabou*, edited by Brenna Bhandar and Jonathan Goldberg-Hiller, 233–63 (Durham, NC: Duke University Press, 2014). An earlier version of part of chapter 8 appeared as "Must We Persist to Continue? Critical Responsiveness beyond the Limits of the Human Species," in *Democracy and Pluralism*, edited by Allen Finlayson, 183–202 (New York: Routledge, 2009).An earlier version of chapter 9 appeared as "Of an Apocalyptic Tone Recently Adopted in Everything: The Anthropocene or Peak Humanity?" *Theory and Event* 18, no. 3 (2015), https://muse.jhu.edu/, accessed December 16, 2018.

INTRODUCTION

The truly apocalyptic view of the world is that things do *not* repeat themselves. It isn't absurd, e.g., to believe that the age of science and technology is the beginning of the end for humanity; that the idea of great progress is a delusion, along with the idea that the truth will ultimately be known; that there is nothing good or desirable about scientific knowledge and that mankind, in seeking it, is falling into a trap. It is by no means obvious that this is not how things are.
—LUDWIG WITTGENSTEIN, *Culture and Value*

In 1992 the Union of Concerned Scientists issued a warning to humanity. According to the union, the current trajectory of development promised "vast human misery . . . and a planet . . . irretrievably mutilated."[1] At this time, the critical areas of concern were the atmosphere, water resources, oceans, soil, forests, living species, and the size of the human population. Each of these areas was identified as a necessary precondition for human survival, with 1,575 scientists joining the public statement. The warning was followed by a set of recommendations said to be within the grasp of all populations of the world: a significant reduction in the destruction of natural resources, sustainable resource management, population stabilization through voluntary family planning, reduction and then elimination of poverty, and sexual equality such that women could determine their own reproductive decisions. To accomplish these goals, the union insisted that investment in and occurrence of violence and war needed to be reduced in order to free up the necessary resources for saving the species. The report estimated that US$1 trillion annually was being directed to the preparation and prosecution of warfare. The starkness of the choice is itself interesting. For the union, in a world of finite resources the species had to choose between war and survival, but it could not choose both.

Twenty-five years later, the warning was issued again, and 15,364 scientists joined the "second notice" to humanity.[2] The group, now renamed Alliance of World Scientists (unions and concerns having fallen out of political favor), found unequivocally that the state of the world is worse than we thought in 1992, and that little if any progress has been made in the intervening years. While no official answer to the 1992 warning was issued, a decision was made. Those in a position to make a decision chose war.

It is not unusual that more than fifteen thousand scientists would agree on something. I imagine millions of scientists agree on other questions, like the basic nature of gravity and the atomic weight of cobalt. Yet it is difficult to imagine the need or interest to issue a public statement about these mere descriptions of fact. What makes this concern worthy of a public address is that the statements issued in 1992 and 2017 are attempts to make a claim on a public, in fact *the* public: the global whole of the human species. The tone of both letters invests the full force of collective scientific expertise, argument making, and powers of persuasion on the case to be made for a threat to the planet. The letters simply assume that if the case is successfully made that humanity faces impending doom, the case for saving humanity will automatically follow as if by some mechanism of logical necessity. Unfortunately, this assumption is not merely off the mark. Global politics for the past five hundred years is proof of the opposite of common sense. There is a centuries-long investment in research, development, and deployment of techniques to ensure that survival is only ever a right for some. This right for some, more often than not, is ensured at the expense of the self-determination and continuation of living for the overwhelming majority of the planet's human population.

Against the banal appeal to a universal humanity or the equally commonplace and catastrophic insistence on an inevitable clash of civilizations, I prefer the idea of "form of life." Not quite race and more than culture or style, this phrase refers to those ways of being in the world—always lived collectively—without which one would no longer be who or what one is. I want to go further than Ludwig Wittgenstein's invocation of form of life as one's particular game of language and gesture—the physiognomy that for him makes one human—into the ways that not just humans but all things creatively striving toward complexity come to make worlds out of their intractable dependence on and contribution to an environment.[3] And beyond Wittgenstein's events of communicative failure, interruptions of these relations and habits threaten existence itself. When efforts are made to wipe out the American bison and buffalo or to militarize borders to interrupt the flow of migrants who follow seasons and crops, it is not just a habit or practice that changes. The interruption of a form of life kills people and frequently cascades into genocides and extinctions. In the case of the buffalo, it was not just the bands and nations of the Great Plains whose precarity was leveraged for the strategic goal of genocide and settlement. The entire prairie ecosystem was targeted, moving on from human inhabitants to predators such as wolves and big cats to make way for leisure hunting and grazing practices that created the dust bowl and the subsequent collapse of riparian habitats throughout the United States.[4]

I take inspiration from Giorgio Agamben's more radical reading of Wittgenstein's form of life in my desire to describe lives that cannot survive being separated from the way they are lived, but like Wittgenstein's linguistic provincialism, I do not accept Agamben's species provincialism that form of life either is what defines the human or is exclusively a human attribute.[5] Quite the opposite, when form of life is seen ecologically, what becomes apparent is how many different species, practices, histories, cosmologies, habitats, and relations come to constitute what we might call a form of life. Form of life is a particular origami in the "fabric of immanent relations" that defines the torsion between the singularity and the interpenetrated relationality of each and every human and nonhuman person.[6] This question will be taken up more substantially in chapter 1, but suffice it to say that form of life, for me, is the current or flow against which we can even identify a change or intervention as violent rather than merely as a change. And geopolitics, the focus of all the following chapters and that which the concerned scientists want to avoid, is the collectively practiced art and science of that violence against other forms of life.

In fact, it is this very geopolitics—nation-states making decisions and wielding power at a global scale—that the scientists want to steer away from war toward saving the planet, which is not premised at some foundational level on a general principle of order or the good. Geopolitics is, at its most fundamental level, a husbandry of global life in which thriving is intimately connected to the particular form of life and the particular lifeworld through which one becomes who one is. Geopolitics is structured to be selective, and to ensure that selectivity by lethal force.

Therefore, to oppose survival to the pursuit of war as a global question for a global audience (as if that audience were empowered or even capable of issuing a global answer) displays a persistent and willful naïveté of how the global was made in the first place. The geopolitical project of planet Earth is a violent pursuit of a form of life at the cost of others—full stop. However, at the same time, with an often zero-sum game over form of life at its center, global war—the presumed opposite of human survival—is not primarily about direct killing. Instead, the violence of geopolitics is an ecological principle of world making that renders some forms of life principle and other forms of life useful or inconsequential. Emmanuel Levinas is quite helpful on this point. In his investigation of the antinomy between philosophy and war, Levinas came to understand the violence of geopolitics and its pursuit of global war to be less a direct material force and more an organizational principle of coercive steering and depriving: "Violence does not consist so much in injuring and annihilating persons as in interrupting their continuity, making them play roles in which they no longer recognize themselves, making them betray not only commitments but their own substance, making them carry out action that will destroy every possibility for action."[7]

The attack on the conditions of life and its formation as a form of life establishes more than a trade-off between the material costs of warfare and the pursuit of the Union of Scientists for planetwide and environmentally sustainable economic and sexual equality. Geopolitics, enacted through global war, is itself a form of life that pursues a *savage ecology*, radically antagonistic to survival as a collective rather than discriminatory goal. Geopolitics, as the organizational matrix of global war, has as its enemy the very pursuit of what the scientists see as a commonsense, pragmatically just planet. Therefore, the line between extreme human misery and just transformation is not practically or impractically out of reach because of a lack of will or misuse of resources. For the majority of the planet, the failure to ensure survival is not about an oversight or bad financial management. Instead, the line between misery and something else is heavily policed and enforced with everything, from odious international debt to hellfire missiles.

Alfred North Whitehead says every science belies a metaphysics, or something we could call more broadly a cosmology.[8] The science in question for *Savage Ecology* is the Euro-American science of geopolitics. I want to understand the cosmology of geopolitics. Thus, this book is an effort to understand how a particular formation of global war, as the slow accretion of a form of life, came to be a dominant form of life cosmologically at odds with the idea of collective thriving. This geopolitical form of life is so caustic, it calls into question if there has ever been anything as universal as a human species to be threatened, much less saved.[9]

Geopolitics or Savage Ecology

> The Anthropocene, the reframing of the Earth in the image of industrial modernity, will be short-lived, a geopolitical instant more than a slow geological era.
> —BENJAMIN H. BRATTON, *Dispute Plan to Prevent Future Luxury Constitution*

Of the various "cenes" of late trying to name what has caused catastrophes at a global scale, no one diagnosis can quite win out. Racism, sexism, settler colonialism, ableism, heteronormativity, speciesism, classism, and technologism are all real forces in the world, and a compelling case for all of them has been made to diagnose the crisis we face. However, each case falters as it tries to close the last loop of its argument such that each of the other forces is somehow subordinate to *this* explanation. Rather than pick a side or stake out new ground on the intersectional axes of destruction, I have opted to contribute to how we understand the state of affairs and the historical conditions that made this state of affairs possible—that is, the means and the ends of our destruction. The motivations behind the state of affairs or master logic in the basement of all things is beyond the scope of this book. I remain interested but agnostic as to what inspires the will to catastrophe. I am less ambivalent about the how of the situation. Geopolitics as a European-led global project of rendering, in the way that fat is rendered into soap, or students are rendered pliable and obedient subjects, is the driver of our epoch and the obstacle to any other version of our world, whether plural or differently unified.

This book is an attempt to make a certain kind of ecological sense out of five hundred years of geopolitics and its warlike means. Here I develop a martial genealogy for what I am calling the *Eurocene*. In this story of development and expansion, geopolitics is not a cause per se, but it is a means that has been elevated and refined into a virtue. It is a means that has become its own ends. Because geopolitics is now a virtue, it succeeds and fails without much consideration for whether it should be abandoned. Those who benefit most from geopolitics shift slightly from time to time—a little more internationalism or a little more unilateralism and back again. The consequences of a geopolitical form of life vary from settler colonial genocide to environmental massacre to strategic interventions into the very rhythms and synaptic terrains of individual human bodies. Yet at each interval of deformation, destruction, failure, renewal, reentry, and invasion, geopolitics persists as the primary operating system of planetary life.

In chapter 1, "The Anthropocene as a Geopolitical Fact," I follow the strange path of Paul Crutzen from his interest in the ozone layer to nuclear winter to climate change to becoming the foremost advocate of the Anthropocene. While Crutzen's early work on the ozone layer and nuclear winter put geopolitics front and center in his scientific analysis, he takes a postpolitical turn after the Cold War by framing climate change as a problem for humanity. In addition to the ways Crutzen's universalist appeal erases the very uneven responsibility for climate damage, the newly depoliticized category of humanity quickly became a justification for great powers to take the lead in geoengineering the planet despite the significant risks for subtropical and tropical inhabitants. In some sense, geopolitics was only a problem for

Crutzen when it threatened the metropoles of the Northern Hemisphere. Now relieved of the possibility of nuclear war between the U.S. and the USSR, power politics is seen as benign, and even transformative. In chapter 7, I return to how this elision of geopolitics informs renewed hope for the future-oriented industrial ecology advocated by Stewart Brand and other ecomodernists.

Moving from the global scale of geopolitics to the hard, martial labor of implementing the geopolitical order, chapter 2, "War as a Form of Life," zeros in on the making of geopolitical bodies and the kinds of corporeal rhythms that inhabit the zones of war and peace in the Eurocene. I ask the question of what it would mean to consider warfare as a form of life, that is, an ordinary practice for many people rather than the ways we often characterize war as an anomalous or rare event that suddenly breaks out. Turning to Maurice Merleau-Ponty's phenomenological account of the body, and Erin Manning's work on dance, I try to theorize what kind of body a human body must be if the extremity of war can *become* normal. Considering warfare as an embodied becoming rather than an abnormal break, I hope, draws our attention to how geopolitical orders are written into the very musculature of our bodies, practices, and communities.

In chapter 3, "From Exhaustion to Annihilation: A Martial Ecology of the Eurocene," I historicize the martial practices of bodies in the ways war speciates into wars of exhaustion, which are primarily reserved for European "peers," and wars of annihilation, which are practiced in settler colonies. Pursuing an ecological approach that looks for relations, heterogeneous actors, things, technics, racializations, territorializations, and practices, the chapter explores how the environment itself—an ecological approach to "New World" ecologies—informed practices of annihilation beginning with the earliest settlement practices in New Spain through to the American war in the Philippines and contemporary practices of counterinsurgency.

In the second part of the book, "Operational Spaces," I take up three different ways that homogenization and war have been operationalized in different ecological orders. Chapter 4, "Bombs: An Insurgency of Things," is a case study on the relationality of improvised explosive devices (IEDs) and the decisive role IEDs played in the U.S. post–September 11 wars in Afghanistan and Iraq. I explore how these variable failures and reinventions played out in the nonhuman character of war through an exploration of the undead war trash of improvised explosive devices. In chapter 5, "Blood: Vital Logistics," the difficult and often contradictory tug-of-war between the metaphors and materialities of blood and race takes center stage through a circuitous history of blood transfusions and their regulation during World War II in the United States, UK, France, and Germany and the ways those policies inform the complexity of enmity and blood use for the U.S. in the global war on terrorism. In chapter 6, "Brains: We Are Not Who We Are," the brain itself becomes a political terrain. The Eurocene, as a neuro-geopolitics, is obsessed with hacking the brain as a new frontier of ecological and martial control. The chapter concludes with a series of questions about whether attempts to weaponize the brain undermine the condition of possibility for agency and freedom. The drive for security and control in the Eurocene comes to devalue the very foundations of autonomy and self-possessed rationality that enlivened the geopolitical drive for homogenization. In chapters 4, 5, and 6, *Savage Ecology* is a story about the aspiration for total control, not control's total victory.

In chapter 7, "Three Images of Transformation as Homogenization," the book moves from the historical making of the Eurocene to its imagined futures. Here I focus on three particularly popular futures espoused as alternatives to the current global catastrophe. Specifically, ecomodernist, Marxist, and U.S. militarist futures all bear the marks of the Eurocene's taste for incorporation and violence. While hoping for transformation, each future remains committed to a project of homogenization at the expense of human animal and nonhuman animal forms of life that are aversive to the smooth transformations each project envisions.

Despite the global scope of homogenization as a geopolitical project, the habitats and ecosystems of the planet still vary by species, by climate, by terrain, and by form of life. I do not want to overstate the success of geopolitics in achieving its dream of a flat planet. However, I also do not want to obscure the increasing intensity and danger, that is, the difference of the contemporary moment of geopolitics. The point of the first two parts is not to, in some sense, declare the kind of "end of history" of the Eurocene or the inevitability that homogenization will prevail. Instead, I want to make as apparent as possible that on every continent—and even the outer reaches of the planet's atmosphere—the technics and waste of geopolitics connect every space to every other space, whether by satellite feed, radioactive isotope, aircraft carrier, unexploded ordinance, sexual trauma, or tragic absence of forced removal. The global network of open wounds, bruises, and scar tissue that runs over the surface of the planet, through its water table and abandoned mine shafts that sprawl out on the vast ocean floor, exceeds the migratory and circulation patterns of any other species or even family of species. Five hundred years of geopolitics has built a global savage ecology.

At the half-millennium of the geopolitical epoch, terms like *biopolitics* seem almost quaint. Life, much less human life, is at best a small sliver of the vast infrastructure of geopolitics. Coastlines, rivers, gravitational fields, and the atmosphere are elements that have been altered in addition to whole populations and individual bodies. The scale of these alterations is not recent. The decimations of continent-wide populations and global temperatures have been in the fabric of geopolitics since its beginnings.

If there is a difference that the contemporary makes, it is that the substance and means of action for change are converging into one substrate for life. Félix Guattari described the world after the cybernetic drive to become the final and total science of all things as a postmedia age.[10] More than the convergence of audio, visual, and data communication described by Friedrich Kittler, Guattari saw Earth itself, along with human consciousness and desire, as converging media. A proliferation of what can be altered is simultaneously paralleled in a flattening of those differences in communication and substance into informatics such that everything becomes at some level plastic in the same way. The sciences of brain plasticity, species plasticity, the plasticity of matter, and the plasticity of the atmosphere are all native to the same historical moment, and understand measurement and change in the same way ontologically. Catherine Malabou's question, "What should we do with our brain?," is now extended to "What should we do with the planet?"[11] The focus on the brain and plasticity as a more general way of thinking about matter as plastic connects the recounting of the past as it is engaged in the first two parts of the book to the vision of the fully plastic future. This future, I argue in chapter 7, is envisioned by forces of industrial liberalism, left and right accelerationism, and the U.S. Department of Defense.

The twentieth century will be remembered as the moment cybernetics truly made humans conscious of themselves and their environment. However, the twenty-first century will be the moment that humans became capable of acting on the processes of that consciousness. The likelihood that this consciousness or these capabilities will serve the new ethic aspired to by the Alliance of Global Scientists seems slim. The postmedia era diagnosed by Guattari looks to be every bit as geopolitical as the eras that preceded it. The benefactors of the world's greatest minds searching for breakthroughs in neuroscience, artificial intelligence (AI), space exploration, and even climate engineering are primarily the martial divisions of the world's governments.

The point of saturation has taken on the feel of an end of history; however, it is not an end. It is something else. The *something else* is the theme of the last part of the book. In this final part, "Must We Persist to Continue?," I describe what I see as the possibility for forms of life other than war and homogenization. Chapter 8, "Apocalypse as a Theory of Change," details an affirmative theory of catastrophe and turbulence. The question of apocalypse *for whom* is thought in parallel with the disruptive and often violent history of geological change. The hope here is not to romanticize Earth's history of mass extinctions but rather to displace the sociocentrism that, in equal parts, ignores the destructive geological power of the planet and the annihilations unleashed on human timescales by very particular humans. Geological and human history are punctuated and mutated by these events. Rather than seeing apocalypses as inevitable, I read them as transformations or bifurcation points where other ways of life become possible.

As a kind of warning against those who would respond to the terror of apocalypses and change with a conservative humanism, the main target of chapter 9, "Freaks, or the Incipience of Other Forms of Life," is Jürgen Habermas and others who fall into a tendency of somatic fundamentalism. I argue that instead of trying to preserve a romantic view of what the human was, we need an agonistic respect and attentiveness for the emergence of freaks, or what we and other lifeforms could become. Rather than fear AI, posthumans, or other emergent forms of life, we should embrace the differentiation of life as preferable to the goals of a recalcitrant humanism or homogenous singularity.

In the book's conclusion, "*Ratio feritas*: From Critical Responsiveness to Making New Forms of Life," I take the idea of speciation and change further into something like a virtue, or what I call feral reason. It offers the possibility of other futures oriented toward creativity and adventure rather than conservation and technological homogenization. In this part, I take apocalypse as a fact but the future as unwritten. To temper the temptation that the future is open to free play, or that any particular grouping of humans truly possesses the determinative agency to make a future, I put forward my best effort to sketch the probable world if it continues along the same sadistic lines of Eurocene geopolitics. In the postvision, which I am calling "The End," I try to imagine the United States of America in the year 2061 if the "great homogenization" continues unabated. The landscape of the story combines the ecological concerns of the book with an emphasis on militarization and the security politics of our contemporary moment.

Savage Ecology is a speculative theory for an ecological approach to global politics. By ecological, I mean a form of analysis characterized by inhuman encounters and deep relational processes across geographical scales rather than a form of political thinking that relies on discreteness, causality, and an exceptional notion of human agency. Hence an ecological approach does not center principally on the environment, what in international relations is called environmental security; nor does it limit global politics to states, international organizations, social movements, or even humans. Instead, I take ecology to mean that all things that make a difference in the vast landscape of geopolitics ought to be included in the geopolitical considerations of contemporary life. The book is populated with Neanderthals, improvised explosive devices, revolutions, brains, dead soldiers, beavers, ideologies, mutants, artificial intelligences, drones, states, and the occasional zombie. The research ranges from sixteenth-century counterinsurgency training manuals to leaked internal Department of Defense reports to the speculative futures of mad scientists like José Delgado. There is no one archive or object of inquiry. For me, all these things and more take part in the catastrophe that many have termed the Anthropocene.

Savage Ecology is also a martial theory of the Anthropocene. Throughout the book I take the idea that we are in a planetary epoch in which the Anthropos is capable of making a "cene" quite seriously. The Anthropocene as a philosophical and political crisis has been too quick to forget the geopolitical arrangements of power and violence that have brought us to this point. Not all of "us" have played an equal part in the making of either the Anthropos or the Anthropocene. In part, the often narrow focus on climate change and the fever pitch of crisis abets the erasure of the U.S.'s role in building and maintaining the current world order. This argument amounts to: "now that everything is broken it is everyone's problem so pointing fingers just gets in the way of a solution." Even critical and posthumanist approaches often lose sight of the role of hegemony and power. This is, in part, because of the efforts of those lines of thought to decenter the human as the sole locus of thinking and action. I am committed to relaxing the focus on human actors in processes of global change; however, I think we can decenter the human without letting go of the very specifically human and often national assemblages that broke this planet.

While there is no global history of industrialized war, capitalism, and ecological destruction, the politics of homogenization as an elite-driven Euro-

American geopolitics of industrialized war and capitalism made ecocide that is now a global historical fact. To put it simply, our shared experience of planetary life has a definitively parochial beginning and present. No anthropogenic, planetary-scale threat faced today—be it nuclear weapons, plastic, climate change, or global war—originated outside the Euro-American circuit of expansion, extractivism, and settlement. As Sylvia Wynter has stated, "we must now collectively undertake a rewriting of knowledge as we know it . . . because the West did change the world, totally."[12] To do this means exiting the Anthropocene as an idea, and collectively—even if not equally—exiting the Eurocene as a failed epoch. I think we should relish Wynter's invitation to consider other "genres of the human."[13] She explains she will not miss the Anthropos because, among so many others, she was never considered human to begin with. We should affirm her lack of nostalgia for the human. To invent a new species is the task that must be undertaken before there can be a "we," an "our," or a "cene" that is more than a requiem for the end.

Unfortunately, for those who want definitive answers, there is no theory provided in this book that puts everything in its right place, predicts the outcome of the next presidential election, or can save us from the now inevitable collapse and reorganization of planetary life. Instead, *Savage Ecology* is a speculative reflection on the depths, nay, fathoms of shit we are in as a community of species. I am certainly not alone in wanting to open up to the sheer magnitude of what confronts the planet. And yet I want to do so without losing sight of the real differences in politics, geography, history, meaning, and cosmology that modulate how each one of us will confront the end of this epoch. In so doing, I hope to emphasize a refrain throughout the book that the end of the world is never the end of everything. An apocalypse is always more and less than an extinction, and whatever makes a life out of the mess we are currently in will depend in some ways on how we come to understand the contemporary condition. Ideas matter even if they cannot save us. Stories, explanations, and philosophical adventures are, in my estimation, the best of what the human estate has to offer. No matter how desperate things get, someone will still ask why this is happening, and we will share in that question the possibility of thinking together. As Bill Connolly often says, "we are not unique; we are merely distinctive," and that distinctiveness is connected to a sense of wonder—even when it is a dark wonder.[14] I want to connect this sense of wonder to a plea for a feral reason. This is a renewed sense of adventure and creativity in pursuit only of itself. Feral is not a way out of all this but rather a way through.

How to Do Stuff with Things

To specify what I mean by nonknowledge: that which results from every proposition when we are looking to go to the fundamental depths of its content, and which makes us uneasy.—GEORGES BATAILLE, *The Unfinished System of Nonknowledge*

We need less by way of context and more by way of concept.
—EDUARDO VIVEIROS DE CASTRO

Something in the world makes us think. How could it be any other way? If thought was its own cause, consciousness that is consonant with the world would be impossible. We live in a world of persistent provocations, and our thinking is at its best when it is along for the ride without trying to steer the course of events. Following Steven Shaviro:

> Things encounter one another aesthetically and not just cognitively or practically. I always feel more of a thing than I actually know of it, and I feel it otherwise than I know it. To the extent that I do know an object, I am able to put it to use, to enumerate its qualities, to break it down into its constituent parts, and to trace the causes that have determined it. But feeling an object involves something else as well. I feel a thing when it affects me or changes me, and what affects me is not just certain qualities of the thing but its total and irreducible existence.[15]

This is another way of saying that all the things of the world should set the agenda for research, as opposed to our anthropocentric image of the world.

If a research agenda is driven by one's presumption of that which is to be studied, then we already find ourselves lost in our imposed telos of the research rather than the object of that research. Take, for instance, the major studies of nuclear weapons. The presumed purpose of a nuclear weapon is to function, to deter, to launch on command, or even to launch on warning. We have many fine studies of how the nuclear arsenal is supposed to work, or more specifically, how we desire it to work. We have theories of nuclear decision-making, game theories of nuclear war fighting, psychological theories, and organizational theories. These studies, from John Steinbruner's *The Cybernetic Theory of Decision* to *Managing Nuclear Operations*, are excellent analyses of hypothetical arsenals in coordination with either definitive human events such as the Cuban missile crisis or equally hypothetical scenarios of nuclear war fighting that "double-click" entirely over the actual process by which six thousand or so weapons get deployed, targeted, launched, and detonated.[16]

The virtue of encounter as the driving force of thought is that it compels us to understand how little we actually describe, much less comprehend, what nuclear weapons, as an assemblage, actually do—that is, not what we would like to do with them but what they are. Delivery vehicles leak coolant, operators lose their minds, code command systems malfunction, and early warning systems misread solar flares, weather balloons, and even geese. Warheads get left armed and flown over "friendly territory."[17] Parts work, break down, and produce algorithmic anomalies. Yet how many great works of security studies or international relations are there on the history of accidents, near misses, organizational confusion, and failed tests? The field has not yet produced a Graham Allison for the arsenal, only a Graham Allison for the presidential decision.

The practical impediment of anthropocentrism—organizing thinking around our projection of the world rather than encountering the world as it is—for good research is how little of the world of geopolitics we spend time thinking about. The vast reality of nuclear weapons finds almost no place in research about nuclear weapons. Despite the occasional consideration of a nuclear accident or an accidental nuclear war, real scholarship on the continent and even planetary-sized assemblages of computers, soldiers, technicians, enriched heavy metals, virtual monitoring and testing, trucks, railways, engineers, underground villages, hollowed mountains, theories of nuclear physics, chain of command, fear, regret, and guilt find almost no place in the theories of international relations. Yet all of it is waiting for us on road trips, with every network dependent on daily ritual, in uranium tailings in Native American reservations and in the cancerous growths of loved ones.[18]

To start with the encounter rather than the application of human-determined purpose directs the researcher to be attentive to how the whole world can be studied rather than picking and choosing the processes that conform to a desired research agenda. I explore what this might look like as a general approach to warfare in chapters 2 and 3, and then in the second part of the book I take on improvised explosive devices, blood, and brains as three specific knots in the filaments of martial ecologies.

Relational Thinking (An Ecology of Things)

The discreteness of objects and actors is a useful but often distracting fiction. If what we want to think through is the problem of geopolitics, then to atomize sectors, objects, and agents of geopolitics will defeat the systemic character of change, behavior, and the emergence of both. By systemic I do not mean structural in the sense of being mechanistic. An ecological approach to security expects a world of highly distributed and complex agencies. Coalitions of agents maintain consistency and contribute to the upheavals that defy the order from which an upheaval emerged. Thus I do not think that ecology is a metaphor for analyzing the world. Instead, relational thinking accretes from empirical scrutiny. Unfortunately, relational thinking is messy because, as John Law says, reality is a mess.[19] The distributed and connective character of change can make things like case selection and variable choice seem arbitrary.

Those who are compelled to pursue positivist analyses of politics via quantitative methods are not likely to find this insight about the world helpful. However, much will be lost to the possibility of analysis if we continue to isolate causally significant variables, or indivisible clusters of variables, from our work. One can, for instance, see how much time has been lost in investigating the relationship between climate change and instability. Thomas Homer-Dixon's *Environmental Scarcity and Global Security* was largely ignored by mainstream international relations theory because of the methodological problems of studying ecological systems in the context of national security crises. Yet who would argue now, more than twenty years later, that we should not have prioritized climate change as a major factor in geopolitics?

So how does one study complex systems rigorously if they, by definition, exceed the mathematical processing powers of our best computer-based tools or the accepted methodologies of the field? I think the answer lies in the rigor and insightfulness of so-called softer approaches. Concept creation when combined with historical analysis and field research can produce scholarship that is insightful beyond our ability to "prove" that it is insightful. Here, I seek to follow Eduardo Viveiros de Castro when he says that "we need a new theory of theory: a generalized theory of theory, one enabling us to think of theoretical activity in radical continuity with practice, that is, as an immanent or constitutive (as opposed to purely regulative) dimension of the intellect embodied in action."[20] This does not mean that quantitative analytic tools or computer-assisted modeling cannot be a vital part of critical work—quite the contrary. Climate modeling, for instance, allows researchers to experience scales of time and space that individual embodied humans cannot. Oral traditions similarly compress and extend time across lifetimes but are too often dismissed because of their nonmodern means of informatic storage and retrieval.[21] Computers, like archives and books, are vital prosthetics in research. They allow us to encounter things in ways that extend our experience beyond ourselves and our native sensory capabilities.

The pack of critical approaches I enjoy traveling with takes issue with the idea that data or modeled outcomes somehow speak for themselves. Rather, data in all forms—from ideas to calculations—are objects of encounter. Data compel us to think but cannot compel us to know. Georges Bataille aptly calls this category of research nonknowledge, "an understanding . . . that borders on knowledge."[22] Data do not transmit information; rather, data provokes further thinking and therefore are not determinative. What modeling, field research, reading, and watching films can do is create the conditions of possibility for encounters not of our own making.

The relational nature of change and emergence means that we must cultivate an attentiveness that might find the most interesting research agenda during a routine check at the airport, or in the repeated failure of your car's GPS near military facilities. The value or rigor of a relational approach that emphasizes the fecundity of encounters is that it marginalizes the capacity of the investigator in favor of the world she investigates. In this sense, undermining anthropocentrism is not just an ethical practice. It also provides a necessary check on observation bias that imposes a telos on the people, things, and systems we encounter, which is a way to pursue the terrifying success and failures of technological interventions into global order. All technical apparatuses from the muskets in chapter 3 to geoengineering discussed in chapter 7 make a difference, but they rarely make the difference that was promised before they were deployed.

Speculation (Scholarship Requires Intervention, Not Proof)

Despite the baggage of international relations, an encounter or empirically driven ecological approach should not need a more sophisticated name than realism. However, to say that things are real does not mean that things are self-evident or easily accessible. It is unfortunate that, for many scholars, things have been reduced to an inert category of rump matter. Things are material *and* they are creative. Things of all kinds possess a quality of plasticity in that they have the capacity to form and be formed. Such formative attributes are variable among different things but importantly are not restricted to language, meaning, or the brain. The constructivist insight about the variability and formative character of the social world should be affirmed but without the unnecessary modifier "social." Rather, we can pursue a speculative description of the construction or process of everything. I do not think such an approach is per se foreign to international relations. For instance, discourse analysis is a process philosophy of sorts, but it is too restricted in what it will consider as the constitutive material of meaning. Some will argue this is because the discursive world is already complex enough. Some will argue that we privilege the discursive because we have privileged access to the world of "our" making. The problem is that such a position often reifies the belief that the world is of our making.

Rather than dismiss attempts at bridging the gap between our world and the world at large as scientism, we can speculate about the creative conjunction of different and differing things, human and otherwise. It is unfortunate that the word *speculation* is much derided in the social sciences. Often to speculate is synonymous with guessing. Following Alfred North Whitehead, I think we should recuperate speculation as the process by which we rigorously intervene in a world that is neither law-driven nor fully accessible to our senses but does resemble what Whitehead called a "doctrine of necessity."[23] In chapter 2, I try to develop an ecological approach to war that can bridge the gap between speculative investigation of the systems of war and the material practices of the body that make those abstractions concrete. Chapter 2 is an intervention into the problem of what war is but it is not a hypothesis about war. Hypothesis testing of various sorts might make sense in a steady-state world where the capacity to test could be up to the task of capturing the system being tested. And discourse analysis alone would make sense if the world were fully withdrawn, or if it were present but meaninglessly inert. However, there are good reasons to believe that neither is the case. Meaningfulness is a construction, but we are not the purveyors of its constructions. Without the blueprints, we have to creatively speculate about the conjunction of heterogeneous actors.

Can Realism Be Critical?

The question often posed, particularly by Marxists, is, What is critical about all of this? Well, it is a plea for a realism whose enemy is common sense. I think we actually have to work quite hard not to be critical. The world insists on its complexity and defies the parsimonious theories we impose on it with such regularity that I do not think the problem is actually how to be critical. The problem is the habits and routines that inure us to the provocations all around to think differently or otherwise than we do.

Such a view of criticism is likely unsatisfactory for those who hope that being critical is synonymous with being normative. For that, I can only offer my condolences, as I do not believe any argument or sufficiently elegant critical theory will deliver to us the ontology we want or think we deserve. Unfortunately, God is very dead, and so if you had hopes that the inner truth of the universe was going to be coincidental with the good, you are out of luck. Ta-Nehisi Coates's letter to his son captures this better than I can: "Struggle is all we have because the god of history is an atheist, and nothing about his world is meant to be. So you must wake up every morning knowing that no promise is unbreakable, least of all the promise of waking up at all. This is not despair. These are the preferences of the universe itself: verbs over nouns, actions over states, struggle over hope. . . . You have to make your peace with the chaos, but you cannot lie."[24]

Thus the continual theological superstition that imbues criticality with redemption or progress is, for me, a dead end. If the horrors of geopolitics are not sufficient to persuade you that there is no providential future for humanity, then no argument or evidence can. Instead, what we have is everything around us, and it is sufficiently creative and weird all by itself. It is also necessary to the task of undermining the petty provincialism that animates geopolitics and a narrow view of humanity. Certainly we can struggle to intervene in those arrangements that are disgusting to our sense of good. Any intervention that is not allied with the world, which is the condition of possibility of sensation and intervention in the first place, will likely fail all the more catastrophically. We can, I think, have a bias for struggle over nihilism, but ultimately realism, or the world, is the greatest enemy against the violence of common sense.

We Need Genre to Be Realists Because Reality Lacks Verisimilitude

Please do not mistake my love of ideas for an escapist retreat into idealism. Quite the contrary: I think the task of theorizing is to invent modes of experiencing the world, even if the route is a circuitous journey that does not lead from fiction to nonfiction but instead from truth to falsity. Fiction in our age of continuous-real-time-captured-by-iPhone news updates is so much more frequently true. The world is real but not easily apparent. There is a world as such but no way of encountering it that is not, as Stanley Cavell says, an interpretation. All encounters are a sensuous process of labor *with* the world and not before the world or after it. Therefore, the fight to see, think, and feel things as they are requires an affirmative sense of genre; CNN is a genre, security reports are a genre, terror alert levels are a genre, and Chomsky-esque truth-telling is a genre, although all of these we are inured to or primed for as a common *sense* of reality.[25] Sometimes we need wilder genres like horror or sci-fi or speculation so that we have the capability to see past what Rudy Rucker calls "consensus reality" into the weird worlds of brain implant experiments, detailed in chapter 6, that have been going on since the 1960s or the emerging freaks of science—explored in chapter 9—that could, if we pay attention, challenge our restrictive normative boundaries of the human.[26]

I take inspiration in Sayak Valencia's work on *gore* or *splatter cinema* as an analytic category for contemporary capitalism to expand the attention of empiricism to include the gore of the real world.[27] Like Valencia finds in the genre of gore, the practices of torture, disappearing, and spectacular violence that suture together the political economy of bodies in the border region between Mexico and the United States are no longer exceptional events but increasingly global practices. The choice of genre is not haphazard. Valencia further distinguishes the sadistic erotics of *snuff* from the specific necro-practices of gore, which produce spectacular forms of extra-state narco violence, the smooth flow of goods and labor necessary for globalization, and the persistence of state sovereign violence all in one stroke.[28] For Valencia, the genre of gore as opposed to other genres of horror and snuff captures these "processes of doubling" and invisibility that characterize the narco-state-capital-death-body machine.[29] Like horror and science fiction more generally, Valencia, like Rucker, describes the "irreal" character of social relations and their reproduction correspondingly requiring a contrarealist genre to make visible what is meant to be ignored or normalized.

Rucker and Valencia practice a kind of transrealism as an *art* form that "deal[s] with the world the way it actually is"[30] because mere description is insufficient to pierce the veil of consensus reality. The endurance of consensus reality as a genre of naïve realism is indebted to an aesthetic but also a corresponding anesthetic that foregrounds a "common sense" in place of an openness to experience of what has not previously been experienced.[31] Consensus or commonsense reality shields us from a world that would otherwise be too real, creating a feeling of the *irreal.* According to Rucker, as long as the evening news feels real, the consensus can continue despite unbelievable contradictions. This is a fact tested well beyond what I thought was darkly possible by the first year of the Donald Trump presidency. As a collective—what Félix Guattari called a machinic unconscious—we tune in and tune out simultaneously.[32] Valencia similarly highlights the degree to which gore capitalism can engage in labor practices and new forms of violence markedly dystopian by any public consensus of a moral life without somehow calling into question the state or globalization.[33] Even catastrophic material contradictions fail to create a legitimacy crisis, and frequently outright fictions mobilize whole nations. There is no better proof of this than the public consensus aided and abetted by thousands of scholars that the greatest threat to humanity is a handful of people called terrorists. Without these new genre-inspired tools for investigation, how else do you make sense of autonomous killer robots and the savage biopolitics of conquistadors, and equally find inspiration in Go-playing AI platforms and nearly annihilated cosmologies resurging against any "realistic" odds?

How do we get from horror to critique? Rucker recommends that we can "turn off the TV (or now ubiquitous internet), eat something, and go for a walk, with infinitely many thoughts and perceptions mingling with infinitely many inputs."[34] Furthermore, artists of all sorts, scholars included, can refuse to allow this "severely limited and reactionary mode condition all of our writing."[35] We can instead employ the tricks of other aesthetic genres and conceptual speculation to expand the sensory capabilities to see the world beyond consensus reality. In this sense, theory can be a kind of dark magic, a destroyer of worlds, an art of sensual experience. We can craft concepts like spells. We can conjure ideas from the virtual in hopes of altering the experience of reality. What comes after that is beyond our control.

To this end, what if the primary goal of studying global politics was not to explain things like laws, rules, and predictions but was rather to broaden how much of the world we could experience and be part of? What if international relations was an empiricism infused with what Cavell calls imagination, such that we can "take the facts in, realize the significance of what is going on, make the behavior real for [ourselves], make a connection"?[36] Cavell says this process of imagination is what Wittgenstein called "interpretation" or "seeing something as something."[37] The failure to see so many things and others as "something" is a plague of much greater significance than any research problem that can be saved by the next methodological breakthrough. And the "seeing something as something" problem is as equally unlikely to be solved by any scientific breakthrough, in the narrow sense. Instead we have to find tactics for making sense of "what is fantastic in our ordinary lives."[38]

Of Mood and Method: Pessimism, Failure, and International Relations

> Of course it is hard for us to think that we are becoming completely wretched! And yet . . .—GEORGES BATAILLE
>
> We're doomed.—EUGENE THACKER

Because I wanted this book to inspire curiosity beyond the boundaries of international relations (IR), I considered ignoring the field altogether, removing all mentions of IR or IR theory. However, upon closer reflection, I have decided to keep these references as I think they are relevant for those outside the discipline and for those who, like myself, often feel alienated within its disciplinary boundaries. In the former case, it is important to know that, unlike some more humble fields, IR has always held itself to be a kind of royal science. Scholarship in IR, particularly in the United States, is half research, and half biding time until you have the prince's ear. The hallowed names in the mainstream of the field are still known because they somehow changed the behavior of their intended clients—those being states, militaries, and international organizations. Therefore, some attention to IR is necessary because it has an all-too-casual relationship with institutional power that directly impacts the lives of real people, and IR is all too often lethal theory.[39] As an American discipline, the political economy of the field is impossible without Department of Defense money, and its semiotic economy would be equally dwarfed without contributory figures like Woodrow Wilson, Henry Kissinger, and Samuel Huntington. The ubiquity of Huntington's "clash of civilizations" thesis and Kissinger's particular brand of *realpolitik* are undeniable throughout the field, as well as the world.[40] Each, in their own way, has saturated the watchwords and nomenclature of geopolitics from an American perspective so thoroughly that both political parties in the United States fight over who gets to claim the heritage of each. Although many other fields such as anthropology and even comparative literature have found themselves in the gravitational pull of geopolitics, international relations is *meant* to be scholarship as statecraft by other means.[41] That is, IR was meant to improve the global order and ensure the place of its guarantor, the United States of America.[42] Having spent the better part of a decade listening to national security analysts and diplomats from the United States, South Korea, Japan, Europe, China, Brazil, and Russia, as well as military strategists around the planet, I found their vocabulary and worldview strikingly homogeneous.

If this seems too general a claim, one should take a peek at John Mearsheimer's essay "Benign Hegemony," which defends the Americanness of the IR field.[43] What is most telling in this essay is not a defense of the U.S. as a benign hegemonic power, which Mearsheimer has done at length elsewhere. Rather, it is his vigorous defense that as a field, IR theory has done well by the world in setting the intellectual agenda for global challenges, and for creating useful theoretical approaches to addressing those problems. For Mearsheimer, the proof that American scholarly hegemony has been benign is that there is nothing important that has been left out. A quick scan of the last ten or twenty International Studies Association conferences would suggest otherwise.

That issues like rape as a weapon of war, postcolonial violence, global racism, and climate change are not squarely in the main of IR demonstrates just how benign American scholarly hegemony is not. As one prominent anthropologist said to me at dinner after touring the ISA conference in 2014, "it was surreal, like a tour through the Cold War. People were giving papers and arguing as if nothing had ever changed." These same provincial scholars aspire and succeed at filling the advisory roles of each successive American presidency. One cannot help but see a connection between the history of the IR field, and the catastrophes of U.S. foreign policy during the twentieth and twenty-first centuries. One could repeat the words of the anthropologist I mentioned to describe the 2016 presidential campaign debates over the future of U.S. foreign policy: it is as if "nothing had ever changed." And yet these old white men still strut around the halls of America's "best" institutions as if they saved us from the Cold War, even as the planet crumbles under the weight of their failed imperial dreams.

If international relations was meant to be the science of making the world something other than what it would be if we were all left to our own worst devices, then it has failed monumentally. The United States is once again in fierce nuclear competition with Russia. We are no closer to any significant action on climate change. We have not met any of the Millennium Development Goals determined by the United Nations on eradicating poverty. War and security are the most significant financial, creative, social, cultural, technological, and political investments of almost every nation-state on Earth. The general intellect is a martial intellect.

Despite all this failure, pessimism does not exist in international relations, at least not on paper. The seething doom of our current predicament thrives at the conference bar and in hushed office conversations but not in our research. In public, the darkness disavowed possesses and inflames the petty cynicisms and hatreds that are often turned outward at tired and predictable scapegoats.

After the fury of three decades of critique, most IR scholars still camp out either on the hill of liberal internationalism or in the dark woods of political realism. Neither offers much that is new by way of answers or even explanations, and each dominant school has failed to account for our current apocalyptic condition. One is left wondering what it is exactly that they think they do. Despite the seeming opposition between the two, one idealistic about the future of international order (liberals) and the other self-satisfied with the tragedy of cycles of war and dominance (realists), both positions are optimists of the positivist variety.

For both warring parties, IR optimism is expressed through a romantic empiricism. For all those who toil away looking for the next theory of international politics, order is out there somewhere, and dutifully recording reality will find it—or at least bring us closer to its discovery. For liberal internationalism, this will bring the long-heralded maturity of Immanuel Kant's perpetual peace. For second-order sociopaths known as offensive realists, crumbs of "useful strategic insight" and the endless details that amplify their epistemophilia for force projection and violence capability represent a potential "advantage," that is, the possibility to move one step forward on the global political board game of snakes and ladders. Still, the cynicism of IR always creeps back in because the world never quite lives up to the empirical findings it is commanded to obey. Disappointment here is not without reason, but we cynically continue to make the same policy recommendations, catastrophe after catastrophe.

I have an idea about where IR's recent malaise comes from. I think it is a moment, just before the awareness of the Anthropocene, after the Cold War and before September 11, when the end of everything was only a hypothetical problem for those of a certain coddled and privileged modern form of life. The catastrophe of the human predicament was that there was no catastrophe, no reason, no generation-defining challenge or war. Now the fate of this form of life is actually imperiled, and it is too much to bear. The weird denial of sexism, racism, climate change, the sixth extinction, and loose nukes, all by a field of scholars tasked with studying geopolitics, is more than irrationalism or ignorance.[44] This animosity toward reality is a deep and corrosive nihilism, a denial of the world. Thus IR as a strategic field is demonstrative of a civilization with nothing left to do, nothing left to destroy. All that is left is to make meaning out of being incapable of undoing the world that Euro-American geopolitics created. Emo geopolitics is not pretty, but it is real. The letdown, the failure, the apocalypse-that-was-not finally arrived, and we are too late.

Still, the United States of America continues to follow the advice of "the best and the brightest," testing the imperial waters, not quite ready to commit out loud to empire but completely unwilling to abandon it. Stuck in between, contemporary geopolitics—as curated by the United States—is in a permanent beta phase. Neuro-torture, algorithmic warfare, drone strikes, and cybernetic nation-building are not means or ends but rather are tests. Can a polis be engineered? Can the human operating system be reformatted? Can violence be modulated until legally invisible while all the more lethal? Each incursion, each new actor or actant, and new terrains from brains to transatlantic cables—all find themselves part of a grand experiment to see if a benign or at least sustainable empire is possible. There is no seeming regard for the fact that each experiment directly competes with Thomas Jefferson's democratic experiment. One wonders if freedom can even exist anywhere other than temporarily on the fringe of some neglected order. Is this some metaphysical condition of freedom, or is the world so supersaturated with martial orders that the ragged edges between imperial orders are all that we have left? It feels like freedom's remains persist only in the ruins of everything else. No space is left that can be truly indifferent to the law, security, or economy. Such is the new life of a human in debt. The social contract has been refinanced as what is owed and nothing more: politics without equity. Inequity without equality.

What about the impending collapse of the post–World War II order, the self-destruction of the United States, the rise of China and a new world order? If humanity lasts long enough for China to put its stamp on the human apocalypse, I will write a new introduction. Until then, we live in the death rattle of *Pax Americana.* While I think the totality of this claim is true, I do not want to rule out that many of us throughout the world still make lives otherwise. Many of us even thrive in spite of it all. And yet, no form of life can be made that escapes the fact that everything can come to a sudden and arbitrary end thanks to the whim of an American drone operator, nuclear catastrophe, or macroeconomic manipulation like sanctions. There are other ways to die and other organized forms of killing outside the control of the United States; however, no other single apparatus can make everyone or anyone die irrespective of citizenship or geographic location. For me, this is the most inescapable philosophical provocation of our moment in time.

The haphazard and seemingly limitless nature of U.S. violence means that even the core principles of the great political realist concepts like order and national interest are being displaced by subterranean violence entrepreneurs that populate transversal battlefields, security corridors, and border zones.[45] Mercenaries, drug lords, chief executive officers, presidents, and sports commissioners are more alike than ever.[46] Doomsayers like Paul Virilio, Lewis Mumford, and Martin Heidegger foretold a kind of terminal and self-annihilating velocity for geopolitics' technological saturation, but even their lack of imagination appears optimistic. American geopolitics does not know totality or finality; it bleeds, mutates, and reforms. Furthermore, the peril of biopolitics seems now almost romantic. To make life live? Perchance to dream. The care and concern for life's productivity is increasingly subsumed by plasticity—forming and reforming without regard to the telos of productivity, division, or normative order.

There are, of course, still orders in our geoplastic age, but they are almost unrecognizable as such. When so many citizens and states are directly invested in sabotaging publicly stated strategic ends, then concepts like national interest seem equally quaint. We are witnessing creative and horrifying experiments in the affirmative production of dying, which also deprive those targeted and in some cases whole populations from the relief of death. To follow Rucker, I want to try to see the world for what it is. We can only say that tragedy is no longer a genre of geopolitics. Tragedy redeems. The occluded character of contemporary geopolitics shoehorned into experience produces the feeling that there is no relief, no reason, no victory, no defeats, and no exit within the confines of national security's constricted world. This is not tragedy: it is horror. We live in an age of horror that, like the victims of gore movies who never quite die so that they can be tortured more, furthers our practice of collective violence and goes on for decades as a kind of sustainable warfare.

A Different Pitch of Failure

Why would I bother with the "night side" of IR theory?[47] In part, I wish to move away from the rationalist fallacy among both defenders and critics of empire. There is a shared belief in the strategic competence of nations like the United States. Even those most vocally critical often see in the covert operations and vast military occupations a kind of purpose or conspiracy. The debate about empire then becomes about its moral virtue rather than the factual question of the strategic competence of imperial states. However, the lives of millions annihilated in Iraq, Yemen, Afghanistan, and now increasingly throughout the continent of Africa do not reflect an amoral strategic competence. The mass murder in pursuit of the war on terrorism and its vision of nation-building is the result of lethal stupidity.[48] In some sense, the investigative journalism of Jeremy Scahill and Glen Greenwald attributes too much reason and order to the catastrophic floundering of the American empire.[49] To see even a dark vision of order in the last thirty years of U.S. policy is itself a form of optimism. No one is in control, there is no conspiracy, and yet the killing continues. A pessimistic reading of U.S. empire and the geopolitical history that precedes it is neither tragedy nor farce. It is a catastrophic banality lacking in any and all history, a pile of nonevents so suffocating that we often hope for a conspiracy, punctuating event, or villain worthy of the scale of violence.[50] For those of us who continually rewatch the reruns of *The Walking Dead* and *Jericho* on our laptops in bed, we are waiting for relief in our privileged but increasingly fragile bubble. I know I am not the only one who finds respite from the weight of politics' "cruel optimism" by watching fantasies of cruel pessimism. A pessimistic understanding of global politics helps explain how we could come to a place where there is a sense of relief in watching everything come to an end.[51]

Failed IR affirms the power of this kind of negative thinking as an alternative to the endless rehearsing of moralizing insights and strategic foresight. The negative is not "against" or reacting to something. Rather, it is the affirmation of a freedom beyond the limits of life and death. That is, it is making a life by continuing to think about the world, even if that thinking is not recuperative, and even if nothing we think can save us. In the face of it all, one celebrates useless thinking, useless scholarship, and useless forms of life at the very moment we are told to throw them all under the bus in the name of survival at all costs. This is a logic referred to lately as hope and it is as cruel as it is anxiety inducing. Hope is a form of extortion. We are told that it is our obligation to bear the weight of making things better while being chided that the failure of our efforts is the result of not believing in the possibility of real change. In such an environment, pessimism is often treated as a form of treason, as if only neoliberals and moral degenerates give up—or so goes the op-ed's insisting upon the renewed possibility of redemption.

In response to these exhortations, pessimism offers a historical atheism, both methodologically and morally. The universe does not bend toward justice. Sometimes the universe bends toward the indifference of gravity wells and black holes. Affirming negativity, inspired by Achille Mbembe, is grounds for freedom, even if that freedom or relief is only fleeting and always insecure. I am not arrogant enough to think a book can attain freedom of this sort, but this book is inspired by refusals of critique as redemption in favor of useless critique and critique for its own sake.

That the pursuit of knowledge without immediate application is so thoroughly useless, even profane, is a diagnosis of our current moment. The neoliberal assault on the university is evidence of this condition, as is the current pitch of American politics. Our indifference as intellectuals to maximizing value has not gone unnoticed. We are still dangerous, worthy of vilification, of attack, sabotage, and derision because we fail so decadently. We are parasites according to Scott Walker, Donald Trump, and the rest. So be it. We are and shall remain irascible irritants to a worldwide assault on thinking that is well underway and facing few obstacles in other jurisdictions.

What would failed scholarship do? Learn to die, learn to live, learn to listen, learn to be together, and learn to be generous. These virtues are useless in that they do not prevent or manage things. They do not translate into learning objectives or metrics. Virtues of this order are selfsame, nontransferable experiences. They are meaningful but not useful. These are luxurious virtues. Like grieving or joy, they are ends unto themselves. But how will these ideas seek extramural grants, contribute to an outcomes-based education system, or become a policy recommendation? They will not, and that is part of their virtue.

Even if there is no straight line to where we are and where we ought to be, I think we should get over the idea that somehow the U.S. project of liberal empire is conflicted, or "more right than it is wrong," or pragmatically preferable to the alternatives. I hope this book can contribute to the urgent necessity to get out of the way by reveling in the catastrophic failure that should inspire humility but instead seems to embolden too many to seek global control yet again. Demolition may be an affirmative act if it means insurgents and others can be better heard. And yet this may fail too. If we can accomplish nothing at all, we can at least, as Ta-Nehisi Coates and other pessimists have said, refuse to suborn the lie of America any longer. Telling the truth, even if it cannot change the outcome of history, is a certain kind of solace. In Coates's words, there is a kind of rapture "when you can no longer be lied to, when you have rejected the dream."[52] Saying the truth out loud brings with it the relief that we are not crazy. Things really are as bad as we think.

If there are those of us who want to break from this one-hundred-year-old race to be the next Henry Kissinger, then why do we continue to seek respect in the form of recognizable standards of excellence? I am not sure where the answer finally lies, but I do know that professionalization will not save us. To appear as normal and recognizably rigorous will not be enough to stave off the neoliberal drive to monetize scholarship, or to demand of us strategically useful insights. The least we can do in the face of such a battle is to find comfort in meaningful ideas and the friendships they build rather than try to perform for those we know are the problem. Some will ask, who is this "we" or is that "they"—where is your evidence? More will know exactly what I am talking about.

The virtues I seek are oriented toward an academy of refuge, a place we can still live, no matter how dire the conditions of the university and the classroom. It is not the think tank, boardroom, or command center. We are, those of us who wish to be included, the last of the philosophers, the last of the lovers of knowledge, the deviants who should revel in what Harney and Moten have called *the undercommons*.[53]

In one of his final lectures, Bataille speaks of the remnants of a different human species, something not quite so doomed, something that wasted its newly discovered consciousness and tool-being on the art that still marks the walls of prehistoric caves.[54] This lingering minor or vestigial heritage is philosophy's beginning. Philosophy survives war, atrocity, famine, and crusades. Thinking matters in a very unusual way. Thinking is not power or emancipation. Thinking matters for a sense of belonging to the world, and for believing in the fecundity of the world despite evidence to the contrary.

How do you get all this from pessimism, from failure? Because willing failure is a temptation, a lure to think otherwise, to think dangerous thoughts. Pessimism is a threat to indifferentism and nihilism in the sense of the phenomenon of Donald Trump. Pessimism is a provocation and an enemy of skepticism, particularly of the metaphysical variety. It is not redemption from these afflictions, but in pessimism there is solace in the real. To put it another way, to study the world as it is means to care for it.

The exhortation that our care or interest should be contingent on how useful the world is and how much of it conforms to our designs is as much opposed to care as it is to empiricism. We can study airports, poetry, endurance races, borders, bombs, plastic, and warfare, and find them all in the world. To consider the depth of their existence can be an invitation to the world rather than a prelude to another policy report. One cannot make a successful political career out of such pursuits, but you might be able to make a life out of it, a life worth repeating even if nothing else happens.

At the end of Jack Halberstam's *The Queer Art of Failure*, we are presented with the Fantastic Mr. Fox's toast as an exemple of something meaningful in these dark times of ours.

> They say all foxes are slightly allergic to linoleum, but it's cool to the paw—try it. They say my tail needs to be dry cleaned twice a month, but now it's fully detachable—see? They say our tree may never grow back, but one day, something will. Yes, these crackles are made of synthetic goose and these giblets come from artificial squab and even these apples look fake—but at least they've got stars on them. I guess my point is, we'll eat tonight, and we'll eat together. And even in this not particularly flattering light, you are without a doubt the five and a half most wonderful wild animals I've ever met in my life. So let's raise our boxes—to our survival.

Halberstam says of this queer moment:

> Not quite a credo, something short of a toast, a little less than a speech, but Mr. Fox gives here one of the best and most moving—both emotionally and in stop-motion terms—addresses in the history of cinema. Unlike *Coraline*, where survival is predicated upon a rejection of the theatrical, the queer, and the improvised, and like *Where the Wild Things Are*, where the disappointment of deliverance must be leavened with the pragmatism of possibility, *Fantastic Mr. Fox* is a queerly animated classic in that it teaches us, as *Finding Nemo*, *Chicken Run*, and so many other revolting animations before it, to believe in detachable tails, fake apples, eating together, adapting to the lighting, risk, sissy sons, and the sheer importance of survival for all those wild souls that the farmers, the teachers, the preachers, and the politicians would like to bury alive.[55]

Although not as much fun as Halberstam's monument to low theory, *Savage Ecology* is for all the other wild animals out there studying global politics. May we be buried alive together.

00. APHORISMS FOR A NEW REALISM

> In this sense of belief in the Devil: that not everything
> that comes to us as an inspiration comes from what is good?
> —LUDWIG WITTGENSTEIN, *Culture and Value*

Inhumanity

When we see the inhuman in something (a great white shark), it is only because we cannot help but feel a connection, a connection we are apt to call human. A good example of the horror of inhumanity is the narrative of dehumanization or othering, or the idea that there is some practice or historical tendency or character difference in certain broken humans—inhumans—which is responsible for genocide, abuse, violence, and so forth. In the fourth section of *The Claim of Reason*, Stanley Cavell argues that this explanation is in fact not accurate. We are able to commit genocide and so on without any othering, and those discourses of dehumanization that follow are in some sense coping mechanisms or opportunities for bad faith to ignore what many people were capable of to begin with. Rather than the zombie scenario whereby a deficit, disease, or supernatural event diminishes our humanity, true horror is that no deficit was needed in the first place. Or maybe to think of it differently, that deficit has always been there. Genocide is fully human, no deficit. Moral tragedy would be seeing the refugees lying on the beach and being unable to do anything to act, being too late. Moral failure would be seeing the bodies and not being able to recognize them as "like us" and therefore being uncompelled to act. Moral horror, the horror of the inhuman as human, is that we could have done something, we did recognize them as like us, and did nothing anyway. We live in a horrifying world, not a tragic one. Dehumanization is a lullaby we sing to each other rather than face the horror that the suffering of others fails to awaken anything inside of us.

Monstrosity

The transition to something else is almost always monstrous precisely because of its weirdness compared with what it is we think we are. How do we think through horror as a genre of political thought rather than as a cinematic or fictional construction? What is the real of horror rather than the Lacanian cop-out of the horror of the real? Government programs like torture and interrogation exceed the normal feedbacks and boundaries of our consensus reality. This is not a metaphor. The gap between reality and perception of that reality is a phenomenological difference that makes the appreciation and observation of events *surreal.* They can only be understood as other genres, other genres of reality. Sci-fi and horror are conceptual and phenomenological necessities rather than "representational styles" or modes of writing. Unlike the often violent flaying of the personality from the prepersonal body found in neuro-torture techniques, exploring the genres of reality holds on to the personality of the artist or thinker. A little personality, a little lingering self, may be sufficient to the task of dilating the modes of perception. Cultivating this tactic we can let in a little of the real rather than being torn asunder.

Taking Liberties

It turns out that freedom as liberty is an ecological doomsday device. The Enlightenment, for all of its self-congratulatory bravado, may in fact end the species. In the realm of downsides, that is a pretty big one. That the freedom experiment is turning out to be a catastrophic failure ought to demand of us something quite dramatic in the revaluation of humanities, economics, politics, and the most basic conceptions of the good. Still waiting.

The Epidemic of Things

Horror's pure virtuality gives rise to anxiety as the permanent state of anticipation. The intensive level of stress created by horror is not relative to horror as its pure virtuality—the always present possibility does not change. Rather, the periodicity and semiotics of its actualization changes how present the virtuality is to our sense of the ordinary or everyday. Frequency then is not quantitative or extensive but an intensive quality. Each successive attack, mass shooting, or explosion is not "one more" but something else entirely. Anxiety is a nonlinear pressure system. Every event, every actualization of horror, leaves us singularly different from the one before. Contrary to Barack Obama's statement after the 2016 Oregon shooting that tragedy was becoming common, these events do not become normal; they change the multiplicity of the normal. What is thinkable as an actualization of horror changes. A soda can blows up a Russian plane en route to Egypt; a street erupts as if itself a bomb; the food we eat, organic, fresh, green, a habitat for lethal *E. coli*; a clever "pod" for laundry detergent kills a child with its concentrated "cleaner" erupting inside . . . And on and on . . . We live in a world of objects, places, and even markers of religious and racial difference that can never return to what they were before.

Nonknowledge

Disbelief is not always the result of a lack of evidence; it is often that the empirical exceeds our capacity for cognition. Horror is such an example. Global nuclear war, for instance, as an event is in the category of what Georges Bataille called nonknowledge. Like the sun before humans existed, global nuclear war is unknowable in that it erases the capability to be witnessed by the condition of possibility for knowledge, that is, the human observer. And yet we cannot help but speculate on the things we cannot know but must think. The horror of realism is a genre or technique to know nonknowledge obliquely.

A New Danger?

Cybernetics—the algorithmic age—is not dangerous the way Fascism is dangerous or Christian Fundamentalism is dangerous. Cybernetics is dangerous the way that Enrico Fermi's self-sustaining fission reaction is dangerous. It is dangerous because it is true and works. And like the atom bomb, we may come to wish we could forget its truth, uninvent it. The horror of cybernetics is that its truth permits the possible future in which we are no longer capable of regretting its truth. We may come to no longer know we ought to regret its coming. This is the problem of what Vilem Flusser calls programming.

Fragility

Fragility is tenuous, negotiable, and generative because it resonates with horror and persistence—radical otherness (nothingness) and the here and now. This is, at this moment in species evolution, exactly where we are. This is a world in which what we are becoming is squarely in the political. What is the politics of the new *dispositifs* of becoming, AI, synthetic biology, Fundamentalism . . . What else could we be?

PART I. THE GREAT HOMOGENIZATION

1. THE ANTHROPOCENE AS A GEOPOLITICAL FACT

In the concluding volume of his *Spheres* trilogy, Peter Sloterdijk says that what marks our current epoch as distinctive is three things: terrorism, product design, and what he calls the "environmental idea."[1] According to Sloterdijk, things are not as I was told in my freshman philosophy class in the 1990s; in fact, "the era of grand narratives" is not over.[2] He reminds us that at the beginning of the twenty-first century, while philosophy might be finished with the grand narrative of modernity, science, geopolitics, and war certainly are not. In fact, for Sloterdijk these sectors are, by the end of World War II, nearly synonymous, and continue to define the ways in which a notion of the planetary has come to define the character of political action. The twentieth century was a Eurocentric project to finish the conversion of places and nature into a kind of dedifferentiated user space. Like the science of ergonomics, modernity was an effort to make things, from continents to seat belts, fit together for the ease of mobility and instrumentality.[3] Global-scale product design, whether canals, nation-building, the nascent weather modification projects now called geoengineering, as well as practices like sustainable development and ur-sciences like cybernetics, require the flattening out and regularization of unruly natures and spaces such that things can be frictionless and useful: everything in its right place, and everything with a name and function. This project of the twentieth century is what he calls explication—a kind of vivisection of ideas and things such that the world could be flayed alive and reconsolidated as a planetary system.[4] The "environmental idea" that emerges from the drive to explication is not the Thoreauvian walk up Katahdin Mountain but is the development of regimes of knowledge necessary to understand how to deprive things of life, from bed bugs to humans. Gas attacks and the rise of aerial bombs in World War I, and the leveling of cities, gas chambers, and atom bombs of World War II, are, for Sloterdijk, the industrialization of the environmental idea. It is a form of war he calls terrorism: "Terrorism suspends the distinction between violence against persons and violence against things from the environmental—it is violence against those human-surrounding ‘things’ without which persons cannot remain persons.”[5]

In the aftermath of two world wars, the environmental project continues, *and* it continues to be martial. Even when combat is not the modus operandi, explication and annihilation are.[6] Despite the excitement by some that wars are coming to an end, and that the global ecological crisis may unite us, the advances of science and the understanding of the ecological crisis that comes to define the transition from the twentieth to the twenty-first century continues to be atmospheric in its means and terrorizing ends. According to Sloterdijk, “air theory and climate technology are not mere sediments of war and post-war knowledge, and *eo ipso* first object of a science of piece that could only arise in the war stress shadow; more than that, they are primarily post-terrorist forms of knowledge.”[7]

Rather than see the attempts to build a global alliance against climate change as a break from twentieth-century geopolitics, I share Sloterdijk’s view that the condition of possibility for climate change as a problem, as well as the attendant suite of political and technological solutions, is consonant with the terrorism of modernity. The hope that global warming could provide a universal ground for the cosmopolitan solidarity as-yet unachieved by other means is dangerously naïve and already often coopted for cynical ends. It would, of course, be equally naïve and dangerous to deny that there is an ecological catastrophe now affecting every region of the planet. However, the danger is the geopolitics of explication and operationalization, or what I am calling homogenization, of which carbon dioxide is just one particularly devastating effect. Therefore, the crisis of what is being called the Anthropocene is intimate with the concept itself. There is a feedback between global thinking, global expansion, and global destruction.

Three Cheers for the Anthropocene

It is worth considering how Paul Crutzen, the progenitor of the Anthropocene as a popular concept, follows Sloterdijk’s vector of martial thinking. Crutzen’s career as an atmospheric chemist has been, since its beginnings, connected to a cosmopolitical vision of global crisis. Before popularizing the term *Anthropocene*, he won the Nobel Prize for work on the significance of the ozone layer as a necessary precondition for human life, as well as emphasizing the significance of global regulations on Freon as a threat to the fragile screen between us, and the sterilizing effect of the sun’s ultraviolet light. Crutzen’s whole career has followed a line of research substantiating the impact of human activity on the Earth system, in particular the breathable layer of that system known as the atmosphere. Alongside work on the ozone layer and the warming effects of high concentrations of carbon dioxide, Crutzen is considered one of the foremost authorities on models that project the environmental effects of nuclear warfare. Further, in both the ozone study and subsequent work on the effects of carbon concentration on the atmosphere, nuclear war figures prominently. Of the four possible threats to the ozone layer identified by Crutzen in his first published article on ozone depletion, which mentions high-altitude planes, chlorofluorocarbon (CFC) foam, and the global production of nitrous oxide, the atmospheric detonations of nuclear weapons resulting in the destruction of one great power is cited as the most significant threat.[8] According to Crutzen, such an attack could destroy as much as 50 percent of the ozone layer as compared to the other threats, which only range from 4 to 12 percent.

Moving from the implicit to the explicit, in 1982 Crutzen and John W. Birks published their haunting article "The Atmosphere after a Nuclear War: Twilight at Noon," where the terror facing the planet was not global warming but what is often called nuclear winter. According to Crutzen and Birks, the burning of forests and cities would block out the sun, destroying enough agriculture and vegetation to threaten the human species, as well as causing cascades of death throughout the larger global web of life. Following the winter, after the dissipation of the reflective postnuclear smog, the flood of unfiltered ultraviolet (UV) radiation would further threaten the possibility of life on the planet, particularly in the Northern Hemisphere, which would be hit hardest by both the nuclear winter and ozone depletion. The article ends with a kind of cautionary assessment of the veracity of the models underpinning Crutzen and Birks's argument. According to the article, the complexity and interdependencies at work in modeling Earth's whole atmosphere make accurate predictions difficult at best. However, the descriptions of weeks of darkness, mass starvation, and later death by solar radiation leave little doubt that we should err on the side of caution.

What is striking in looking back on Crutzen's career is the degree to which his first "Anthropocene" was one in which power politics would alter the geological record of the planet. Furthermore, the risk of human extinction came from the unpredictable consequences of cooling the earth and the chaos of decreasing as well as increasing solar radiation. The greatest threats to the planet for the first thirty years of Crutzen's career were the ways that Cold War bipolar competition—geopolitics—might disrupt or even destroy the cycles of planetary life. However, these same insights and models, as well as a particular way of thinking about humans and the geological record, came to form the basis of Crutzen's now outspoken advocacy for intentionally cooling the planet through geoengineering. Crutzen spent the first two-thirds of his career trying to prevent nuclear winter and the last third trying to figure out how to replicate nuclear winter's effects in a way that could be survived by most people.

Rather than see these two career trajectories as opposed, I think Crutzen's thinking displays a continuous concern for the Northern Hemisphere and a particular cartography, rather than a geography, of human survival.[9] Crutzen, as well as the concept of the Anthropocene itself, cannot escape preceding geopolitical conceptions of the Earth. Crutzen and others who rush so quickly to the necessity to transition efforts from climate abatement to climate modification are unsurprisingly *not* moved by claims that artificial cooling will likely cause droughts and famines in the tropics and subtropical zones of the global south; nor are they moved by how such plans may accelerate ocean acidification.[10] The utilitarian risk calculus that favors the greatest good for the greatest number has no geographical or historical sensibility of how unequally aggregate conceptions of the good are distributed around the planet.

Global thinking, even in its scientific and seemingly universalist claims to an atmosphere that "we" all share, belies the geopolitics that enlivens scientific concern, as well as the global public policy agenda of geoengineering that seeks to act on behalf of it. Saving humanity as an aggregate, whether from nuclear war, Styrofoam, or climate turbulence, has never meant an egalitarian distribution of survivors and sacrifices. Instead, our new cosmopolitanism—the global environment—follows almost exactly the drawn lines, that is, the cartography or racialized and selective solidarities and zones of indifference that characterize economic development, the selective application of combat, and, before that, the zones of settlement and colonization. More than a result of contemporary white supremacy or lingering white privilege, the territorialization of who lives and who dies, who matters and who must be left behind for the sake of humanity, represents a five-hundred-year geopolitical tradition of conquest, colonization, extraction, and the martial forms of life that made them all possible through war and through more subtle and languid forms of organized killing.

I am not suggesting that Crutzen and others are part of a vast conspiracy; rather, I want to outline how climate change, species loss, slavery, the elimination of native peoples, and the globalization of extractive capitalism are all part of the same global ordering. That is, all of these crises are geopolitical. The particular geopolitical arrangement of what others have called the *longue durée*, and what I am calling the Eurocene, is geologically significant but is not universally part of "human activity" despite the false syllogism at the heart of popular ecological thinking that a global threat to humanity must be shared in cause and crisis by *all* of humanity.[11]

Departing from Sloterdijk, I am hesitant to so easily locate modernity or explication as the root or cause of the global catastrophe. No single strategy, war, act of colonization, technological breakthrough, or worldview fully explains the apocalypse before us. However, there is something like what Gilles Deleuze and Félix Guattari call a refrain that holds the vast assemblage together, a geopolitical melody hummed along with the global expansion of a form of life characterized by homogenization rather than diversification. Accordingly, if we are to make some sense of such a vast world that is, even for Crutzen and Birks, "quite complex and difficult to model," I think we must consider the particular refrain of geopolitics that is capable of, by scientific as well as more humbly embodied standards, destroying worlds along with *the* world.[12] To eschew geopolitics simply because, as a refrain, it is too big, too grand, or too universal would ignore the conditions of possibility for nuclear weapons, power politics, and carbon-based globalization, and would greatly impoverish the explanatory capability of even the best climate models. So maybe it is not so strange that Crutzen and others' attention to the nuclear threat of great powers has all but disappeared despite the fact that Russia and the United States still possess thousands of nuclear weapons, and as of late have been all too vocal about using them. Instead, the Anthropocene, as envisioned by Crutzen as a universal concern, requires with it a depoliticization of the causes of that concern.

Therefore, Crutzen's fascination with nuclear winter is geopolitical not because it is about nuclear weapons—although that does not hurt. Rather, Crutzen's attention to nuclear winter is geopolitical because it is an image of the Earth system as a system with particular beneficiaries animating that interest. Sloterdijk's diagnosis of what I am terming the Eurocene, or the space of what he calls European "earth-users," is present in the very cybernetic understanding of the planet as a spatial and substantive whole.[13] In the cases of both nuclear winter and climate change, the atmosphere is a model, or more accurately, the last model. The whole Earth becomes a single integer in a larger set of planet systems rather than a set of habitats, zones, or locales. The Earth is merely another system isomorphic as a unit of analysis with Mars or the exoplanet TRAPPIST-1f. The shift in scale from place to the planetary is much more than a pulling back from the ground upward. The integrated Earth as the representation of a system *and* as an actual material system is aided by a process of integration, proceeded by a few hundred years of Sloterdijk's conception of explication where each part of each environment is disaggregated, described, and then reassembled to explain the whole. The process of integration is not merely a metaphoric or metaphysical geopolitics. It is the condition of possibility to understand the planetary as being political, as well as the condition of possibility for its charting as an economic and military cartography. Unlike the *weltanschauung* of Heidegger's world image, the planetary "user space" requires five hundred years of conquest, fossil fuel extraction and exploitation, settlement, hundreds of expert fields from geography to chemistry to ecology, and the normative consolidation of cosmopolitanism as a right to the freedom of movement at least for those capable of the feat.[14] The worldview or world image alone is a necessary but insufficient cause. The practices that habituated, expanded, and intensified that worldview are what is critical to its emergence. In this sense, the Anthropocene, like Crutzen's award-winning models of climate change and nuclear winter, is much more than an explanatory model. These models are the outcome of five centuries of integration and homogenization such that the infrastructure capable of making the Earth as a system knowable could be built, and the circulation of knowledge and data could be amassed to even make the diagnosis of a geological epoch in the first place.[15]

Properly accounting for the origins of our ecological crisis is vital. No political project oriented toward the many possible futures stretching out before us can consider the questions of ecology and justice on a global, much less geological, scale unless we first take on the unfortunate historical generality of the Anthropocene. The continuing project of Europeanization, now led by U.S. imperial power (although perhaps not for much longer), is central to how the planet got to this point. Understanding this is essential for how any "we" worthy of the plurality of the planet can invent something less nasty and brutish than what currently counts as global order. A consideration of the Eurocene, a geological history and name that foregrounds the geopolitical confrontation that stands in the way of any such future, is required in order to take the scale of our predicament seriously, while also confronting the power politics that made that scale possible.

What Is in a Name?

The argument for renaming the last five hundred years of the Holocene is based on two claims. The first is that there is significant material evidence of human-induced change to the climate system on a global scale. The second is that renaming the Holocene is essential to raising awareness that climate change and environmental change are more generally anthropogenic. Accuracy and consciousness raising are the twin urges for renaming. On both counts, we should reconsider what we mean by human if we want to call this the Anthropocene.

First, the "human" footprint is much more complex than just CO_2. We should do more than acknowledge the vast debates over the various contributions to the geological record, and at the very least consider that on a geological timescale, CO_2 concentration is relatively dwarfed by radioactivity in its uniqueness. It is comparable to the modern waste product par excellence—plastic—not to mention the layers upon layers of human-made objects of all sorts of other materials.[16] Furthermore, if the claim is that the Anthropocene is meant to name the scale of human effects on the planet, it should include the ability to warm and cool the Earth, as the project of Europeanization has done both at remarkable levels of intensity.

Beginning in 1610, a mini ice age took hold of the planet. The explanation for this, although debated, is that some 20 million people killed by the European invasion of the Americas resulted in vast reforestation of the North and South American continents.[17] The providence that conquistadors spoke of was not the blessing of God but syphilis, influenza, and a number of other nonhuman animal species that went along for the ride. The first waves of death were in some sense without malice; even if the conquistadors had been "friendly," they still would have been contagious.[18] However, the well-armed explorers and settlers that leveraged the apocalypse for their own gain leave no doubt about whether the genocide and terraforming of the Americas was European in cause and intent. There is no way to know how many languages, cities, ideas, cosmologies, and ways of inhabiting the world were lost during these first waves of mass death.[19] Yet we can observe their material absence as a trace in the Orbis spike, or period of cooling, that took off after 1610 when a wilder arboreal nature took back what had been inhabited land.[20] This was a register of the altered sociotechnical order that followed Euro-bacterial imperialism.

However, rather than see the condition of the Eurocene as a problem of encounter or first contact gone awry, it would be more accurate to mark it as a transformation in what constituted European conquest, the emergence of a particular pathway of modernity. The tragedy of the Americas was not inevitable. After all, there was no real gap in human relations or species difference that could support a before and an after contact. Waves of explorers have been arriving in the Americas for 130,000 years and even European Vikings managed to come and go without laying waste to the continents.[21] Global trade and exploration around the planet has been a continuous practice extended well beyond recorded history. What makes the Eurocene different from earlier moments of encounter—the geopolitics that characterize our epoch—is what Sloterdijk calls the making of "operational space." For Sloterdijk, the change after 1492 was first and foremost an "operativistic revolution," by which he means that imperial expansion by Europe and subsequent colonization was "an opening of extended operational space."[22] The particular mechanistic worldview of European conquest combined with the capital-resource feedbacks of European political economy flattened places into dedifferentiated spaces. The planet became the map rather than the earlier ecological endeavors to map the diversity of the planet.

The history of nuclear weapons is more recent but no less demonstrative of a geological footprint. There is now a distinctive radioactive glow in the layers of earth since July 16, 1945. The bombings of the civilian populations in Hiroshima and Nagasaki are not the end of that story. In the years that followed, more than two thousand nuclear weapons have been tested. Of those weapons, 97.5 percent have been detonated by European powers. Akin to Sloterdijk's martial reading of scientific practice, detonations do not appear as tests from the perspective of the Marshallese, Western Shoshone, or the thousands of "downwinders" who experienced the aftermath of radioactivity carried by the shifting air pressures of the atmosphere. A sixty-year nuclear war in the form of nuclear testing has spread cancer, incinerated sacred lands, and made other space uninhabitable on a temporal scale several orders of magnitude more significant than the ten-thousand-year lifespan of atmospheric CO_2.[23] And what does the future hold? The nuclear powers of the Eurocene—United States, Russia, United Kingdom, France, and Israel—still maintain 97 percent of the 15,913 nuclear weapons on alert around the planet. Self-annihilation is still a very real possibility despite what would be inferred by the beleaguered state of the arms control agenda.

As for plastic, the Texas-sized trash gyres that swirl in the world's oceans are another reminder of what a cosmology of disposability and synthetic chemistry has wrought. Plastic may not have quite the longevity of CO_2 and irradiated earth, but for hundreds, maybe thousands, of years, it will continue to circulate, wreaking havoc throughout the food chain. It is hard to imagine the world that now squeezes the last few cents out of the poor with single-serving plastic shampoo pouches and bottled water that is needed, because nearby lakes and aquifers have been sold to Coca-Cola, without the accompanying post–World War II European project of development that followed the land grabs of the twentieth century.

With the survival of the human race at stake, what difference can a concern for these other causes, much less its name, possibly serve? A recounting of the distinctively European history of our geological era is much more than polemical. The Eurocene names a practical problem not captured by the Anthropocene. Eurocentrism, more than a worldview, is a five-hundred-year project of violent terraforming and atmospheric engineering.

This is why I am not merely interested in the explanatory power of a better model. Instead, I am after the politicalization of geology, or the linking of ecological catastrophe with the power politics of the Euro-American global order. I think it is becoming increasingly evident that contemporary debates on how to "save the planet" are still infected by the geopolitical attachments to power and privilege responsible for the crisis. It is all too convenient that the demands of scientists and others to forgo international cooperation or large-scale reductions in industrial ways of life in favor of unilateral climate modification amount to saying "only a hegemon can save us now." I hope instead to show how that geopolitical order that makes hegemony possible is at the very core of the crisis. Despite efforts to distance discussions of politics and the environment from one another in favor of a highly functionalist, low politics of cooperation, we cannot escape the fact that ecology is historical and history is geopolitical.

Yet the double movement of politicizing ecology and historicizing geopolitics is insufficient, as it can become too easily and too narrowly anthropocentric—"we" make nature and therefore "we" make history. Instead, "we" humans, as defined as late moderns, are in desperate need of an ecological approach to geopolitics as well. By ecological, I mean a form of analysis characterized by multispecies encounters and deep relational processes across geographical scales rather than a form of political thinking that relies on discreteness, causality, and human agency. Hence an ecological approach does not center principally on the environment, what in international relations is called environmental security; nor does it limit global politics to states, international organizations, social movements, or even humans. Instead, I take ecology to mean that all things that make a difference in the vast landscape of global security ought to be included in the geopolitical considerations of contemporary life.

From this ecological perspective, geopolitics has culminated in a planetary epoch in which a particular Anthropos is capable of making a "cene."[24] However, I think the Anthropocene as a philosophical and political crisis has been too quick to forget the geopolitical arrangements of power and violence that have brought us to this point. Not all of "us" have played an equal part in the making of either the Anthropos or the Anthropocene. In part, the often narrow focus on climate change and the fever pitch of the contemporary crisis erases the Euro-American role in building and maintaining the current world order. The argument often advanced by great powers and environmentalists alike amounts to something like this: now that everything is broken, it is everyone's problem, so pointing fingers only gets in the way of a solution. Even critical and posthumanist approaches often lose sight of the role of hegemony and power. This is, in part, because the effort of those lines of thought to decenter the human as the sole locus of thinking and action is also a necessary but insufficient maneuver. This chapter attempts to relax the focus on a narrow human world while holding on to the very specifically human and often national assemblages that broke this planet.

Reprising Geopolitics

In popular and even academic discourse, geopolitics is often used interchangeably with any kind of statecraft. In some cases the choice of the words may connote a kind of realist bent of national interest. However, from its beginning, geopolitics, as a way of thinking about global politics, carried with it preceding iterations of geography and ecology well beyond narrow conceptions of the state. Raymond Aron, one of geopolitics' more adroit theoreticians, argues that essential to geopolitics is the multidimensional character of planetary life in which the politician finds themself.

> How much concrete reality does the geopolitician retain in the designs of the stage and of diplomatic-strategic actors? The conduct of foreign affairs appears *instrumental* to the geopolitician, the use of certain means towards certain ends. Resources—men, tools, weapons—are mobilized by states with a view to security or expansion. Yet lines of expansion, like threats to security, are indicated in advance of the world map if, at least, the geographer can fix his attention on the natural data on which the prosperity and power of nations depends. Geopolitics combines a geographical schematization of diplomatic-strategic relations with a geographic-economic analysis of resources, with an interpretation of diplomatic attitudes as a result of the way of life and of the environment (sedentary, nomadic, agricultural, seafaring).[25]

Following Aron, geopolitics and ecology are intimate. To complete the schema of global politics means understanding the intersection between forms of life and their particular relationship to the habitats they live in. Aron's sketch of geopolitics is at once anthropological, geographical, and ecological in its account of how interest can be pursued strategically and successfully. As a science, geopolitics emerges out of the confluence of German geographers, ecologists like Alexander von Humboldt, and German nationalist historians like Leopold von Ranke.[26] The *rooted* ontology of Nazism's blood and soil resonance with geopolitics is not an aberration in its biological-geographical conjunction of racism but only in its resolve and catastrophic scale.

Geopolitics, even in later iterations that attempt to create distance from the echoes of National Socialism, presumes a strategically significant difference in the relationship between the environment and the forms of life that inhabit it. The author of the term, Rudolf Kjellén, coined it as an extension of what he called *leibens politik,* or biopolitics for which geopolitics was its planetary pursuit.[27] The sources of competition (threat) as well as the possibility of success (domination) required a kind of biological imperialism alongside the political defeat of different forms of life.[28] In this sense, the rationalization of imperialism and global politics after the period of formal colonization still contained within it a necessity of homogenization, that is, a war on ecological as well as human difference.[29] It is not surprising then that even Aron's less overtly jingoistic rendering of geopolitics as statecraft would suggest the potential necessity of geoengineering, as early as 1962, as part of a robust strategic pursuit of hegemony.[30] For Aron, like the geopoliticians who preceded him, the only way to not be constrained by the material/ecological context of one's power was to gain the ability to alter, manage, and even create one's ecological order.[31]

It is not a coincidence that the U.S. and Europe finally started giving in a little on the exceptionalism of its existence about the time it became clear we broke the planet. Now it is "our" planet. In the aftermath of the geometric project of world making—the geography of geopolitics—the Malthusian science of human population studies takes over to sew together the thin layer of human life that inhabits the world in order to make a species correspondent to the planet. Semiautonomous from the national projects of biopolitics described by Michel Foucault (*History of Sexuality*, vol. 1), global thinkers like Julian Huxley conceived of a global biopolitics, a biopolitics of species rather than national populations. According to Alison Bashford, Huxley extended the ecological project of Humboldt's geography to a planetary ecology of a planetary vision of the human animal. The ecological character of Earth—feedbacks, migrations, "breeding storms"—all affected the species as a planetary species rather than specific or isolated habitats or populations.[32] A moral economy of responsibility resembling the geometric sphere means that the decision of *life* comes to impact every other person and therefore enters the domain of politics—geopolitics—for every other person too.

For Bashford, political ecology congeals as a field of study around a global object of inquiry such that a "truly scientific eugenics" is possible.[33] The milieu now planetary and its species of concern now significantly homogenized as a species enables a species thinking as a "planetary consciousness" that is formed like earlier iterations of global thinking geopolitically.[34] The problems of populations, diseases, migrations, and resource scarcity are not present as such; they each become problems through the *discovery* of new trade routes, forced population displacements, disruptive primitive accumulation, and enforced deprivation that makes species-scale population changes visible. Geopolitics makes species and species-scale problems in one stroke. Global biopolitics is more than an episteme; it is a project of terraforming turned eugenic, which requires a new episteme in order to govern what has been made into one conquered territory, one standardized trade route, one expropriated resource, one extinguished language at a time.

Bashford identifies the convergence of food aid policies and contraception during the Lyndon B. Johnson administration as the coming to fruition of the eugenic vision of the species. In Bashford's words, "geopolitical problems were solved by biopolitical solutions."[35] For me, on a slightly longer cut of history from terraforming to species making, geopolitics and biopolitics were synonyms all along. Savage ecologies with slightly different substances to be homogenized—one spatial, the other biological—are each now indistinguishable.

With this history in mind, it is apparent that the imbrications of the so-called Anthropocene and geopolitics are much older than the naming of the concept. As a form of politics, the pursuit of an Anthropocene was already changing the conditions of life before geoengineering was possible.[36] Humans did not stumble into this predicament. Geopolitics as a practice of statecraft is bent on expansion and homogenization, and the Anthropocene is an epoch of globally significant activity concomitant with geological feedbacks. The cooling and warming of the planet made conquest possible in the first place, with the globalization of diseases and the European biome wiping out millions of Native Americans. This made a mini ice age possible, and the resources and wealth plundered from the "new worlds" fueled the carbon liberation explosion of industrialization, and on and on. To say that the Anthropocene is thoroughly geopolitical or vice versa is almost a tautology, and yet the practitioners and thinkers of each of these silos of global thinking often ignore one another. Therefore, there is no choice to make between the Foucauldian epistemic history of a system of thought and the materialist geophilosophy of the Deleuze and Guattari concept. Entangled, geopolitics and the Anthropocene emerged together politically, geologically, discursively, and violently. Concepts make worlds and planets make concepts. And the crazy assembly of technics, persons, habits, places, spaces, microbes, meaning, sensibilities, and economies, all known as politics, savagely distributes "who gets what, when, how."[37]

Rather than detail the ways humans have left a geologically significant (to human geographers) mark on the planet, I am going to try to push the debate over the Anthropocene toward the Eurocene, or the ways in which the Anthropocene is being made and by whom. The "anthro" is both too specific and not specific enough to explain where we are. The European project of the Anthropocene is a multispecies, technic-extended, elite story more aptly characterized by a relentless expansion of a state of war than a slow diffusion and integration of peoples and markets. Thus rather than focus on the imprint that defines the Anthropocene such as carbon or nuclear fallout, I am more interested in the "process history" that made the consequences of what is called the Anthropocene possible. There are two reasons for this. First, carbon centrism vastly underestimates the scale of the crisis. Second, carbon centrism lends itself to an operationalist logic consonant with, rather than in opposition to, the making of a dying planet. For me, the Anthropocene is always a geo-biopolitical concept. Its Malthusian past is also a Malthusian future in which making particular forms of life live comes about directly through a necropolitical administration of murder and authoritarian abandonment.

I do not want to suggest that any use of the term *Anthropocene* is tantamount to the terminal necro-geopolitics that made the geological epoch. The word itself is useful in its dramatization of scale and significance, but the continued value of that dramatization will be determined by how it collides with the political landscape that constrains and enables what is thinkable at the scales dramatized by the term. So far, the political outlook is not great. Outside critical considerations in the academy advanced by thinkers such as Joanna Zylinska, William Connolly, Timothy Morton, Anna Tsing, and Roy Scranton, the political projects taking up the Anthropocene often lack the nuance and reflexivity of such thinkers.[38] Of the many tribes of the Anthropocene, those coalescing politically at a global scale are limited to those who think we can make "deep cuts" to growth and carbon emissions and live more modestly, those who think we can engineer the climate system to avoid the necessity of transformation, and those who think we can ignore the problem and simply manage via security politics the tidal wave of displaced people, eruptive agricultural collapses, and surging sea-level rise. These tribes are doing more than pursuing war by other means; in the final instance, they are waging war. All these proposals require writing off the vast majority of *Homo sapiens* in one way or another. That some tribes intend or even celebrate that outcome more than others means little for those who will be locked out of modernist images of a sustainable future.

Much of this critique may seem familiar, and even dated, but it bears repeating. Contemporary theoretical debates often treat concepts like modernity and Eurocentrism as dead, because too few in the academy put up a fight in their defense. This does not mean the political project shaped by modernity and Eurocentrism gave any ground whatsoever. The interdisciplinary excitement over the Anthropocene and its global consequences has served in many cases as a way to rhetorically and institutionally move on from questions of Eurocentrism and its settler colonial present before much of anything has been done to address it.[39] We are not finished with Eurocentrism, we are not finished with modernity, and there is no human "we" that can make the decision to move on or set aside history in favor of the emergency we face.

The emergency politics of the Anthropocene, particularly when contracted into the last decade of political stalemate and neoauthoritarian retrenchment, also resonates with a particular geopolitics in the sense that it favors the power politics of the same states and the same practices of statecraft that made the Anthropocene. For 90 percent of the planet, this is a five-hundred-year emergency with catastrophic punctuations of disease, famine, and warfare. Insomuch as there is a "we," we do not live in a contemporary emergency of decades but a centuries-long present of slow violence.[40] In this sense, we are not finished with the tools of rhetorical analysis as an essential way to make sense of how we have come to understand our current moment. And yet that framing would not be possible without the eruptions of hurricanes, insurgencies, disappearing megafauna, and spectacular accidents. However, the lag between the scale of violence and catastrophe and the recognition of the crisis says something about hegemony, and who can speak and what is legible or sensible.

I do not think I am alone in wanting to open up to the global magnitude of what confronts the planet. Yet in this chapter, I want to do so without losing sight of the real differences in politics, geography, history, meaning, and cosmology that modulate how each one of us will confront the end of this epoch. In so doing, I hope to emphasize a refrain that the end of the world is never the end of everything. An apocalypse is always more and less than an extinction, and whatever makes a life out of the mess we are currently in will depend in some ways on how we come to understand the contemporary condition. Ideas matter even if they cannot save us. Stories, explanations, and philosophical adventures are the best of what the human estate has to offer. No matter how desperate things get, someone will still ask why this is happening and we will share in that question the possibility of thinking together.

As we explore the dark fascination with the futures of our species, the catastrophic inadequacy of our dominate form of life becomes more and more apparent. The dominant forms of planetary life display an obsession with warfare and order—part technological hubris, part ecological sabotage—which have ripped their way through every continent on the planet, making a geological mark. The making of the Eurocene has been created by no single class or nation, much less by a clearly defined agenda. An aggregating and heterogeneous collection of people, things, perspectives, hatreds, malignancies, and creeping global expansions has unleashed our contemporary condition. We live in a moment imperiled by an immature giganticism. All of us experience this moment differently, but a rare few can escape even for a moment the degree to which a weight impinges upon us all. We live in an apocalyptic era unequally created by a minority bent on the accumulation of wealth and a self-interested regenerating political order. However, the "we" that will bear the burden of this five-hundred-year project of rationalized exploitation is much vaster, and includes bumblebees; humpback whales; poison arrow frogs; wolves; Hawaiians; Micronesians; African Americans; the inhabitants of Flint, Michigan; Syrians; Mayans; Queers; Christians; Muslims; Atheists; transhumanists; hipsters; shamans; entrepreneurs; homeless veterans; war orphans; albatrosses; elephants . . .

Unfortunately there is no high ground from which the entire moving arrangement can be seen. Every perspective obscures and reveals some larger or smaller part of the story of how "we" broke the world. The scale—when? where? magnitude? how long?—and the connection between each unfolding catastrophe or history of venal will-to-power and presumed superiority share a connection but not an identifiable cause or choke point that can be isolated and targeted from the heights of rational abstraction. My perspective, my point of view, is from the United States of America, maybe the second-to-last-empire. The U.S. is where I find myself and it is also where I belong, despite my best efforts to gain distance from such a horrifically destructive and arrogant form of life. To be American is not merely to be a citizen of the United States; it is, rather, to be part of a precarious mixture of European industrial and demographic expansion, a homegrown sense of Christian providence, liberal institutional development, and a ruthless martial art of extermination and settlement that has continued unabated since its founding. It is in this context that I will try to explore what I think is the character of contemporary geopolitics. For me, in the dying light of the American empire, we face a last great planetary struggle for homogenization.

The Character of the Global Crisis

I would sum up my fear about the future in one word: boring. And that's my one fear: that everything has happened; nothing exciting or new or interesting is ever going to happen again . . . the future is just going to be a vast, conforming suburb of the soul.
—J. G. BALLARD, Interview, *RE/Search*

In a world that encourages uniformity, that judges values by their utility, perhaps these animals like so many of their kind, also, are doomed to disappear in favor of some more commercially useful species. Yet, I cannot avoid a bitter sense of loss that, we, born to a world that still held these creatures, are being robbed of a priceless inheritance, a life that welcomes diversity not sameness, that treasures astonishment and wonder instead of boredom.—JACQUES COUSTEAU

Every day we are told things are worse than we thought.[41] Sea-level rise is happening faster than we thought, species are disappearing faster than we thought, and the possibilities for reversal are slimmer and slimmer. The proposals for human survival gaining traction, including geoengineering, the centrally managed supercities of Stewart Brand and others' "Ecomodernist Manifesto," space colonization, and becoming digital beings all resemble the wonders of thriving planetary life less and less.[42] On April 20, 2016, the *Washington Post* headline read, "And Then We Wept." The news was in and it was not good. The Great Barrier Reef, the Amazon rainforest of the world's oceans, is 93 percent bleached. The coral foundation of its vast ecosystem is dead or dying.[43] A year to the day before this announcement, we were told that the northern white rhino was extinct.[44] The last white rhino, a male named Sudan, is being kept under guard twenty-four hours a day from poachers.[45] No army or protection is sufficient for survival as there is no mate remaining. The young men carrying machine guns are Sudan's only company as he waits to complete his species extinction, a task thoroughly accelerated by human male desires for the aphrodisiac qualities of rhino horns.

Each event—a dying global reef system in Australia, the loss of a singular species in central Kenya, a slow shift in ocean levels—exists in an interregnum between the brutal facts of existence through which all things must pass. The crisis of our contemporary moment is that the cycle of passing and renewal has been interrupted by the metabolic rift of modern human animals. Which trajectory we are facing is unclear. Is the sixth great extinction upon us? The difficulty in classifying extinctions is differentiating a normal rise, decline, and extinction of species against which to compare and periodize "events" of catastrophic and lethal acceleration. Even the five great extinctions took place over unfathomable periods of time.[46] In all the great extinctions, "events" are hundreds of thousands of years long. Furthermore, the incomplete nature of the fossil record makes population sampling very difficult. One has to figure out ways to reliably distinguish between whether the absence of evidence is indeed evidence or is merely the absence of evidence. After extensive review of excavations worldwide over at least 150 years of research, one can estimate what is called the "background" extinction rate. This is the expected rate of species loss over a given period of time. This rate is not definitive. At best, it is a kind of working rule of thumb. That being said, the academic debates over whether the current rate of extinction exceeds any version of the background rate is like two people on the Empire State Building bickering over whether it is the fall that kills you or the certain impact at the bottom.[47] Even conservative estimates put the loss of species across the plant and animal kingdom at thousands of times the background rate from earlier human and prehuman eras. To put it another way, even if the most conservative estimates are right, we are in real trouble. To take just one example, thanks to habitat loss and the chytrid fungus, the amphibian extinction rate is forty-five thousand times higher than the background rate. Amphibians survived four of the five great extinction events in Earth's history, yet one generation of human travel has spiked amphibian extinction rates above what was caused by multiple asteroid impacts, supervolcanoes, cataclysmic climate oscillations, and a collision with a comet.[48] In an irony only humans will appreciate, the current apocalypse is marked by a noticeable lack of raining frogs.

Amphibians are not alone in the race to extinction. As recounted by Elizabeth Kolbert, one-third of all reef-building corals, one-third of all freshwater mollusks, one-third of sharks and rays, one-fourth of all mammals, one-fifth of all reptiles, and one-sixth of all bird species are disappearing.[49] What makes this particular era of disappearances unique is not just the rate of extinction but also the distribution. The entire ocean is facing unprecedented instability.[50]

Furthermore, extinctions are occurring globally, even in those areas spared by heavy industrialization and development.

While climate change is unlikely to help, the current amphibian apocalypse is driven almost entirely by the human-induced movement of people and things around the planet.[51] The chytrid fungus now affecting the majority of the planet is responsible for mass die-offs of amphibians, depriving them of oxygen and causing heart attacks. While climate change should certainly be important to any global political agenda, the already occurring sixth great extinction calls into question more than just the dependence on fossil fuels. From the perspective of those forms of life being wiped off the planet, the entire rhythm and circulation of just-in-time globalization enforced by great power navies—one of the most defining characteristics of the Eurocene—is threatening extinction.[52] Insofar as an environmental agenda has gained political currency in the past two decades, no political party or significant constituency takes seriously the proposition that global travel should come to an end. Freedom of movement is almost unquestionably championed by liberal societies. Those who do challenge it are often reactionaries and xenophobes, not environmentalists.

Since the first slow and then accelerating egress from Africa, humans have spread to every continent on the planet. That movement once resembled something like the linearity of diffusion but has reached, for some in the elite, terminal velocity. There are now humans who live in constant motion on permanent-residence cruise ships to avoid taxes, and there is a global class of anxious airport-hopping business elites who reside in no place in particular.[53] The latter are so allergic to friction slowing their circulation that even in this age of security and checkpoints they have been granted special routes and forms of identification to avoid the coagulation of administration now managing planetary circuits.[54] This is just one example of how liberal practices come up against McKenzie Wark's reworking of what Marx calls metabolic rift. For Wark, following Marx, the advent of labor that freed humans from the animal world also put humans out of synch with natural processes. The result is that humans, to be human, require too much food, water, and energy for natural cycles to fulfill.[55] From this perspective, there is no version of the contemporary order that can be egalitarian and sustainable. Disposable consumer-based economies cannot scale for any length of time. So in some sense, Wark and Marx are right. The cycles of Earth and much of its inhabitants are out of synch with humans and their love of labor. For Wark in particular, this leaves little else to do but accept that any viable human project will have to embrace geoengineering and even space colonization alongside other efforts to build a "post-scarcity society." However, such concepts should be made more precise in identifying the particular forms of life that are at odds with or exceed multispecies ecological feedbacks. If humanity is to find itself in another dark age, rather than a unified global project for environmental management, there are many possible ways of living that could be sustained within the dynamic equilibrium of Earth systems. But the point stands. If we remain within the currently restricted vision of the future of global culture—an America for everyone—the adaptive character of even large Earth systems, such as the hydrologic cycle, will collapse or enter periods of extreme turbulence.[56] To put this another way, the ought of the cosmopolitanism "good" as currently conceived and the ecological are not consonant.

However you feel about transnational capitalism, it is indisputable that the uninterrupted movement of things and people around the planet comes at an extraordinarily high cost to human and nonhuman animals alike. This is at times difficult to discern as the human population steadily increases and the world seems suffuse with living things. Therefore, the problem of the current crisis is not reducible solely to some aggregate of living biomass. What is being lost is the diversity of life that inspires wonder. Apocalypses are not primarily about extinction—they are irreversible transformations.

The often misguided debates over climate change capture this problem quite acutely. In fact, despite how difficult it is to admit that the deniers of anthropogenic climate change may be half-right, they are correct that fluctuations are a normal part of Earth's history.[57] However, what sustains the conservative bent of this claim is the sense of providence that the full argument entails. Those who champion adaptation and "natural" fluctuation trade on the presumption that Earth adapts and fluctuates for *us*. Fluctuations will occur and creatures will adapt, but in the past that has meant everything from a world of only single-celled anaerobic bacteria to vast seas of virtually nothing but trilobites. Climate denialism is, ironically, no less anthropocentric than many of its scientifically validated opponents.

The Peril of Similarity, or, The Great Homogenization

In addition to extinction-level events, Earth has also experienced a number of monoculture events, that is, epochs of great homogenization. Whether by reptiles, plants, or humans, domination by one species has resulted in collapses *and* explosions in creature diversity. It is not without precedent that one form of life could predominate and even spawn a new earthly order, as in the Cambrian explosion 540 million years ago, which is considered by most geologists the most innovative period of evolution. The great transformation of the planet by photosynthesis provides another salient example. However, the terraforming accomplished by plants is not likely to be repeated by humans unless an incipient form of life that thrives in a carbon-rich, hot, radioactive, dioxin-saturated environment comes to take over the planet. Even then, it is not just the warmer temperature or toxic nature of the planet that is dangerous to life. Periods of rapid warming and novel additions to the atmosphere have often caused violent feedbacks such as rapid cooling or, in some cases, ocean stagnation from the loss of ocean currents and upwelling. In such cases, the cascading die-offs of creatures great and small can themselves tweak and shift vast planetary cycles in new directions of amplifying and intensifying destructiveness or creativity, depending on the inheritors of the new dynamic equilibrium.[58]

The problem is also that humans are not innovating or undergoing speciation to fill the gaps left by other forms of life, as dinosaurs once did. Diversity is collapsing within the human species as well. Most languages and most ways of life outside the narrow scope of Euro-America are disappearing at an accelerating rate. According to linguist David Harrison as well as a number of other linguists working at the United Nations Educational, Scientific and Cultural Organization (UNESCO), of the 6,912 languages currently spoken worldwide, less than half of them will survive the twenty-first century.[59]

Language extinction is not just the loss of words. According to Harrison, each language contains a different cognitive map of the human brain. This claim cannot be overstated. In an example from Harrison's research among the Urarina people of Peru, some languages, although very few, place the object of the sentence at the beginning. The action and subject are grammatically organized by the object. According to Harrison:

> Urarina places the direct object first, the verb second, and the subject last. . . . Were it not for Urarina and a few other Amazonian languages, scientists might not even suspect it were possible. They would be free to hypothesize—falsely—that O-V-S word order was cognitively impossible, that the human brain could not process it. Each new grammar pattern we find sheds light on how the human brain creates language. The loss of even one language may forever close the door to a full understanding of human cognitive capacity.[60]

Given the bloody philosophical wars that have been waged over the relationship between humans and objects in the external world they encounter for the entire history of recorded thought, linguistic worlds such as that of the Urarina represent possibilities that decades of critique may not be grammatically equipped to produce. Given how bound up our current political and ecological disasters are with the problem of objectification, or why we treat objects so badly, this might be important.

In order to consider Harrison's provocation fully, we have to give up on the idea that there is some kind of formal isomorphism in the basement of all languages. There is no metalanguage. Instead, Harrison says, "languages are self-organizing systems that evolve complex nested structures and rules for how to put the parts of words or sentences together."[61] Rather than think of language as the way that humans master the world, Harrison explains, it is language "that has colonized our brains."[62] After a life spent trying to record and hold on to as many of the disappearing languages around the world as possible, Harrison argues that every language is a singular "accretion of many centuries of human thinking about time, seasons, sea creatures, reindeer, flowers, mathematics, landscapes, myths, music, infinity, cyclicity, the unknown, and the everyday."[63]

Furthermore, the loss of languages is not an issue of "multiculturalism." The loss is not just one of a way of life, like being an activist or an academic: it is the extinction of a form of life. With each language that dies we lose a glimpse of the cosmos never to be repeated. As Agamben has said of the form of life, it is a set of practices and conditions of being that is inseparable from being biologically alive.[64] Few cases capture the inextricable relationship between life and living like those groups that have survived five hundred years of colonial expansion intact in the forest of Brazil.[65] As they have successfully postponed the virulence of the European world of disease, exposure to "us" (global culture) will mean certain death. With no inherited immunity, these groups will return to the soil with their cosmic perspective. The primary cause of the displacement of uncontacted peoples in Brazil is logging and drug violence, both part of globalization.

I should be clear about what I mean by perspective. A perspective is not a "point of view" in the postmodern trivial sense, as if there is no truth and only an "opinion of the truth." This kind of consumerist commonsense postmodernism is a dead end. By perspective, I mean what Eduardo Viveiros de Castro calls radical perspectivism, whereby the selves of a host of different entities—jaguars, rocks, uncontacted peoples, plants—all experience and theorize the world in heterogeneous alliances not reducible to each other, much less as something like ideology or belief. According to Viveiros de Castro, what we find in comparative cosmologies are the possibilities of human–nature relations that are no less real or material than Western scientific observations but that organize the world around *feritas* ("wildness") rather than *humanitas* ("culture, humanity").[66]

Given how self-destructive and inevitable Euro-American anthropocentrism often feels in contemporary modern life, forms of life organized otherwise are more than just curiosities. Instead, other cosmologies and the languages that dwell in them offer the possibility of radical mutation. For Tristan Garcia, this mutation is an adventure in philosophy and metaphysics that refuses to accept subject/object and human/nonhuman binaries as inevitable problems of cognition. Instead, Garcia traces what some have called a flat ontology, or a way of being where each human is in an egalitarian give-and-take relationship with things, animals, and other humans for creating meaning about the world. The superiority and sovereignty of self-consciousness for making meaning in the world is ditched to explore something else entirely.[67] Garcia's work draws on a minor continental tradition of philosophy, but it is difficult to imagine the inspired escape from "the metaphysics of access" in favor of the dignity of things without the cosmologies of Amerindians, or without Viveiros de Castro's role as a kind of intercosmology diplomat to inspire it.[68] Consequently, as the linguistic and cosmological differences of the world flatten and merge, it is not just "background" loss or functional survival of the fittest that is taking place. Humans as the sole inheritors of the hominid legacy are experiencing catastrophic loss, a kind of internal hollowing out. The ecological crisis reaches deep into our material and mental constitutions.

The destruction of perspectives—whether it is those of poison dart frogs, sawfish, Navajo speakers, mpingo trees, bluefin tuna, isolated people of the Brazilian rain forest whose names belong to them alone, or artists and philosophers forced to abandon their creativity in favor of brain-dulling precarious labor—leaves this world less interesting and less complex than it was before. With each loss of these forms of life we lose not just a diversity of opinions about the universe but distinctive practices of tilling the earth, water management, creativity, revolutionary thinking, aquaculture, human–animal ecologies, as well as political and ethical practices.[69] As more than mere "points of view," forms of life carry with them means for inhabiting Earth that in some cases far exceed the mono-technological thinking of contemporary global development. Therefore, homogenization entails a restriction of our sociotechnical horizons. To be clear, these vital practices are not restricted to the human estate. They include the North American beaver's river management practices and their ability to combat soil erosion, the duties of megafauna and apex predators to keep grazing creatures on the move and thus prevent overconsumption in prairie ecologies, and so on.[70] The expanse of possible human/nonhuman alliances lost in the singularity of our current apocalypse is unknowable in an unusual way. Each lost alliance or form of life means a future that can no longer come about. The geopolitical advance of homogenization is killing futures as it strangles the present.

2. WAR AS A FORM OF LIFE

Even if we still decide we want to retain the Anthropocene as the name of our current predicament, it is worth pausing and spending some time trying to determine what contribution war makes to the making of this epoch. Whether we call the geological present the Anthropocene or not, it is important to consider what *makes* the current epoch nameable at all, that is, the material formations of the contemporary. As Paul Rabinow often repeats, the question of inquiry into the contemporary is the pursuit of "what difference today introduces with respect to yesterday."[1] And as this is a book about the martial character of homogenization, it should come as no surprise that I think warfare, or homogenization by organized violence, plays a central role in the making of the contemporary global system. Significantly, warfare, as a driver of mutation and change, is largely left out of the contemporary debates about the Anthropocene. And, even if geological significance is the cause for naming, I suspect that the fossils left behind by this era will more often than not be implements and impressions of war.[2] From the atom bomb to the untold billions of martial artifacts, including shells, planes, and fallen soldiers, the last five hundred years will certainly be characterized by an accelerating rate of organized and disorganized murder and violence.

War as a Concept

War as defined by classical war studies suggests a distinct class of actors, interests, aims, and expertise. As a result, the study of war as well as much of the social mobilization of war presumes an exteriority of war and warfare from other sectors and institutions like the economy or the state. For those who study military history, war—in this limited understanding—can certainly be decisive in the rise and fall of nation-states and even transformations of the global system when that system is only indexed by the states that populate it, but the pursuit of these histories still presumes a kind of exceptional character of war. War following this line of thought is a cataclysmic event that interrupts the otherwise *normal* character of daily life. For others, particularly in the field of strategic studies, warfare is a tool, an instrument whereby states and sometimes organizations pursue ends beyond the limits of politics and persuasion. War compels and determines a course of action as an orchestral direction of overwhelming force. For those who hope to abolish war, a parallel exteriority animates their theorizing about war. War, according to these thinkers, is reducible to the self-interest of hegemonic states, the militarism of soldiers, and the self-amplifying loop of profit and power. Presidents, generals, CEOs, arms dealers, and patriots come together to pursue war as an end in itself. Again, those actors and those pursuits are treated as outside the *normal* realm of human social relations. But what if war is history? What if the very form of life that created, was reinforced by, mutated with, and emerged from the Eurocene is warlike? State-making, territorialization, expansion, annihilation, settlement, and globalization are all warlike relations. I want to consider the possibility of war and warlike relations as processes of making a form of life in which warfare is normal. And what I mean by normal is much more than what we mean when we use concepts like ideology or legitimacy or discipline. By normal, I mean the very fabric of relations that makes a form of life and a world: a war body, a war assemblage, a war ecology.

I am not suggesting that war is the only form of life. There are surviving forms of life interior and exterior to the Eurocene. No process of annihilation succeeds without leaving at least a trace.[3] However, the *normal* workings of daily global life are a state of war. Rather than think of state of war in the juridical or theoretical sense, which distinguish war from peace on the grounds of declarations or measures of order, I want to consider war as an ecology endemic to the Eurocene. So by *state of war* I mean state in the sense that physicists or chemists think about states of matter. Every state of matter is an order, and despite that order, every state of matter has some elements of other states. A state of matter exhibits properties like solidity, liquidity, gaseousness, or the full-on freak-out of plasma but is not entirely made up of that state. And yet the state still has an effect despite that heterogeneity. So to say that we live in a global state of war, and that the making of the Eurocene was that making of a global state of war, is to say that war intensifies the field of relations that make the world what it is right now, not that it exhausts the possibility of what the world can become. Instead, the practices and organizations—from resource extraction, enclosure, carbon liberation, racialization, mass incarceration, border enforcement, policing and security practices, primitive accumulation by dispossession, targeted strikes, to all-out combat—are relations of war rather than merely correlates or opportunities for a war metaphor. To put it a bit more bluntly, politics, colonialism, settlement, capitalism, ecological destruction, racism, and misogynies are not wars by other means—they are war. War is not a metaphor; it is an intensive fabric of relations making the Eurocene.

To make this claim requires rethinking—somewhat bombastically—the meaning of war. If war has such a wide application, it would seem to mean nothing. In talks, roundtables, and casual conversations, colleagues have often suggested that such an expansive definition of war is polemical or even absurd. Others have said that spreading war so thin cheapens the sacrifices and tragedies of those who have experienced "real war." It is curious to me that many of the same people have no difficulty assigning similar *base* or *structuring* characteristics to capitalism, settlement, or patriarchy. I do not see war as a replacement or a displacement of those structuring structures. Instead, war is like those other complicated, heterogeneous, abstract machines but interrelated and importantly semiautonomous in the making of the world.

The importance of shifting the point of emphasis or break between war and other "big processes" is to emphasize the way collectively making death comes to be its own organizing ecology rather than just an instrumental means for other ecologies, such as racism or sexism or capitalism, that are often more obviously invested in ordering—subordinating orders—than destruction. Furthermore, I do not think, given the extreme level of violence and deprivation necessary to create the global ecology we now inhabit, that it is "a stretch" to call war the constitutive fabric of planetary relations. Instead, war as an intensive difference takes possession of other categories, at which point phase shifts take place in categories like racism or economics. What was the slow, lethal burn of postslavery policing escalates into the fury of outright combat in the streets, a race war in the streets of 1921 Tulsa or the 2015 streets of Baltimore. Even in our sacred texts of democratic theory, the pulsing tributaries of war run throughout descriptions of political formation. John Locke argued with little dispute that slavery was the institutionalization of war.[4] And W. E. B. Du Bois said of the process of reconstruction after slavery that war had begun again, and in fact had never ended.[5] Do we think that the same could not be said for the vast carceral project directed at black people described so well by Michelle Alexander or Loïc Wacquant?[6]

The common retort is to ask whether this line of thinking means all forms of killing should be considered war. I think that is a reasonable question. However, I believe that retort actually demonstrates the problem of war in the Eurocene. Take, for instance, the lynching of African Americans after the American Civil War or the routine murder of Native Americans by settlers before the arrival of the American cavalry during the period of westward expansion. Were these killings disparate acts of murder or strategically valuable microevents of war? How would patriarchy's sadistic continuity and heterogeneous creativity across geography and time survive without the nearly viral practices of domestic violence that suture norms to bruises, scars, and corpses? To exclude the way each micropractice of war—murder—aggregates over time and space into continental scale, slow-motion warfare would significantly impoverish our understanding of the role war plays in the making of global systems. Any one act of brutality could be dismissed on idiosyncratic grounds and attributed to lazy claims about human nature. Or these acts of brutality can be brought into a conceptual jurisdiction of war such that we can get a glimpse of how these seemingly disparate practices of violence resonate, congeal, and order the global system.

The historical granularity necessary to *prove* this seemingly absurd proposition is well beyond the scope of this book. Instead, this chapter and the larger aim of the book try to identify operators or machines in the emergence of war that confound the tried and true questions of sovereignty, security dilemmas, and the increasingly apparent absurdity of circumscribing global political change to the behavior of "great powers" in a rationalist and state-centric sense. The commitment to a world run by the causal agents of states, regimes, and norms (whether thick or thin) appears to me like the Velveteen Rabbit: toys whose straw stuffing is beginning to poke through worn skin and whose button eyes have been long since lost. Unlike the mythic rabbit, there is nothing *real* for these fetishes to become.[7] The first step to understand war as a form of life that is world forming requires building a community of concepts capable of capturing the diversity of relations at work in the making of war. It is a messy inquiry.

In recoil from this indiscernible mess, there are those who would rather "un–black box" phenomena by exhaustively identifying all actants or detailing complex processes of social and material change, but these approaches are ill-suited to describing the broader ecology of war. The problem with the desire to explain or un–black box phenomena is that it seduces us to focus on those phenomena on which we think we have the best chance of imposing an artificial autonomy or separation of things and events. Those phenomena that seem to give themselves up to analyzable bits and pieces are prized for publication and research funding. This cult of discreteness has not gotten us very far. Reality does not, as some realist philosophers say in a creepy analogy to butchering, "cut at the joints."[8] Positivism, even the complexity and systems variety, wants an object of inquiry at an instant. In a particularly

Whiteheadian moment, Merleau-Ponty responds to the desire for discreteness with disdain:

> In human existence, then, there is a principle of indetermination, and this indetermination does not merely exist for us . . . from some imperfection in our knowledge. . . . Existence is indeterminate in itself because of its fundamental structure: insofar as existence is the very operation by which something that had no sense takes on sense. . . . Existence has no fortuitous attributes and no content that does not contribute to giving it its form, it does not admit any pure facts in themselves, because it is *the movement* by which facts are taken up.[9]

In the attempt to vivisect existence—to reduce to an instant what is process and movement, rhythm and relation—those more subtle connections or resonances whose effects are felt but not discrete are overshadowed by those relationships we can chart and measure. Such an approach circumscribes our thinking rather than allowing it to remain open to the emergence of thought as provoked by a wild world.

Domesticated notions of complexity are stand-ins for the merely complicated. Real complexity suggests that there is novelty in the world rather than thinking that novelty is an effect of a complicated process. This is what Merleau-Ponty means when he says that the indeterminate is the "fundamental structure of the world." Those who see in complexity the mere character of complicatedness invest in that image of complexity the hope for more sophisticated predictive models and desire a mechanistic universe defined by initial conditions.[10] For those striving to create positivism 2.0, chance exists as an endangered species to be extinguished once all the data are in. I think this diminishes the creative and chaotic elements of becoming characteristic of ecologies like war, and reinvests the desire for order and control with a false and dangerous telos. The aleatory is reduced to a question of epistemology rather than being seen as a generative principle of the cosmos, giving new confidence to those who would see to make useful predictions or pronouncements upon their models of the world.

I admit it is tempting when new scientific discoveries verify our own theoretical belief in thinking that somehow science has finally gotten it right. However, one need only look at what happens when the image of thought founded on this new faux-empiricism combines with the shabby categories of interest and security to see how quickly scientific *facts* about complexity or quantum physics or networks or sociobiology can be put to use for preventive war and social control.[11] In international relations, even insights about uncertainty can become a predictive social scientific method.[12] The postcolonies and pockets of peoples surviving settlement are still finding that the "study" of their cultures' complexities serves the interests of those who would obliterate difference rather than those who would insist on a new pluralism. Anthropologists, sociologists, and economists—all armed with the latest in social-actor-network theory, complexity equations, and advanced social media scraping algorithms—are deployed as part of the subsequent revisions of the Human Terrain System in Iraq, Afghanistan, Chicago, Los Angeles, the Dakota Access Pipeline, Yemen, Mexico, Mali, and back again. Knowledge separated from its ethical considerations is readily weaponized.[13] Rapidly scoping from the micro of culture to the macro of the planetary scale, the Eurocene is already transitioning from territorial warfare to geopolitically motivated geoengineering and regional terraforming to institutionalize hegemony geologically while still refining the microscopic collection of data on threats as small as a single person.[14]

Inspired by the work in early chaos and catastrophe theory, Deleuze and Guattari suggest another path for the pursuit of complexity. Rather than ignore the overwhelming complexity of the world, scientific inquiry can be a way to cope with chaos "defined not so much by its disorder as by the infinite speed with which every form taking shape in it vanishes."[15] Creative sciences attack the problem of chaos by attributing functions to chaos so that its shifting patterns, orders, and relations can be thought. Functions are a kind of "fantastic *slowing down*, and it is by slowing down that matter, as well as the scientific thought able to penetrate it with propositions, is actualized."[16] I want to develop war as something between a function and a concept. I am trying to draw together consistencies by extending out, "building bridges," and occupying larger zones of components and functions that attempt to slow down, distinguish, and make actual indexible territories of interest, which only appear whole because investigation as a science has imposed a kind of temporary and hesitant "freeze frame."[17] One could become worried that such provisional moments could get mistaken for reality itself, however, with close attention, care for the world, such mistakes are hard to maintain. The relational (in movement-process) and substantive character of the world defies reductionism. Novelist and philosophical tinkerer Tristan Garcia explains the neither/nor of concept object-relations as follows:

> A thing is nothing other than the difference between that which is in this thing and that in which this thing is. Unless one guarantees this double sense, there are no thinkable things. Every reductionist who

claims to deduce that which this or that thing is from that which composes this or that thing only succeeds in dis-solving the very thing that they claim to account for. We attempt to accomplish the exact opposite of this: to guarantee things as invaluable differences embedded in the distribution channels of being of the world. To complete our task, we set out to discover the meaning which circulates among things, between that which composes them and that which they compose, inside or outside us, with or without us.[18]

War in particular demands this double aspect of things to capture war's territorializing and deterritorializing tendencies to make and unmake things and be a thing all at the same time.

Although throughout the book I take inspiration from new scientific research—examples from neuroscience, physics, evolutionary biology, artificial intelligence, and experimental psychiatry—that inspiration will not be used to build some new, more stable, method of inquiry, but instead I will try to trouble a still pervasive image of the world as law-governed. Therefore, war serves a double movement: it presents itself as an ordering principle or form of ecology despite the frequent conflation between war and chaos, while also undermining the image of the world as one ruled by the laws of a singular transcendental order.

Even many postpositivists, particularly interpretivists, of various kinds are unsettled by this kind of open-ended or experimental thought, absurdly big claims, oriented around assemblages, resonances, or systems of thought particularly in the context of war and violence because it means letting go, at least for a moment, of the desire to ascribe blame or culpability—from my perspective, consonant with causality—to particular individuals in time and space. This is a deficit, but it comes with the benefit of elucidating, even if only vaguely, the operators in the generation of technics, affects, peoples, and weak or novel connections in the savage ecologies that often determine, or at least circumscribe, the incipient possibilities of action by the individuals we so desperately want to hold accountable for their failures or vices. Remember Walt Whitman's words that began this book, "The war—that war so bloody and grim—the war I will henceforth forget—was you and me." The question in the context of this research is: Why is there so consistently an arrangement of racial supremacy, power, ethics, and violence to prosecute or distribute war and technical-martial logic to support and distribute it? While it is valuable to investigate how leaders can function to amplify conflict, what is striking is how quickly any one leader can be replaced and how little the trajectory of war changes.

Following Sloterdijk's pronouncement on the becoming atmospheric of contemporary warfare, collective violence is saturating every corner of the Earth system, but like carbon, neither the distributors nor distribution of violence is equitable. Saturation is no excuse for universality. There will never be a *we* that is human as such. And yet there is also no tribe, history of proper names, or nation-state that can bear the responsibility in any meaningful way. Twenty or thirty generations of malicious and sadistic decisions cannot amount to the collective effect of the heterogeneous relations that produced the Eurocene but neither should we let go of the particular forms of life that congealed around an instrumental approach to collective violence that swallowed and then organized peoples, nonhuman peoples, and things throughout Europe and then the regions those people, nonhuman peoples, and things settled. Mapping something like the totality of those actors and relations is impossible and maybe even counterproductive, but tracing the lineages of warfare that came to enable the expansion of the Europeans until they became a "cene," a geologically and geopolitically significant order, may gather up a swarm of conceptual machines still buzzing through our contemporary moment.

Consequently, I am less interested in why once such an institution or assemblage is in place, a leader at a given moment succeeds in making actual the already present virtual tendencies of war. Consider how difficult it is to reconcile our lost hope for Barack Obama with the expected failure of George W. Bush or how quickly the terror over Donald Trump was normalized once the adults from the military stepped in, precisely because foreign policies of each arrangement of leadership are in many ways indistinguishable, particularly from the perspectives of their victims.[19]

It is not surprising to me, then, that sovereigns make war, or that they take advantage of democratic paradoxes to do so. The problematic that drives this section is how such a complex, mobile, and global ecology of war so closely aligns and adheres to such a seemingly local decision as a sovereign act of violence or declaration of war.

One might take a lesson from the electrification of sound. In order to amplify or magnify a sound and preserve the fidelity of a particular harmonic arrangement, one cannot simply "turn up" the volume. It requires a certain interface between the means of amplification, the ambient qualities of the room, the number of people present, and the resonant capabilities of those people, the furniture, the walls, the floor, and the ceiling. Similarly, political decrees or decisions to produce effects must reverberate and interface with complex assemblages of institutions, economies, ethical dispositions, affective discourses, and other machinic operators. From this perspective, sovereign "decisions," whether by presidents or suicide bombers, appear to be on both sides of the razor's edge between cause and effect. Such an approach requires, as Deleuze writes, "not so much . . . convincing, as being open about things. Being open is setting out the 'facts' not only of a situation, but of a problem. Making visible things that would otherwise remain hidden."[20] So we have a world full of sovereign violence, but the place of a given sovereign in the distribution of that violence remains obscure.

What follows is an attempt to flush out the contemporary milieu or ecology of war in its mutational, material, and global tendencies to open up the landscape for something that requires a history. In the following chapter, I return to the specific historical mutations of war that contributed war's contemporary becoming as a Euro-cum-American facialization of the Eurocene.

The sections that follow look for these "loose associations" or different assemblages that may not yet have converged or connected but have tendencies and internal resonances that I think organize war's various becomings. In particular, I consider the ecology of annihilation that characterizes American practices of warfare now dominant among "great powers." These parts do not follow one from the other. They are often disjointed in their locations and themes. The hope is that the consistency that does connect them is one of possibility or chance rather than necessity so that the sense of inevitability that often characterizes the analysis of global security is not so easily territorialized for the capture of war. Some connections will appear more obvious than others, but, like other kinds of ecologies, sometimes the connections are merely the happenstance of coinciding; contingency can produce novelty and novelty can be catastrophic and horrifying. From this perspective, I ask what it would mean to consider the driving force of the Eurocene, war, ecologically.

War Is Ecological

What's this war in the heart of nature? Why does nature vie with itself? The land contend with the sea? Is there an avenging power in nature? Not one power, but two? —PRIVATE EDWARD P. TRAIN, character in *The Thin Red Line*

To what date is it agreed to ascribe the appearance of man on earth? To the period when the first weapons were made.—HENRI BERGSON, *Creative Evolution*

Every conception of culture, identity, ethics, or thinking contains an image of nature. —WILLIAM E. CONNOLLY, *Neuropolitics*

According to Raymond Williams, *ecology* was not used outside science until the twentieth century.[21] Its common usage means the relationship between plants, animals, and their habitat, or, put simply by Gregory Bateson, organism plus environment.[22] In the natural sciences, ecology most commonly denoted Lamarckian notions of evolution. In Bateson's case, it is a theory of evolution that is creative and participatory at multiple levels of complexity and organization—species, populations, individual organism, and assemblages of living and nonliving things. This approach provides a kind of minor tradition against the grain of reductionist Darwinians who located the engine of history in the survival of certain populations because of the random fact of a particular genetic variation. As Alfred North Whitehead remarks in *The Function of Reason*, the explanation for creativity in the world cannot simply be the survival of the fittest. If survival is the only measure of success, then the rocks win. They certainly have lasted and will last the longest.[23] To put it differently, life and complexity run contrary to the principle of survival and stability. To put it more bluntly, what many people call vitalism is for the purposes of this investigation the organizing forces that form negentropic islands of order contrary to the chaos that surrounds and in many cases provides sustenance for them. Vitalism is meant to represent what Erwin Schrödinger called negative entropy, or the capacity for some organizational processes to feed on the free energy of the system in order to militate against the compulsion toward equilibrium.[24] The vitalism problem is normally fought out at the scale of the organism and is generally resolved by either an acceptance or a rejection of what Aristotle called entelechy.

However, the problem with giving a name to a single vital force is that it does not get us very far. Even if you accept a vital principle, you still find yourself falling down a series of philosophical rabbit holes from fights over panpsychism, mechanism versus holism, and neo-Platonic claims of hylomorphism ad infinitum. From an ecological point of view, the resolution as well as those who reject the resolution seem to only kick the can down the road. Once relieved of the demands for an entelechy, you still have the problem of the organism or what Georges Canguilhem called the "living and its milieu."[25] On what grounds do we distinguish the cell or the microbial colonies living in a body (for which the body, say a human, is their ecology) from the place that humans live and eat (which is their ecology) from the larger systems like swamps or even larger systems like oceans from even larger systems like whole planets? And what about the organelles inside cells—are they alive? After all, things like mitochondria were once independent organisms. And what of the atomic and subatomic particles that constitute the cell organelles? When do the physical-material components cease to be alive and be mere components? The problem of vitality, which seems primarily a way to explain the liveliness of a single individual, melts into the relationality that extends for many at least to the upper atmosphere and beyond as well to the basement of quantum weirdness. And what if we go bigger? The sun and its relation to the planet or the shielding of Earth from sterilizing rays by the tilt and rotation of the solar system and galaxy is a vital part of Earth's ecology. At what scale do we even ask after the problem of the vital?

Rather than try to resolve the fight that has raged across every eon and every cosmology, an ecological perspective could simply accept that there is vitality without naming the origin, source, or singularity of a vital force while also following the entangled scales and geographies of relations where they make a difference in one's zone of concern. For me, investigating war directs me to the phenomena of a planetary concern and processes that are vital. War is vital, which is to say ecological. Whether war possesses the holism or consistency of an organism is not answerable from an ecological perspective as the emphasis on relationality rather than singularities makes the boundaries between bodies and systems reflections of position and bodily limitation—that is, I am a human asking the question. From an ecological perspective, "wholes" and "singularities" are not ontologically real. Only relations and processes are real; wholes and singularities are at best fleeting and nodal like knots in a string.

As ecological researchers, we do not play with readymade objects; instead we make real cuts into a real fabric of immanent relations to form concepts that we can talk about and understand to more and lesser degrees. Concepts are real too but they are not Platonic forms or Whiteheadian eternal objects waiting to be discovered. Instead, like stars and planets and other things, they are made from collisions and encounters. In this case, the concept—as opposed to a planet—is made from the collision of the world with me and my milieu of interlocutors (human and otherwise than human) and vice-versa. Concepts like ecology or war do not represent things; they are things that the language attempts to represent so that we can share our conceptual encounters with our friends. And even then our sharing is not quite a representation or at the very least is not a representation in the sense of a correspondence to something outside itself. Our sharing requires as much tone, rhythm, metaphor, shared history, and context that we can muster to draw others into the virtual archives of experience we draw upon to fashion our worlds of words we call books. Such a set of relations is also an ecology too but an ecology of a different order than those we directly investigate and yet one that cannot ever be fully disaggregated from all the others either.

If that is not tautological enough, I want to specifically address ecological relations, which rather than a redundancy is meant to dramatize the creativity and variable interinvolvement from top to bottom, cosmos to microorganism in all things. So an ecology of things describes an indefinite set of more and less interpenetrated force fields, following Timothy Morton's insistence that "the ecological thought permits no distance . . . all beings are related to each other . . . in an open system without center or edge."[26] Even as Morton at times is repelled by the airlessness and objectlessness of such an ontology, his hyperobjects and other formulations cannot help but thrive in such an ecological thinking. As a way of thinking, ecology comprises all the systems causally relevant to one another even when that relevance fluctuates with changes in the intensity of interinvolvement.[27] Therefore, all systems are actual and present in every other system, but the significance of the presence fluctuates.[28] The differences in fluctuations are the differences that provoke us to notice and think. To give some space between what is present and what is potentially present in the everything that is the ecological, Deleuze, like Canguilhem, often referred to the broader ecological category of all actual things in their pluripotential relations as the milieu, ranging from the milieu of a particular songbird to the cosmic milieu of the current epoch.[29] Even the milieu that comprises all the milieus, the set of all sets, if you will, is not a closed system for Deleuze. It is made up of many overlapping and intercalated assemblages, each influenced by the varying intensities of force fields that bear on their multiplicity, that is, the structure of their "space of possibility" of what we mean by difference in itself or novelty that differs.[30] Channeling Morton again, "the very question of inside and outside is what ecology undermines or makes thick and weird."[31] Sometimes those things that provoke or encounter us are territorialized before we encounter them, and in the case of concepts, I think, we participate in their territorialization. War, as a concept, is a participatory territorialization; its definiteness is lent to it by our interest, but war is receptive and resonates without our interest because it is a real fabric of immanent relations making and being made by milieus.[32]

To move from the abstract to the concrete, war, as a particular kind of ecology, may be populated by martial assemblages such as soldiers, tanks, uniforms, gas masks, rations, and bullets. These assemblages are then crosscut by the mobile and nearly instantaneous arrival of the sex trade, black markets that provide a missed brand of toothpaste, a claymore mine, a valuable medicine, and food for civilians whose support infrastructure has been devastated by the indiscriminate aerial bombing of their cities of residence.[33] There can be nonhuman animal assemblages such as great apes and forest habitats that fight back against civil wars, or swarms of disease-carrying mosquitoes that alter the internal ecosystem of warm-blooded mammals.[34] These overlapping assemblages may then be further amplified by a drought or famine that raises tensions between civilians and soldiers due to the relatively constant supply of food and water that passes by hungry or thirsty civilians on the way to a well-stocked military barracks.[35] All these assemblages feel the weight of gravity, and even the potentially mutagenic effects of cosmic rays; but on the temporal register of human daily routine, on any one of maybe twenty-six thousand or so days in an average Western lifespan, these force fields of the cosmic variety will remain incipient.[36]

But maybe not. At a critical juncture of command, or a heightened moment of tension, say, between two superpowers on the brink of war, the intervention of an asteroid like the one that hit Tunguska on June 30, 1908, could cascade into global nuclear war. In what must be an infinitesimal slice of time in the life of an asteroid or a comet, what if the Tunguska event had taken place fifty-four years later during the events of October 1962?[37] Would the Soviet Union have been able to distinguish the nearly thirty-megaton blast caused "naturally" by the cosmic object from a nuclear ground burst?[38] What about forward-deployed forces like those of Saddam Hussein's in the First Gulf War who were instructed to launch all chemical and biological weapons if radio contact with Baghdad was disrupted, with that disruption being the effect of a particularly intense solar storm?[39] Warfare is chancy, more and less than its command structure or troop training, because it is ecological.

Does this mean that wars never happen because of "first-move advantages" or offense/defense theory? Is the security dilemma irrelevant?[40] No, and these features certainly inhabit the assemblages discussed above. However, we make too much of them precisely because causal stories are within the jurisdiction of human leaders and experts. The presence of something like "easy conquest" may be observable post hoc, but it would not be the *cause* of a war in its totality. In the cosmology of global politics I am trying to build, events have a complex natural history, an ecology that is radically empirical but antagonistically antipositivist in that it does not conform to the metaphysical religion of nature's uniformity. As E. H. Carr remarked after his engagement with early complexity mathematician Henri Poincare, the propositions of scholarship must abandon Isaac Newton for hypotheses to "crystalize and organize further thinking . . . advancing simultaneously 'towards variety and complexity' and 'towards unity and simplicity.'"[41] Such a double articulation—not a dialectic—eschews parsimonious theories of international politics in favor of a deep intimacy with the processes of the world or what Brian Massumi following William James calls *extreme realism*.[42]

Thus warfare in practice and war as concept are not instrumental or successfully restrained by cold calculations of national interest. Instead, war names the mutational rhythm or machinic character in warfare that is not reducible to warfare. This is why we should refer to war as an axiomatic or abstract machine. War is a concept, a slice of chaos, to describe what remains consistent enough to demand a concept, and provocative and mercurial enough to explain the periodic breaks, innovations, and catastrophes of martial life.

Therefore, the question of war is not one that is to be answered by a primary cause or a number of factors in which each can be subsumed under a law, becoming more clear as the more detailed description of its inner workings is finely vivisected and categorized. Nor is war timeless. What has been categorized under the heading of war ranges from interspecies combat between early hominids to the dropping of atom bombs. Still, there is not a continuous line that can be drawn from Paleolithic tool use to the ballistic missile Minuteman III. Instead, there is a changing ecology of relations out of which the wielding of force and combat is organized and mobilized and endures over a slice of time and space we can name the Eurocene. To begin with a category of war presumed to be ahistorical that is then used to strain historical events that are in some way recognizable as war tells us very little about any particular practice or event and even less about the relations that connect those events through time and space such that a concept *makes sense* to create a new kind of science of war.

Becoming War

In our contemporary era of networks, counterinsurgencies, and indiscernible zones of peace and conflict, war enters the battlefield more obviously at odds with sovereign warfare. It may help to work in reverse, as the contemporary conflicts demonstrate quite starkly how "rigorous" definitions of war fall apart, before working our way through some of the slower, more subtle historical attractors of war such as annihilation. The examples in the contemporary memory are not hard to find. Consider the soldiers of My Lai who did their jobs horrifically too well, and the absent without official leave (AWOL) soldiers who refuse to fight in Israel or elsewhere. The tragic irony of the global war on terrorism cannot be understood until we grasp the relationship of war to the state of affairs rather than the relationship between warfare and the state of which much ink has been spilled. For instance, war is not initially apparent in the seeming strategic deployment of the Mujahedin—mercenaries armed and trained by the Central Intelligence Agency (CIA) against the Soviet army—until the Mujahedin's character of war exceeds and escapes the state apparatus's strategic proxy warfare to return as civil wars in the Congo or the networked logistics of Al Qaeda. The U.S. attempt to break the deadlock of bipolar deterrence via nonstate actors did not fail to disrupt the bipolar balance of power. Instead, it worked too well, unleashing a new mode of organization for violence and warfare. Instead of the nitpicking debates over personalities and financial connections that try to prove or disprove that the CIA "created" Osama bin Laden, we should map the ways new organizations of violence were let loose, imitated, reinvented, and then echoed across the planet.[43] Whether by conspiracy or imitation, the Mujahedin reterritorialized in the post–Cold War, creating a veritable franchise of warlords throughout Africa and Central Asia, and not just with the name brands of Al Qaeda, Boko Haram, and the Islamic State in Iraq and Syria (ISIS) but the numbers of other novel forms of warfare taking place in conflict far outside the civilizational drama of the war on terrorism. Drug cartels in Mexico, neo-Nazis, Christian militias, and even neo-Nazi sympathizers within the German military are using the decentralized structures of information sharing, improvised precision weapons, social media recruiting, weapons development, and on and on—a new order congealed out of a previous order already containing the vestigial structures necessary for what would come next.

Similar to the globalization of Mujahedin-like organizational types and techniques, the U.S. invasion of Iraq in 2003 could not be contained or instrumentally directed despite the great power status and traditional state form of the United States. The funding of opposition militias, both Sunni and Shia, against the Revolutionary Guard, combined with the dramatic assault of shock and awe, succeeded in the successful overthrow of the Baathist regime of Saddam Hussein. Yet the fighting did not stop there. It continued and it multiplied. The new assemblage of opposition created by the U.S. invasion, while not unified or even organized, successfully forced the U.S. to give up on permanent military bases and, for a time, enter a state of withdrawal asymptotically approaching zero. However, U.S.-led warfare returned and at the time this book is going to press is gaining momentum. In writing this manuscript over a period of six years, I have wondered whether I would need the past tense to describe the on-again, off-again conflict in Iraq and Afghanistan. So far the past tense seems indefinitely postponed.

While it is impossible to identify strict causal relations between events, the actions of the United States in Iraq severely undermined its ability to gain support for its military intervention in Libya, Afghanistan, Yemen, and beyond, and the organizational types set loose during the Cold War have entered into fecund relations for the multiplication of species of war making. The frequent riots in Mosul against continued U.S. drone violence and special ops assassinations reverberates throughout the on-again, off-again uprisings and state captures of Arab uprisings are now almost forgotten outside the region. The wars in Iraq and Afghanistan continue to return home to the U.S. territory in the form of debt, unemployment, and hundreds of thousands of soldiers with traumatic brain injuries, each further amplifying the reach of war long after any particular temporal segment of warfare has ended.[44]

The ecology of war also produces new bodies that emerge from the preparation of warfare and combat, both friend and enemy, as deterritorialized and reterritorialized bodies. In the case of the U.S. soldier, the body is subjected to intensive training and discipline, given amphetamines for response time and alertness, new eyes for night vision and multidirection sensation, then drenched in affect overdrive forms of right-wing media, as well as lingering religious or communal or national zeal brought with each soldier that is then worked up into a battle-ready lather.[45] Insurgents and bombers are also prepared in a number of ways with the elixirs of fear, hate, black market drugs like the ISIS amphetamine of choice Captagon, revenge, duty, religion, ideology, and reconfigured communications technologies.[46] All result in alterations in brain chemistry and perception.[47] The body in war abandons its organs, the rectum, the eyes, the brain, a limb, to house the bomb or weapons platform that is its new organ, its machine that will alter the war space that it will soon be plugged into.[48] Hence the productive capacity of bodies-in-war to produce more fear, anger, revenge, hate, sorrow, and frustration is limited only by the creativity and plasticity of the relations that create them. There is no moral order or normative boundary to what a body can do.

War Is a Creative Force

War is what escapes and deterritorializes constraints. Contrary to those like Martin Heidegger or Theodor Adorno who see war as a pure instrumentality, the ignoble novelty of violence demonstrates that the enframing of war is never complete. The standing reserve of warfare (soldiers, bullets, bombs, civilians to be protected, totals of enemies to be killed, etc.) is never wholly subsumed as a resource or instrument. The fog of war, unforeseen escalation, *levée en masse*, low-tech assemblages, blind allies, ad hoc militias, and the defensive advantage of weakness express the creative elements of war as they have been named; however, this list is not exhaustive. It is not an exaggeration to say that no historical development of humanity has entirely escaped the gravitational pull of war. It would be more accurate to say that war has organized a common and highly dispersed martial form of life that thrives in the Eurocene.

We do not like to ask the question, but what if ever-present violence has become a way of being, a form of life? What would it mean to speak of war as a "fabric of immanent relations" rather than merely a regrettable means for politics?[49] As scholars of war and geopolitics, in my estimation, we do not take seriously enough the metamorphoses of war. War is often relegated by analysts to the status of an effect or an object rather than a concept unto itself. So wars are declared; wars are waged; wars are ended; war is even outlawed. People do not speak of war as they do the political, the ethical, and so on. We need to consider what centuries of wars do. What is war's analogue to the political as the political is to politics? This is important if we want to consider that just maybe the martial is more central than the political or ethical to the forms of life that thrive and expand in the Eurocene. Like politics or ethics, each tactile act in warfare is another comportment of the body, a technique of musculature, posture, style, gait, each with its own possibility not just to survive war but to live war.[50] And the life that emerges, spreads.

This raises a series of questions ignored by normative investigation of how we ought to fight or what would constitute a just war. For example: How long can peace be absent before the body finds its satisfaction in an assemblage of war rather than in the "beauty" or justness of peace? When does a body or collective find the transition to peace to be as abrupt and violent as the outbreak of conflict? And are we really so convinced of a future "pacific" human society, as Kant was in his *Theses for a Universal History with a Cosmopolitan Intent* (1784), that we cannot imagine warfare ceasing to be an aberration? Hasn't it already for the powers that organize the international order? The stakes here are not trivial; there is no providence that guarantees we are meant to live in peace, which means there are also no natural laws to prevent us from fully becoming war. And how many centuries has war come to define the expansion, integration, and annihilating homogenization of planetary relations? How many generations of bodies since 1492 have been created by a state of war?

To answer these questions, the body in war has to be taken up as a body rather than a rational agent weighing the costs and benefits of conforming to norms or following international legal codes. There is no agent before the body. And the swing of a machete and the sight of a gun may be no different from the infra-assemblage of our bodies than the swing of a bat or the sighting of a jump shot. As Erin Manning puts it, this is "the body, more assemblage than form, more associated milieu than Being."[51] What differs is a minimal difference in affect rather than a transcendental moral calling. Manning again states, "Affect promises nothing. It creates across and beyond good and evil."[52] Therefore, the body is charged no less by hate, anger, rage, and fear than it is by joy, pleasure, and generosity. It is just charged differently.[53] The swing of an arm can be the opening of a dance or the mad plunge of a bayonet. For Merleau-Ponty, this is what characterizes the human, the body, and contingency made form: "Man is an historical idea, not a natural species. In other words, there is no unconditioned possession in human existence, and yet neither is there any fortuitous attribute. Human existence will lead us to revisit our usual notion of necessity and of contingency, because human existence is the change of contingency into necessity through the act of taking up."[54]

The affective field, the habit of the muscles, and the encounter by which each of these is engaged alters the effect of the movement, which is what is signified by the making of "meaning," or what Merleau-Ponty calls the "taking up." The effect is made sensible—bodily and habitual—and retrospectively we call this meaning. The movement itself is indivisible and intentional rather than instrumental and willful.[55] At first glance, this appears as a contradiction, intentional but not willful; however, Merleau-Ponty has a process in mind that requires us (intention) but is not reducible to our interiority (will). He describes the making of a new body as a process of meaning-making whereby "the new meaningful intention only knows itself by donning already available significations, which are the results of previous acts of expression. The available significations suddenly intertwine according to an unknown law, and once and for all a new cultural being has begun to exist."[56]

In war, new bodies are made, but not *by* war any more than war is made by bodies. Instead, there is what Massumi describes as an "instantaneous back-and-forthing between now and the future, and between disparate domains of activity. . . . The strike of paradox renders the gesture inventively 'undecidable'—in addition to being true."[57] In the slight difference between axe swings—one to chop wood, the other to sever arms—is the incipient possibility of different "cultural beings" and the longue durée of different forms of life. This incipience is contained not in the decision but in between the world and the body and then subsequently in the worlds those bodies make.

The in-between of incipience is also not restricted to the human. A slight difference in urban development—one high-rise to promote downtown living and another to contain racial difference—may just as easily extend from the indiscernibility of reality and enable or amplify a particular becoming, or make it more durable or more contagious. Similarly, we should puncture the myth that the preparation for war and the prosecution of war are different in kind. Instead, we should treat the body and the body of war as an "ecology of process."[58] In preparation—the becomings of war—it is not just the habit of sighting and pulling triggers, innovating new strategies and the means to eliminate populations like the cities of Dresden or Hiroshima. It is the affective mood, the technological and urban regulators and amplifiers of the flows that slow and congeal into new habits—like dropping a bomb from thirty thousand feet and other incipient possibilities, pre-adaptations, that might at another moment find expression, like refusing to launch nuclear weapons in a time of crisis.[59] War makes worlds and worlds make war.

Shifting our interests from events and acts to processes and habits directs our attention to how the outbreak of war may be subterranean in habitual activities that are not seemingly warlike. We should not be fooled by the common sense that because things are not always at the fever pitch of war, war is not working behind the scenes in our imaginative, judging, and bodily faculties as well as our ports, freeways, internet connections, satellite feed, and toxic runoff, emotional and molecular. Some preparations for war move too slow to be seen.[60] If we want to attend to war's invitation to be thought, we must make more vivid those preparations for warfare that often get lost in political realist discussions of armaments and troop movement, like the slow accretion of carbon over a century or two, the reorganization of waste, the resignification of belonging, and other more gradual processes in the conversion of lifeworlds to "operational atmospheres" for past and future wars, exposing what Sloterdijk calls "new surfaces of vulnerability."[61] The consequences of a life of war far exceed any particular battle or even world war.

3. FROM EXHAUSTION TO ANNIHILATION

A MARTIAL ECOLOGY OF THE EUROCENE

Like all ecological systems, the ecology of war goes through periods of relative stability, suddenly punctuated by intense changes in which some organizations of violence and technical assemblages predominate over others. According to Hans Delbrück (the inheritor of Carl von Clausewitz's project to organize the history and analysis of war on political and socioeconomic lines), the nineteenth century was one such moment of bifurcation between two relatively distinct modes of warfare. For Delbrück, this period was a contest between Frederick the Great's wars of exhaustion and Napoleon Bonaparte's wars of annihilation. Importantly, warfare in the twentieth and twenty-first centuries, and the geopolitics of homogenization discussed in the previous chapters, reflect the outcome of this contest. One could, following Foucault and others, imagine this break as one between epistemes. However, we need to go further than the epistemic to make sense of how the transition took place and became, in some sense, path dependent despite the horizontal relations between people, politics, things, martial logic, and bodily habit. If Foucault thinks, as Deleuze says, that every technical object is social before it is technical, then we have to leave the episteme for another time.[1] If we think that there is an incipient inhuman character to Foucault's return, time and time again, to the "government of things," then we can simply shift the weight of emphasis.[2] I am not that invested in resolving the tension between these two readings of the episteme. In part, I think the tension between the "social" character of the diagram and the "smallest things" of the security apparatus is an ambivalence in Foucault's thought rather than a purely textual-hermeneutic dispute that we can get right. In the end, rather than resolve the ambivalence, I want to situate ecology and the shifts in ecological orders in the middle of Foucault's tension: an ecological history and a history of ecology.

Therefore, these two strategies of war, exhaustion and annihilation, are not *just* ideas but are *also* ideas. Each represents a historically specific confluence of race, class, training, nationalism, tradition, technological advance, timing, and the creative flux of selection, mutation, intervention, jungles, deserts, cites, transportation, and beasts of burden—all of which participate variably in war's creative evolution.[3] For instance, the primacy of annihilation owes no small debt to the explosive tsunami of nationalism that swept Europe. Without the democratization of weapons and advances in logistics and communication, the horde of France would have been more of a liability than an asset. Likewise, an understanding of race as biological difference whereby enmity could be understood in terms of pure survival is unimaginable without the proto-wars and ecologies of extermination in the "New World" that premediated those in Europe. Therefore, we should be wary of arguments that rely on the inevitability of strict contingency in which once the *people* were mobilized, war had to follow. Whether demographic arguments about youth bulge like those advanced by William McNeill or, alternatively, constructivist claims about identity and identification, or technologically determinist arguments about ballistic weapons, we lose the ecological character of change when we privilege one of these processes over all others.

Historians with a bent toward the nationalism of the nation-state as the bifurcation point in European history, such as Foucault and Delbrück, make too much out of the continental origins of wars of annihilation by always returning to the Napoleonic wars as a point of origin. As Roxane Dunbar-Ortiz, William McNeill, Siba N. Grovogui, and Michael Shapiro remind us, the origins of annihilation, the pitting of whole populations or races against one another, do not run in a straight line from Paris to the atom bomb.[4] Mass war in the form of total mobilization finds its object of annihilation in earlier civilizational conflicts, for instance the European crusades, the imperial wars of conquest,[5] and then colonial wars for permanent occupation in the Americas, the Caribbean, and Africa. These conflicts were taking place for two centuries before France's revolutionary *lévee en masse*.[6]

Yet each of the trajectories—colonial and metropole—does feed back into an amplifying and intensifying technological milieu of industrialization and credit-financed state expansion, whether for commercial or territorial expansion.[7] This is often a distinction without a difference. Each mode of commercial and territorial expansion feeds off the other for support and their raison d'être. It is too easy to pick sides between Vladimir Lenin's *Imperialism, the Highest Stage of Capitalism* or the inverse, Jared Diamond's *Guns, Germs, and Steel.* Neither the mode of production argument nor the claims of technological determinism are sufficient to fully explain the other, more pluralistic imperial models of capitalism such as the Ottoman Empire, the annihilating forms of imperial Soviet state communism, or the vicious forms of neoliberal communism China now pursues as a geopolitical model inside and outside its borders.[8] The variances between them suggest that modes of production, like modes of war, are more like abstract machines than structures. That is, as forms or attractors, capital and war intensify tendencies rather than constricting possibilities for mutation and change. The Eurocene emerges out of a particular, contingent, and singular ecology rather than as the inevitable outcome of superior technology, or some banal argument about human nature and the will-to-power.

The changes that enabled annihilation to become the dominant mode of the state of war, and in fact a geologically significant geopolitics, are most apparent in (very different from caused by) the changing composition of armies, and the new assemblages that populated them over the course of the sixteenth through nineteenth centuries. Major-General J. F. C. Fuller argues somewhat too romantically that over these four centuries, there was a shift from chivalry to science and industry.[9] However, his insight is not unfounded. Once war as a multiplied duel gives way to force compositions like pike and cavalry, which then disappear to give way to new democratized modes of warfare, something else emerges for which the metaphor of the duel seems not only romantic but completely inadequate. The transition between modes of warfare, while not the entire story of how we came to live in the martial Eurocene, is a significant part of the story of how the contemporary ecology of war emerged as a global ecology. There is no Eurocene without the capability to lay waste to whole civilizations, there is no settlement without extermination, and there is no globalization without homogenization and expansion backed by navies and freelance violence entrepreneurs. While I would not characterize war before gunpowder and national mobilization as chivalrous, there are decisive changes in the practices and organizations of war that fundamentally alter the still lingering image of war as a contest in favor of something more brutal, more adept at homogenization and extermination.

A mutation takes place in the martial arts of European states and the U.S. state. Great powers shift from the privileging of individual skill on the battlefield to a quantity theory of war in which numbers, at least for a time, matter more than tactical execution.[10] In specific historical terms, one can see the difference in the episteme of skill versus the quantitative mode of warfare in the way leaders of the Gauls were chosen by performance in combat. One's right to lead among the Gauls came from one's ability to survive and excel in combat. The leader often actually led, in the physical sense, the troops into battle. The leaders were the front line or tip of the *cannae* or wedge shape that fighters formed as they raced toward their enemy. The metaphorical meaning of "leading" a battle in which verbal and later cybernetic commands or leadership replaces the physical presence of commanders on the field of battle tracks with the deskilling, depersonalization, and democratization of warfare into a quantitative enterprise.[11] The faith in overwhelming force over strategic or limited ends explains a great deal about contemporary practices of combat wreaking havoc with little success around the globe. The quantitative and then annihilative mode of warfare also says something about the creeping entanglements between war and other ecological orders such as the economy, resource extraction, urbanization, and scientific innovation.

These seemingly distinct modes of organization and relations making are increasingly bound up with and to war precisely because a quantitative approach to annihilation requires—as a contest—the full mobilization of production, whether demographic, economic, technological, or affective. Not surprisingly, each of these *sectors* becomes a target of war, resulting in an inseparable entanglement of arsenal and form of life. Using the example of gas warfare, Peter Sloterdijk argues that the environment itself became the operational space of warfare. Unlike targeting or fighting, gas changed the atmosphere of the lifeworld into a death world. The transition for Sloterdijk was a logical outcome of a form of war for which annihilation was analytically and materially practiced, and therefore, habituated as a preference, war became technological: "technology militarily encapsulates the nature of enmity: it is nothing other than the will to annihilate one's opponent. Enmity made technologically explicit is exterminism. This explains why the mature style of warfare in the twentieth century was annihilation-oriented."[12] Therefore, for Sloterdijk, poison gas for crops, poison gas for troops, and poison gas for victims of the Holocaust, while not being indistinguishable, share a common heritage in the explication of the ecological order. The habitats of bugs, soldiers, Jews, queers, Roma, Communists, and other dissidents share an atmosphere; therefore, the annihilation of each can be accomplished by taking that atmosphere away. Rather than seeing gas as an aberration from other forms of war, Sloterdijk sees it as an extension of war becoming ecologically aware.[13] War as a technological assemblage of making each relation more and more explicit, whether economy, combatant, or resource, also made each a target and each relation that sustained that target a potential weapon. Over four centuries of European development and expansion, violence capability and participation have undergone a process of radical democratization for both citizens and things. Everything is a target and the dependence of every target implies a weapon, a "new surface of vulnerability." According to Sloterdijk, "no other epoch displayed such advanced expertise in the art of annihilating."[14]

Put powerfully by British major-general and historian J. F. C. Fuller, "with the advent of firearms we do not merely turn over another page in human history; instead we open a new volume, the title of which is The Will to Power."[15] What is often referred to as the "gunpowder revolution" in discussions of revolutions in military affairs is five hundred years later still without equal in terms of the transformative effect it has had upon the distribution of global violence. Nuclear weapons may possess more destructive power, but the majority of combat and noncombat killings still take place across the barrel of a gun. In 1495, shortly after the adoption of gunpowder in Europe, Charles VIII (according to Niccolò Machiavelli) was able to take over Italy "chalk in hand," overcoming the strong presumption in favor of defensive warfare dependent on fortified cities and castles.[16] Charles VIII accomplished this feat because of a regional monopoly of artillery. He could mark on a map with chalk the target zone of conquest and it was done. Military preeminence at this scale had never been possible with so few soldiers. The gunpowder era was marked by intensifications of the rate of change as much as an intensification of destruction. Fuller notes that only twenty-six years after Charles VIII's reign of terror, which ended with the transition from moats, walls, and towers to wet ditches, ramparts, and bastions covered by heavy guns, war reverted again in favor of the defense.[17] In just twenty-six years, an entire organization of dwelling and fighting was ended and replaced by another, which again shifted the balance of what violence could accomplish.

However, on the battlefield, the expense of artillery limited the initial impact of gunpowder. Therefore the democratizing power of gunpowder lagged in terms of the structure of military units. Artillery could deal a decisive shock to the enemy, but accuracy, speed of reloading, and range were still very limited. As improvements were made, the change was reflected in a marked transition from cavalry to infantry. For a hundred years, war was a mechanical assemblage of artillery, pike and musket.

The affective fallout was also significant. What it meant to go to war and be in war no longer resembled the heroic duel between equals. War as an aristocratic pursuit was lost to the technical leveling of ability and accessibility with regard to modes of destruction. Blaise de Monluc (1502–77), the marshal of France, decried the firearm as the "'Devil's invention' because it distributed the means of killing to the common people."[18] Beyond the military classes, Miguel de Cervantes and John Milton decried the death of honor and courage because of the "base cowardly hand" of a "chance bullet, coming from nobody knows how or from whence."[19]

Like many technological transformations, others saw promise in the mechanical order of a more advanced form of warfare. Technical complexity and the standardization of arms and capabilities were equated with progress. W. H. Lecky declared one hundred years later a corresponding decline in battlefield casualties. According to, Lecky, the "triumph of barbarism" was impossible thanks to "military machinery."[20] Not unlike the proclamations about the end of war after the advent of nuclear weapons, many in the sixteenth century presumed that firearms and the science of mechanical war had established a kind of permanent Western order. However, that optimism rang hollow in a matter of decades, not centuries. The assemblage of musket and pike warfare met its match on the shores of newly discovered lands in which the optimistic glow of technological equilibrium lost its footing on the path to colonial domination. I do not want to imply that European martial technology failed in the pursuit of conquest, and then settlement, but rather that the very different ends from intra-European religious and territorial disputes fundamentally altered the evolution and speciation of lethal technology. As will be discussed later in this chapter, the martial technological speciation brought with it a new apparatus of martial governance that was often indistinguishable from annihilation and genocide.

Yet the resonance between modes of war and modes of governance and control began before the imperial encounters in the Americas, Africa, and Asia. According to J. F. C. Fuller, the competitive advantage of artillery and its technological children also marked a turning point in the predominance of the state form in Europe.[21] The cost of artillery, the essential component of early gunpowder warfare, shifted power from the nobility and other city-state and commercial leagues to centralized autocratic states, which were more organizationally effective at collecting taxes and compelling the mining of iron and other metals essential to the production of artillery.[22]

Lewis Mumford goes so far as to call early states "mining states" because of the essential role played by the ability to amass the raw materials necessary for the new mode of warfare.[23] The expense of artillery was compounded by the necessity to hire, train, and employ larger numbers of musketeers. The state form, therefore, came to dominate both the nobility and the church as an organizational unit precisely because of its resonance with new modes of technical warfare. According to Fuller, war became a "political instrument" and ceased being a "moral trial."[24] Following McNeill, even commercial state development was driven by the search for materials to forge the new weapons of war. The beginnings of capitalism can be found, according to McNeill, in the demand of states for copper, zinc, and tin to make bronze and brass, and the inspiration of those proto-capitalist tendencies of early states was martial, not financial.[25]

Heavy artillery also made states capable of geopolitical aspirations, but the usefulness of artillery alone did not last forever. As has often been the case in war, portability and mobility run into conflict in this period with destructiveness. In 1703 the invention of the bayonet "ended the first lap of the gunpowder age by wedding the medieval to it; for the spear was now mated with the musket," replacing the need for pikemen.[26] According to McNeill, the bayonet and flintlock musket spurred one of the first modern arms races, as it quickly became standard in all European armies.[27] The lethality and effectiveness of the infantry, and therefore the demand to replace aristocratic cavalry with democratized troops, was increasingly apparent as reliable guns could now be converted to a saber weapon via the bayonet when the pace of battle no longer permitted the still laborious task of reloading.

The shift to guns and infantry further increased the need for supplies and the production of more weapons. The effect, according to Mumford and Fuller, was an increasing demand on industry to standardize arms.[28] Mass production and an emphasis on numbers and repetition, rather than singularity and skill, takes place on the battlefield as much as or more than the early factories of commercial industrialization. According to Fuller, "The quality idea, upon which eighteenth-century fighting power was based, was, in the last lap of that century, steadily giving way to the quantity idea."[29] Again, here there was a martial logic for which mercantilism and proto-capitalism and industrialization were means rather than independent drivers. However, there was also a positive feedback between the demand for mass production, the resulting economic and resource demands on the state, the expansion of overseas conquest, and, therefore, war. Even if there was a martial inception, once global networks and states locked into place, the difference between war for economic advantage and gaining economic advantage for war quickly became indistinguishable. Two kinds of war emerge from the vicious circle of mercantilist states. European states fought other European states that were in competition for colonial exploitation, and they fought colonial wars of occupation and expropriation to pursue primitive accumulation and, later, settlement.

Critical machines of capitalism played a vital role in the becoming global of this phenomena. The drive to institutionalize the mass death set off by first conquest in places like the Americas required new levels of speed and profit only made possible by finance and credit. According to Mumford:

> One abstraction reenforced the other. Time was money: money was power: power required the furtherance of trade and production: production was diverted from the channels of direct use into those of remote trade, toward the acquisition of larger profits, with a larger margin for new capital expenditures for wars, foreign conquest, mines . . . more money and more power. . . . Money grew in part out of the increasing mobility of late medieval society. Landed wealth etc. was difficult to transport, whereas money could be transported by a simple algebraic operation on one side or another of the ledger.[30]

Mumford's "ledger" took the form of the Bank of England in 1694, and was further accelerated by Thomas Savery's steam pump in 1698. As a result, "war was endemic, because its main object was to extract wealth from other nations."[31] The machinic and amplifying relations between ever more portable forms of warfare, primitive accumulation, finance, credit to enable investments (before the return on those investments had created new capital), and the demand for security in newly formed settlements created terraforms and institutionalized a new kind of planet—that of the Eurocene.

The next major spike in the drive to war by quantity came as human mobilization was affectively nationalized to take advantage of the democratization of violence capability. The steam engine and the credit system, as well as the steady stream of resources from the colonies, could now be put to use on a scale that was no longer limited by the cost of mercenaries or the self-limiting nature of militaries whose ranks were still mostly dominated by the aristocracy. The French Revolution and its subsequent capture by Napoleon resulted in a French army that "quadrupled in size, and battles in slaughter, power being sought through multiplication."[32] To give a sense of the scale, on March 18, 1793, French forces, which totaled 14,000, were pushed back and defeated by the Austrians. A year after the Revolutionary Convention's decree of universal service, an army of 770,000 men stood ready on January 1, 1794.[33]

There were important material changes that accompanied the democratization of the French army. The intense and often elaborate resource demands of the aristocratic officer class were replaced by more modest mobile supply caravans. More soldiers with lower expectations for food and shelter could be placed in the field with fewer supplies.[34]

According to McNeill, the supply of soldiers and money for war often outpaced the range and mobility of warfare.[35] The emergence of national identification reached its fever pitch after the French Revolution when the willingness of every man to fight was not translatable in the moving of every man to the fight much less maintaining those forces once deployed. Martial national affect as a technology of mobilization exceeded the mechanical innovation of transportation and communication, but that asymmetry created enormous demands for further innovation.

The substance of the new mode of warfare—the increasing reliance on gunpowder—also demanded new kinds of industrial and manufacturing capacities that, while improving in terms of scale of output, were not portable. Unlike earlier modes of warfare in which weapons could be acquired or even made near the field of battle, such as pikes, the chemistry of gunpowder warfare could not be improvised. Also, there were lingering restraints from earlier modes of warfare, such as cavalry. Thousands of men meant thousands of horses that had to be fed and maintained along with the increasing numbers of troops. According to McNeill, the movement of war from one location to another continued, at best, "by slow, sporadic stages."[36] The flow of war was therefore constricted and complicated by the disjunction between the lust to fight and the means of production to supply that lust. The contradiction between desire and supply came to a head for Napoleon in the Peninsular Wars, in which Spanish peasants, freebooters, and mercenaries poached and sabotaged traditional horse caravans en route to resupply French forces.[37] One of the most damaging blows to Napoleon's campaign before the final defeat at Waterloo was not delivered by a competing state at all, as the Spanish monarchy was in complete disarray. Instead, the vulnerability and dependence on long-distance supply lines enabled untrained mobs and sabotage to accomplish in slow motion what the Prussian army at its peak could not do.

Even the success that Napoleon was able to have with his unprecedented military horde was a bit of an accident. In an aside, Delbrück insists that the potato had more to do with the expansion in size, intensity, and mobility of war than any particular advance in the transport of food. Had it not been for the spread of potatoes and particularly large crop yields starting in the last half of the eighteenth century across Europe, the military experiment in democratization may well have failed. It was the surplus of food created by the New World import that made it possible for large numbers of troops to survive, through the raiding of local crops, without exhausting local stores such that popular uprisings occurred against the invading army.[38] This fact only further extends the significance of the ecological feedbacks from the New World.

Transportation, supply, and communication were no less a problem on the battlefield. As Fuller, Delbrück, and McNeill note in their descriptions of early mass warfare, communication had its own material limits. The tactical formations of war, whether columns or lines, required, in addition to training and discipline, methodical deployment and command of soldiers. However, as McNeill points out, "even when bugles supplemented shouted commands, [orders] could not reach beyond the battalion level, i.e., 300–600 men."[39] Therefore, a demand to expand communication emerged alongside a relaxing of command and discipline on the battlefield. The strict column or line formation that had characterized successful eighteenth- and early nineteenth-century armies became not only a tactical failure when confronted by the size and chaos of Napoleon's army; now the structure was also impossible to implement given the new scale of popular war. Thus attention to any one soldier's capabilities diminished significantly.[40]

According to McNeill, beginning with the Crimean War, the production and rapid movement of supplies began to catch up with the demands for larger and larger numbers of troops. He notes that military innovation at the level of armaments continued apace, but the outcome was determined by advances in supply and transportation. This worked hand in hand with the French Revolution's *levée en masse.* Inseparable from the experience of nationalism with its affective/ideational dimensions was the possibility of supplying and moving every soldier to the battlefield. Without advances in transportation and supply, the stirring of nationalist spirit in the nineteenth century would not have stirred the nation's feet. It is this intensification of movement or flow that enabled the spread of popular, gunpowder warfare to the rest of the globe. The result, according to Fuller, was that "both armies and factories increasingly ceased to be the servants of the people to become their masters. Henceforth, the mass struggle dominated life (1859, *The Origin of Species*), work (1867, *Das Kapital*), and war (1832, *Von Kriege*), Darwin, Marx and Clausewitz becoming the ruling trinity of the nineteenth and twentieth centuries."[41]

The struggle of the masses also changed the possibility of governance. The general staff became a distant bureaucracy because there were too many people on the field to command directly, which led to the rise of battlefield CEOs (which are of course missing from the battlefield as compared to the Barbarian Huns, who were led by their strongest, highest-ranking warrior). Leadership and command linked to new modes of communication technology and were never the same. According to Fuller, "Though the general can still plot and plan . . . he can no longer lead or command because the masses are too vast for his grasp. Command now passes to the General Staff, its foremost problem being the development of fire power. Thus the phalangial order—shock by bullets and shells—once again becomes supreme."[42]

The quantity thesis of warfare prevailed as the appetite for it could be supplied, moved, armed, mobilized, and commanded. Once the whole population can fight, something like a biopolitical view comes to the fore to organize decision-making joining the proto-biopolitics of nationalist sentiment. A macro view of the field from above—from a distance—replaces the telluric view of the field commander. Different orders of knowledge divide the situation room from the battlefield. By 1866 European armies became "mass armies. . . . Quality is ousted by quantity and war becomes the affair of the 'average man.' . . . It was the quantity idea underlying the conception of the nation in arms, and more still the quantity needs of such a society, that fostered industry."[43] And annihilation? This component of the quantity theory of war is essential as the modalities of war creep further and further into civilian life. It is not just the biopolitical framing of one nation against the other but the quantity theory of war that translates every person, every farm field, every resource into potentiality for the prosecution of war. Consequently, everything and everyone becomes a potential target. The images of total war "raise war from the cockpit of gladiatorial armies to the grand amphitheater of contending nations."[44] Not surprisingly, economic instruments such as blockades, war debts, and sanctions become means of warfare too. During this period, the first multilateral sanctions regime enforced through periodic naval blockade is deployed by the United States, England, and France against Haiti in the immediate aftermath of the Haitian Revolution as a means to contain a war France could not win.[45]

The effort to totalize war gains another technical component in World War I with unrestricted submarine warfare. However, in addition to matching the blockade as a means of isolating the enemy, the submarine could disrupt flow at any point along the shipping routes without warning or the need for the defensive patrols of a blockade. The submarine as an approach to warfare is a precursor to the IED in terms of the way that this form of warfare related to the general topos of war as supply chains and communication infrastructures became a widespread necessity and vulnerability. The unbounded nature of war takes off further as airpower enacts the view from above already imagined by the chateau generals who commanded via telegraphy from afar. The blurring of the lines between civilian and military, combatant and noncombatant had, of course, already taken place in the ecology of the preparation and mobilization of war. However, the increasing technical capacity to extend speed, reach, and destructive power closed the strategic loop between the preparation for war and warfare, as the destruction of the opponents was a condition of possibility for preparing for war.

Also, World War I altered the balance of the profits of war by one more class of decision makers removed from the field of battle, shifting "plunder by the generals and troops to the gains made by financiers, war contractors and manufacturers."[46] The result was Fuller's prescient statement regarding Clausewitz, and later to be rediscovered by Foucault: "Thus it came about that the vanquished nations inverted Clausewitz's famous saying that 'war is a continuation of peace policy.' They replaced it by 'peace is a continuation of war policy.'"[47]

By the time Ford's CEO, Robert McNamara, entered the war, the connection between war as output of an abstract quantity of bullets and an even more abstract mobilization of the whole economy is the standard practice of all great powers in Europe.[48] Therefore, when Robert McNamara joined the War Department during World War II, the introduction of post-Fordist and proto-cybernetic thinking about the rationalization of industry was already incipient and immediately resonated with the already material imperatives of mass warfare.[49] Consequently, tactics and battlefield strategy slid down the scale of attention among those counting bullets and bombs. An input/outcome rationality took hold at the level of armaments, the mobilization of the economy, and now there was plenty of data to feed the operationalization of economy and war as well as microlevel interventions into the bodies of soldiers to increase firing ratios and consequently kill ratios. Old forms of training were no longer useful if the goal was to put more bullets on the battlefield rather than earlier eras' premiums on accuracy and tactics. According to Lt. Col. David Grossman, the main crisis of World War II was S. L. A. Marshal's discovery that 85 percent of individual riflemen did not fire their guns, even when ordered to do so or when confronted by an enemy combatant.[50] Output and capacity became ends in themselves, like military objectives and the taking of territory or honor in battle. The quantity theory became axiomatic in modern warfare.

From Annihilation to Quantity

In parallel and continuous resonance with the transformation of war on the European continent, forms of war by annihilation were developing toward a quantitative mode of war rather than the other way around. Settlement and colonial expansion, from their beginning, negotiate a modulation between extermination and counterinsurgency. Rather than the narrative often casually employed whereby great wars were replaced by savage counterinsurgencies somewhere in the middle of the twentieth century, wars of annihilation as proto-counterinsurgencies run in tandem historically with the transformation of "great wars." The differential between classical war and degraded insurgency warfare is geographical rather than temporal. As can be found in the writings of Christopher Columbus, all through the subsequent waves of arrival on the shores of the Caribbean and the Americas, forms of warfare were being adapted and innovated to suit the ecology of conquest and the native habitat that was to be transformed for European operations.[51]

New World, New Wars

> He brought light to the colorless region,
> With fierce Mars, with Minerva, and Apollo.
> Don Bernardo de Vargas, for only to him
> Was granted the authority of heaven.
> Nature was made better.
>
> LÁZARO LUIS IRANZO, in Vargas Machuca,
> *The Indian Militia and Description of the Indies*

In 1610 Captain Bernardo de Vargas Machuca published the first counterinsurgency field manual. While there are many earlier extant manuals for war going back to ancient China and Greece, or earlier if you include the extensive Assyrian wall carvings on the art of war, military historian Geoffrey Parker claims that Vargas's text is the first to be published that extensively deals with the long tradition of putting down rebellions and fighting guerrilla wars.[52] Unlike Sun Tzu's *The Art of War*, which seems to presume any or every possible enemy, or the many accounts of war stories from Thucydides or Herodotus, which merely retell the heroic and tragic tales of particular battles, Vargas's text is devoted to one specific target and its environment. The bulk of the manual reads more like Alexander von Humboldt's *Cosmos* or an early anthropology text like Immanuel Kant's *Geography* than it does a discourse on the martial arts.[53] Vargas does wax poetic about traditional martial questions like leadership and valor, but the longest sections of the manual detail the belief systems, hunting and housing practices, and rituals of the Indians as well as the flora and fauna that Vargas sees as essential to the way of life he confronted. As a result, at least part of the difference between Vargas's text and other military manuals is an ambivalence regarding the problem of enmity. For Vargas, the goal of successful combat is not the unequivocal defeat of the enemy but the long-term effort to settle conflict sufficiently so that conquest can be replaced by settlement and a normalized system of financial and political tribute between the Spanish and the Indians. In this way, Vargas's text is closer to Machiavelli's *The Prince* and its lengthy discussions regarding how best to rule conquered peoples than it is to classical martial texts.[54]

However, contra Machiavelli, Vargas's description of material context or contingency, what Machiavelli calls *fortuna*, takes on a more scientific character.[55] For Vargas, material context can countermine one's aims and even overwhelm one's skills, and can be known in much greater detail than Machiavelli's anthropomorphic vengeful nature-as-woman. The longest book of Vargas's manual, titled "Hydrography and Geography," includes extensive descriptions of weather patterns, seasonal differences, edible and dangerous plants, as well as what animals and natural resources can be put to use in the efforts to quell rebellion. In one of many examples of the co-species character of empire and settlement, Vargas goes on at length about the value of dogs in the pursuit of the enemy.[56] For Vargas, dogs provide necessary companionship along with the extension of human sensory capabilities. A loyal dog acts as a kind of early warning system for nighttime raids as well as ambushes. The dog, through its sense of loyalty for its master, seeks to protect the soldier even in the height of combat, attacking Indians and pulling them from horses to the advantage of the soldier. Because Vargas identifies the "indigenous" advantage of Indians in their intimate knowledge of their own forests and other environmental "high grounds," dogs are valued for their ability to overcome the inferiority of settler knowledge about the environment for martial ends.[57]

Indigenous knowledge of the environment recurs as a threat throughout the manual. Dangerous plants and animals are a kind of stockpile of potential weapons the Indians can turn against the Spanish.[58] Vargas sees knowledge of not only the poisons used in battle but the source of the poisons as essential to the art of counterinsurgency. Furthermore, the martial significance of the environment does not end with the weapons of nature but also includes the bounty of nature. Wild and cultivated sources of food are extensively described as they represent potential resources for the enemy as well as, with the knowledge of what can and cannot be eaten, a resource for one's own troops. Long before the creep of warfare into the "lifeworld" of twentieth-century Europe described by Sloterdijk, the *umwelt* of the Indian was the space of war, its weapons, and the target for successful settlement and governance through war. The horror with which commentators like Sven Lindqvist describe the targeting of cities, factories, and agricultural resources during twentieth-century colonialism and World War I and II is already explicitly present in the martial arts of sixteenth- and seventeenth-century settlement practices.[59]

The "origin story" of Vargas is significant not so much because it came earlier than the descriptions of Sloterdijk and Lindqvist but because the ecological mode of warfare in the Americas is coemergent with the practices of colonial settlement and governance rather than the effect of settlement. In fact, war is the practice of colonial settlement and governance. Following the order of Vargas's manual, the descriptions of the environment and fighting in that environment transition seamlessly into what Vargas sees as the fundamental schema for those forming militias, which is what comes after the direct use of combat. He breaks up the project of martial conquest into four parts: (1) how peace is settled, (2) how a city is formed, (3) how land is to be distributed, and (4) the kind treatment owed to the Indian with reward for conquerors and settlers.[60] Despite the expected reverence for great military leaders such as Alexander and Caesar throughout the first section of the text, Vargas says in this section, "One who settles and conquers deserves great renown, but one who holds on and moves settlement forward deserves more."[61] The art of war, for Vargas, is exemplified, with the most valor, in the art of settlement. Land tenure and the management of Indian populations includes the continued use of violence and even annihilation but toward the goal of functionalism in the creation of the vast user space for Spain rather than in victory over a particular enemy. The Indian is a figure of enmity only insofar as the Indians' form of life extensively described by Vargas as lazy and evil presents an obstacle to the making of New Spain as a flourishing urban economy. At times Vargas describes the hope for peaceful coexistence, which he even insists at one point requires that Indians be able to continue their form of life without Spanish interference, but only because he believes that allowing for independent Indian "markets" will produce more gold and tribute for the Spanish than outright slavery. If and when the experiment of independent Indian economies fails, the alternative is readily offered to return to war and the institutionalization of warlike relations in the form of direct "vassalage and obedience."[62]

The last piece of the counterinsurgency program described by Vargas comes from what the Indians owe the Spanish. Here, the more recognizable paternalism of the colonial project is visible. Vargas writes an entire appendix later added to the manual devoted to disputing Bartolomé de Las Casas's polemic against Spanish colonization. For Vargas, the greatest gift given by the war waged by himself and others is the gift of *política* or civility. According to Vargas, *pace* Las Casas, the Indians were inclined to cunning and unnatural acts, including the murder and rape of their own families. What could be taught by war and governance-as-war as well as the taming of nature and the transformation of the wilderness into useful cities and farms was civility. For this, Vargas says, referring to St. Augustine, "thankfulness must be as great as the benefit received."[63] Those Indians not sufficiently thankful were not the enemy on political grounds but were evil and could not be tolerated on metaphysical grounds. Therefore the decision by the Indians to reject the militia as state-to-come was tantamount to a declaration of war. Any cruelty or violence following from that declaration was the responsibility of the Indians rather than the Spanish. This logic, according to Vargas, disproved the whole of Las Casas's philosophical treatise on the soul of Indians.[64] Whether or not Las Casas won the debate on analytic grounds, the project of state making as war making that attempts to finish the job of annihilation started by the conquistadors affirms Vargas as the debate's practical victor. The process by which the Americas became part of the Eurocene is one of ecological counterinsurgency. Historian Alfred Crosby refers to the Americas, as distinct from Africa and Asia, as "neo-Europes" because of the near total terraforming of the environment by European plants and animals and a result of European ways of life.[65] There is no state formation, no colonization, in the Americas that is not a practice of ecological counterinsurgency. The regimes of knowledge about the weather, peoples, plants, and nonhuman animals is significantly imbricated in the very development of war practices. War in the Americas was ecological from its beginnings in more obvious ways than the practices of war during the same period on the European continent, and with that deeper relationally in the making of targets and violence, war was also significantly more brutal. So contra Clausewitz and Foucault, to say that war is politics by other means or that politics is war by other means would be misleading. In fact, settlement was war. In the Americas, at least, European politics was a martial art.

Furthermore, the informal and pervasive character of war in which conceptually the practices of war bleed into other established categories is explicitly present in Vargas's manual. And therefore, the ways in which war organizes and makes a world at least somewhat autonomously of states is not unique to the era of so-called failed states after the Cold War. Instead, the making of the Americas resonates with the proliferating categories of combatants, styles, and intensities of warfare that confound theorists in the contemporary period. Alongside ISIS fighters and those who fight them, Blackwater's civilization mission on behalf of white Christianity, the Janjaweed warlords of

Sudan, Mujahedin, or the war laborers of Sierra Leone described by Danny Hoffman, one can place Vargas's militias in a kind of connective morphology with these species of war-making organizational types even if there is not a developmentalist story of necessity that can be told through history. Vargas was, like these other mutations of warfare, a violence entrepreneur par excellence.[66]

Like Machiavelli's *Prince*, Vargas's manual was written as a kind of job application in hopes that official employment could be gained from the king of Spain. According to Vargas's editor Kris Lane, Vargas was part of class that emerged after the end of "great" conquistadors: "By 1599 hundreds of bands of mixed Spanish, creole, mestizo, African-descended, and indigenous paramilitaries roamed the American backcountry from New Mexico to Chile participating in what were often called *castigos*, or 'punishments,' privately financed police actions against alleged indigenous rebels, thieves, and fugitives. . . . They increasingly saw themselves as a professional class . . . [who] wrote the king repeatedly requesting compensation in the form of government posts and pensions."[67]

Unlike mercenaries, these violence entrepreneurs sought out and defined their own aims and conflicts in hopes that the pursuit would be recognized worthy of a service after the fact. Also, unlike privateers or freebooters who moonlighted as soldiers when not fully engaged in piracy, those like Vargas saw themselves as virtuous and in the spirit of the sovereign even when not directly in the service of the sovereign. It is hard not to see resonances between Vargas and the other bands of militias as incipient organizations to the Ku Klux Klan or Minutemen of the contemporary U.S. in which the innovation of new forms of violence is enlivened by reverence for a cause or order neither incidental nor coincidental with the state.[68] Although Vargas represents the informal means of settlement and colonization pursued by the Spanish, the linkages to the use of American Civil War veterans to act as roaming and available forms of force on behalf of settlers and early towns in the westward expansion pursued primarily by Anglo-American interests are strikingly apparent. Like Vargas, the U.S. cavalry and other deputized settlers given the sovereign right to kill were essential to the practices of annihilation and state-building.[69]

Rather than see these seemingly "freelance" or disparate acts of killing as distinct from war, I want to include them in a category of warlike relations that practically and conceptually are normalized as a means of state warfare as the regularization and capability of state forces comes to displace or engulf the bands of violence entrepreneurs that preceded them. Deleuze and Guattari describe a similar process in the relationship between the nomad-cum-war machine and the state form.[70] However, unlike the war machine, the violence entrepreneur is not nomadic in the molecular sense described in *A Thousand Plateaus*. Instead, the aims and organizational *refrain* of Vargas and other bands doling out *castigos*—as well as the later and contemporary iterations of violence entrepreneurs—already possess the economy of desire internal to the state that Foucault, in the introduction to *Anti-Oedipus*, called *micro-fascism*.[71] The violence entrepreneur in the smooth space of the frontier was a force of territorialization vis-à-vis a state to come and a force of sadistic deterritorialization to the indigenous orders that a future state would need to eradicate in the effort to make "the West" a user space for settlers, commerce, and sovereign order.

What begins, at least textually, with Vargas is a trajectory of martial development parallel to the shift from wars of exhaustion to wars of annihilation taking place on the European continent. The American trajectory is deeply indebted to an ecological understanding of settlement and transformation rather than conceptions of sovereign territoriality or the balance of power. Unlike the purely quantitative tradition heading toward annihilation on the cleanly Euclidean battlefield of Europe, the practice of ecological counterinsurgency as settlement and development in the Americas fills out Sloterdijk's claim that the European project of modernity is a kind of operational revolution in which peoples, places, and things are sufficiently named and "explicated" such that the disaggregation of all things into useful and discrete parts can produce the globe as a homogenous user space rather than the bumpy and singular collection of places that preceded the modernist project.

At the end of Vargas's manual, we do not yet have the direct calls for extermination found in later American military thinkers like George A. Custer or Alexander Hays, nor is there an explicit proposal for geoengineering or the extreme homogenization demanded by ecomodernists.[72] However, in Vargas we have a predecessor in which a process of cultivation at the expense of other forms of life is essential to the martial arts and a conception of a *good* ecology or prosperous nature. In an effort to understand an ecology of war and its attendant form of life, counterinsurgency constitutes a vital lineage of the Eurocene and must be extended throughout the subsequent centuries after Vargas rather than periodized to the point of idiosyncrasy. A series of striking resemblances even if not an explicit continuity is visible throughout the sixteenth through twenty-first centuries, all of which run contrary to the presentist effort to locate "irregular warfare" or "small wars" as particular and minor components of the mid-twentieth and early twenty-first centuries.

Contrary to and effacing the legacy of ecological counterinsurgency as nation building, there is a popularized version of the American way of war that follows the *Counterinsurgency Field Manual* that emerged alongside General David Petraeus's meteoric rise to command during Operation Enduring Freedom. In this version of events, the U.S. experimented with counterinsurgency in Vietnam but, in the wake of the failure of this war, left it behind. According to Sarah Sewall's introduction to the academic version of the *Field Manual*, counterinsurgency after Vietnam was "relegated to U.S. Special Forces."[73] Writings on counterinsurgency by military historians and practitioners reinforce this view, focusing on counterinsurgency as mostly being modern and European in origin. In particular, the attempt to "reinvent" counterinsurgency as an art of war in the case of the U.S. *Field Manual* required revisiting the British in Malay and the French in Algeria as the classical origins of population-centric warfare, eschewing the much longer history of U.S. "irregular warfare" at home and abroad.

That counterinsurgency techniques and strategies reach back to the beginnings of liberal thought and remain undiminished in the so-called progressive eras of Europe and the U.S. suggests that there is much work to be done in reconstructing the lineage between what Foucault calls governmentality and the practices of warfare and military intervention practiced in parallel. For me, it is impossible to capture the degree to which counterinsurgency is at the heart of the U.S. paradigm of state making and war fighting without placing the Indian Wars and subsequent efforts at American Indian population concentration and pacification at the center of U.S. political development rather than relegating Indian removal to an anomalous practice in the periphery.[74]

The "renaissance" of the study of counterinsurgency, for both critics and practitioners, has thus far missed the opportunity to take seriously the role genocide plays in state formation. While this claim is frequently made, the gravity of its history is rarely appreciated. To stage what I mean as a kind of thought experiment would require a world where the population-centric efforts and the celebration of counter insurgency (COIN) more generally would be immediately compared to practices of warfare that culminated in slaughters like those that took place at Wounded Knee, in which unarmed civilians were rounded up and murdered in mass. That counterinsurgency and pacification campaigns can be considered the "softer" side of war by contemporary commentators requires a near total amnesia of what wide-scale pacification and counterinsurgency looked like in the U.S. territories between the end of the Civil War and the late 1890s. To put it more succinctly, if scholars are not placing General Philip Sheridan alongside David Galula and other gurus of COIN, then we are contributing to a historical erasure that beyond its dubious character ethically also impoverishes our understanding of the transversal character of geopolitics.[75] Ecological warfare is not exterior to *normal* politics; it is constitutive of it. So even when some scholars mention in passing the Indian Wars as proto-counterinsurgency, little attention is given to the managerial and ecological character of clearing land, eliminating nonhuman animal species such as buffalo, or urbanization as extensions of the project of annihilation to continue long after the official end of the Indian Wars and extend to the present violence most recently displayed at the Standing Rock reservation in North and South Dakota, where advanced counterinsurgency techniques as well as so-called nonlethal weapons of population control were deployed carelessly.[76] Counterterrorism experts such as TigerSwan provided deep surveillance and network analytics for disrupting activists. Supporting technology such as infrared surveillance equipment as well as sonic weapons such as long-range acoustic devices (LRADs) were used as part of the effort to disrupt those defending the land.[77]

Among the many collections on the return of counterinsurgency since September 11, 2001, there is one exception to the general trend of treating the Indian Wars as an artifact rather than a continuing project of the U.S. state. However, even Dustin Wax's brief description of ethnographic research conducted by the Bureau of Indian Affairs (BIA) after the 1934 Indian Reorganization Act does not connect the ethnographic "turn" at the BIA directly to the military techniques of counterinsurgency innovated for the 150 years preceding the act and the subsequent years of surveillance most famously in raids supported by the counterintelligence program (COINTELPRO) on the Pine Ridge Indian Reservation that culminated in the imprisonment of Leonard Peltier.[78]

Specifically, as researchers of security practices, we are ignoring a vast historical resource for understanding the distinctive character and practice of the U.S. military, as well as domestic social governance carried out by police forces that have practiced counterinsurgency for as long as the United States has had a "race problem." A notable exception to this historical erasure and contra Foucauldian accounts of the emergence of security, Patricia Owens targets the history of liberalism and the domestic in particular during the Progressive Era, to account for the character of U.S. war fighting in the Philippines. Owens describes "policies of attraction," whereby military campaigns in the Philippines used a dual focus on extreme violence, in some cases killing every male over ten years old, and the comparatively "attractive" offer of civilian camps with enforced sanitation and schooling for children. It is in the camps, or civilian zones, that Owens finds what she calls "homologies" with the agenda of the U.S. Progressive Era. Here Owens asks, "Given that civilizing missions involve the domestication of dominated others, how did these Progressive Era policies represent a distinctly social form of modern domesticity?"[79] While I quite like Owens's concept of homology, as it allows for distinct origins to converge on particular forms of governance, I do not share her recognition of similarity. For me, the family resemblance to the Philippines began at least sixty years before the Progressive Era, when Secretary of War John C. Calhoun created the Bureau of Indian Affairs. In the context of American political development, it is significant that one of the first federal bureaucracies with jurisdiction over the home and social issues was created by and administered by the U.S. War Department before later being moved to the Department of the Interior.

Despite the numerous comparative histories of welfare states, and the many histories of the U.S. welfare state, no history that I know of includes the Bureau of Indian Affairs in its account of U.S. administrative development. Despite this fact, the BIA was and continues to be an administrative agency dually charged with the pacification of indigenous people via the administration of life. In the history of the United States, I know of no earlier agency organized around what Foucault called biopolitics. And the development of biopolitical management was at work in the federal governance of American Indians long before it was extended to the "domestic" population via the efforts of the Progressive Era, and the institutionalization of welfare during the administration of Franklin Delano Roosevelt.

At least as early as the 1830s, American Indian populations were being relocated and concentrated on reservations as a means of pacification and elimination. Resettlement came with the promise of services, as confinement often prevented normal means for acquiring food through hunting, fishing, and agriculture. The process of concentration created the need and dependency upon which biopolitical governmentality emerged. Uprisings, because of the frequent failure to fulfill those basic needs for food and shelter, combined with many American Indians refusing to be resettled, further intensified the linkage between the Indian Wars of extermination and the biopolitical management of reservations. The central focus on pacification also intensified the institutionalization of Christian missionary schools into federal boarding schools. Like Vargas and his violence entrepreneurs, religious schools were captured by the U.S. state as new organs of control. The rudimentary diagram and practice of education as pacification follows from its precursor, the moral imperative of conversion. In the missionary and federal model, the organ of the school was a machine for ending a form of life, a machine for homogenization.

Christian missionary schools date back to the earliest days of settlement; however, this technique of pacification was not institutionalized by the federal government until the 1850s. The federal boarding school practices of separating children from their families and Indian ways of life, as well as beliefs and language, were an essential part of the process of pacification-as-homogenization.[80] These techniques, which link education with security and domestication, are significantly more homologous to the practices in the Philippines than the social work of the Progressive Era, which included little or no direct connection with military planning or state expansion.

Federal boarding schools for American Indians were often housed on military bases, as was the case of the school opened at Fort Spokane in Washington State. This history undermines even Owens's attempt to push the practices of counterinsurgency back as far as President William McKinley's mission in the Philippines, "to protect the natives" and replace "tribalism with individualism," to the history of the Progressive Era.[81] At the very least, McKinley's words invoke a more compelling homology to American Indian policy and management than to the nearly anachronistic forces of Progressivism in the U.S. that gained their traction more after the turn of the twentieth century than before it. While I agree that in all three—the BIA, the Progressive Era, and the Philippines—education was a core component of governing the social, the more explicit process of "domestication" as "pacification" is much less apparent in the Progressive Era. Certainly racial politics were implicated in the Progressive Era's interest in eugenics, birth control, and the panic over Southern and Eastern European immigrants, but the connection between the BIA and the Philippines suggests that long-standing practices of American warfare against native peoples are a more significant homology particularly if our aim is making sense of the continuation of U.S. geopolitics in the Eurocene.

However, it is worth foregrounding the Philippines in a genealogy of the Eurocene, particularly if we look closely at the military leadership and troop composition of the U.S. occupying force in the Philippines. The Philippines was under the command of General Elwell Stephen Otis, who was infamous for the intensity and cruelty with which he pursued the occupation. The Philippines was not the first place Otis had used these tactics and therefore not the laboratory for Otis's particularly vicious practices of counterinsurgency. Nearly twenty years before the invasion of the Philippines, Otis was deployed to Montana after the defeat of Lt. Colonel Custer at the Battle of the Little Bighorn. Otis continued to move up the ranks through his victories, not in foreign wars but in his extreme cruelty he deployed in the defeat and pacification of the Sioux Nation. Otis's career as a soldier and commander was almost entirely devoted to the Indian Wars. In fact, the Philippines was Otis's last command before retiring.

Many of Otis's commanders would have also had experience fighting on the frontier before being deployed to the Philippines. The Indian Wars had been used to soak up vast numbers of unemployed veterans after the U.S. Civil War, and the continued management of reservations and pursuit of noncompliant bands of Indians accounts for nearly all the use of U.S. military force until the Spanish-American War and Philippine-American War.[82] Therefore, the practice of concentrating women and children in camps, or reservations to be educated, while "adult men" (a broad interpretation generally including boys ten years and older) were viciously murdered, is not a homology so much as it is a direct continuation of U.S. frontier policy in a more distant context.[83]

The lineage to the U.S. Indian Wars does not end in the Philippines only to return in Afghanistan or the second Iraq War. In Vietnam, the Strategic Hamlet Program, as well as Operation Phoenix (which sought to "reeducate" or "attract" potential assets among the North Vietnamese Army (NVA), and assassinate and torture others in leadership positions to up the "attraction" of reeducation), follows the same insights of the BIA and U.S. cavalry in the Western Territories. One should also view the strategic alliance the United States developed with the Hmong people of Vietnam as a continuation of counterinsurgency tactics developed in the pacification of the U.S. and one strikingly similar to the mixed-race bands of irregulars represented by Vargas and also by Vargas's tactical advice to employ Indians loyal to the crown. The practice of employing native informants was common outside the Spanish context. In the mid-eighteenth century, particularly during the French and Indian War through to the final stages of eradication at the end of the 1890s, the development of "indigenous" allies was seen as an essential part of "knowing the enemy" as well as developing local knowledge of the environment. These practices are not merely homologous but are directly linked to the counterinsurgency practices in Vietnam and the Human Terrain System deployed in Iraq and Afghanistan. Emblematic of this tradition is the special unit of Native American trackers created in 1972 known as the Shadow Wolves. The mission of the Shadow Wolves is to provide the United States and its allies with traditional forms of knowledge critical to U.S. national security. Now under the jurisdiction of the Department of Homeland Security (DHS), the Shadow Wolves are included as part of the war on terrorism: "The Shadow Wolves have traveled to Kazakhstan, Uzbekistan, Latvia, Lithuania and Estonia to teach ancient tracking skills to customs officials, border guards, and national police in those countries in order to detect and follow the tracks of people who may be transporting components of weapons of mass destruction."[84] In particular, the Shadow Wolves, like Vargas's allies, are prized for their unparalleled indigenous knowledge of the U.S. Southwestern deserts. And like Vargas's ecological understanding of tactical knowledge, the DHS explains that the name of the unit "refers to the way the unit hunts, like a wolf pack. When one wolf finds prey, it calls in the rest of the wolf pack."[85]

Without doing too much damage to historical specificity, one can draw a line from the attempt to pit tribal groups against one another or weaponize indigenous knowledge during the seventeenth century through to the present, all the way to the efforts in Iraq to turn the tide by employing the Sons of Iraq and other militias in the so-called Anbar Awakening.[86] There is no historical gap whereby we could use a phrase like "return" or "revisit."

If we follow this historical trajectory from Thanksgiving to Waziristan, what we see is that there is a consistent strategy of concentrating women and children, the forced reeducation of "friendlies," the indiscriminate slaughter of potential male combatants, and the development and use of local or indigenous ecological knowledge. Furthermore, since the creation of the BIA, stabilization or order making has been premised on reeducation and domestic development. The U.S. has more recently referred to this part of counterinsurgency as nation-building, but it differs little from the ways the BIA sabotaged and managed tribal councils in the various arrangements of the trust doctrine after the era of treaties.[87] Despite the ways that Progressive Era and later mid-twentieth-century theories of development may have renamed these tactics, the practices on the ground have changed little since the Bureau of Indian Affairs was created in 1824. Therefore we should not be surprised or dismiss as rhetorical flourish that the Kunar Province in Afghanistan was commonly referred to as "Indian Country," or that Osama bin Ladin was code-named Geronimo.[88]

Furthermore, it would be a mistake to consider the history of U.S. Indian removal and eradication as merely additive to the larger story about biopolitics and war. Rather, the absence of this history is habituated by the case studies that are chosen by counterinsurgency specialists and historians alike. We need to take seriously the constitutive role of the total war and then ecological counterinsurgency waged by the U.S. on native peoples throughout the American continent.

Historical narratives matter, and the conspicuous absence of North America from the literature on counterinsurgency and the story of wars expansion by Delbrück, Fuller, McNeill, and others is an essential component of restoring the practice to the innocuous status of a mere tactic rather than as the cornerstone of the settler colonialism that built one of the most powerful and destructive nation-states in history. If the architect of the post–World War II international order is the United States, then the architecture of order and security was invented on the Western Plains, and the victims of these experiments had names like Sitting Bull and Geronimo. Given the pivotal role of the U.S. in the history of the twentieth century, any genealogy of the social or the domestic that starts and ends in Europe will be missing core components of what distinguishes settler-colonial statecraft from colonial statecraft. In the former, there is no possibility of decolonization or amends. There is no version of political order not indebted to counterinsurgency. The existence of such states is ontologically violent in perpetuity with respect to form of life and the catastrophe of homogenization. What history the citizens of those states, myself included, make of that debt is yet to be seen but so far is not so good. For these reasons, one cannot homogenize the origins of liberalism or counterinsurgency. Settler states are exceptional, and American exceptionalism is at the forefront of this difference. The U.S. is the violence entrepreneur par excellence in the making of the Eurocene.

For me, the United States and other settler colonial states that remain in the world today are not states that sometimes employ counterinsurgency, like their contemporary British and French peers. The U.S. was created through counterinsurgency and will always be a counterinsurgency state, whether those techniques are being practiced in the streets of Ferguson, Missouri, the Pine Ridge Indian Reservation, or the drone routes of the Federal Administered Tribal Areas of Waziristan.

Is It All War Now? The Ecology of War Revisited

In light of the total ecological wars of U.S. settlement and the total consummation of technological advance and warfare, there is a temptation to use Fuller's claim about the *opening* of the conceptual boundaries between war, economy, life, and politics at the end of the nineteenth century to allow war as a technological Western art to be the final word on the possibilities of life. Many others have joined Fuller in this claim. For Marx, civil society is a kind of congealed war.[89] For Du Bois, the Reconstruction era was war continued.[90] For Foucault, politics was war by other means.[91] Foucauldians and those who follow Agamben make similar claims in the expansion of Foucault's inversion to a general description of global politics.[92] It is this, Foucault's formulation taken globally, that has most caught fire among critical scholars of international politics and elsewhere. Foucault's account of the reversal of war and politics rests on a distinction between the sovereign or juridical orders of the sixteenth to seventeenth centuries and the rise of a security *dispositif* beginning at the end of the eighteenth century.[93] However, this epistemic shift from sovereign war to governmentalized policing is insufficient to explain the formation of what Foucault elsewhere called counterconducts, that is, the emergence of new practices and relations of war that are not derivative of or contingent on the particular carceral logic of the city and its milieu. Again, if we follow this logic, we fall victim to a historical periodization that reinforces the provincial Eurocentrism that all too often confines genealogical analysis to the European continent as the source of change. To repeat the critical maneuvers of Foucault rather than merely follow his extant texts requires attention to the material formations of war (particularly those outside the city and beyond the European continent) that follow their own organizational and emergent logics, and which far exceed the inside/outside distinction Foucault draws between the rise of the police and what he characterizes as a lingering form of classical war at home in the European balance of power.[94]

By turning Clausewitz's words inside out, what we are left with is little more than war as an instrument; it is just that now war is waged on more fronts, both internal and external. This does nothing to complicate the concept of war. J. F. C. Fuller's materialist account of the slugfest between what he calls the *constant tactical factor* and the *quantity theory of war* comes to a similar conclusion as Foucault, but the inversion of Clausewitz and the targeting of whole populations provokes an image quite different from that of biopolitics. Instead, quoting Fuller from 1942: "War ceasing to be a struggle between life values becomes a blind destructive force, like an earthquake, a volcanic eruption or a typhoon. Whole populations are now attacked, wiped out, enslaved or herded from one country into another like cattle. . . . the entire life of the enemy state comes to be the object of attack." Fuller continues quoting Ely Culbertson, "From time immemorial, men fought against men, and weapons were but accessories; in this and in future wars, machines fight against machines and men are all but auxiliaries . . . engines of destruction which devour their substance.[95]

This process occurs because war is an assembly of things set in motion and held together by the racial refrain of settler colonization, the imperial and postimperial transforms territory into operational space, and, as an assemblage, war is also transformed by those mutations and expansions. Following the ecological account developed in the last chapter, war is not a tool of the state or a failure of order. War is a phylum of organizing principles, refrains, and protocols unto itself. War drags along with it the whole of the population—its vitality, industry, inventiveness, movement, rhythm, and affect—and attests to the human and nonhuman character of the population and violently metabolizes other forms of life in its path.

If we want to call war biopolitical, then bios must extend beyond the human and certainly the thin European conception of humanism, and even the nonhuman animal, into the creative anime of all things, and into the formative and energetic forces of war from its beginnings. Mass war was set loose by the French Revolution's *levée en masse* as much as it was made possible by gunpowder, supply lines, radio, interchangeable parts, mass production, and the increasing speed and efficiency of transportation and communication of information, whether linguistic or otherwise, as well as new terrains of racialized enmity. All these feedbacks point to a milieu or ecology of war that, while overlapping with the market or the epistemophilia of nineteenth-century humanizing sciences, lured these assemblages into connection at least as much as Foucault's account of late capitalism or humanist governmentality.

Therefore, the *constant tactical* factor—war as mutative creativity—is not just an axiomatic or abstract machine of war. It is a global abstract machine traversing the war/nonwar divide; war individuates specific forms of life. Hence, we can talk about warlike relations and governance as a martial logic without having to find combat present in either of these conceptual expansions. This statement would seem consistent with those inspired by Antonio Negri and Michael Hardt, who seek to supplement Foucault by rebranding the global system an empire. However, while Negri describes the veritable deterritorialization of war, politics, economy, and life, he too quickly subsumes these deterritorializations under the logic of capitalism. For him, war is little more than a supplement or effect of capital: "War seems the only possible solution. . . . With the disappearance of the internal criteria that allow that self-regulation and self-valorization of development, it is the violence of the strongest that creates the norm. . . . it guarantees the smooth running of society and widens the terms of the market."[96] This is too simplistic, and perhaps this formulation should also be flipped on its head. War ought to be seen from the other side *too* as a martial machine with capitalist biopolitical effects, not just a capitalist axiomatic with martial effects. This would shift the emphasis from an anthropocentric preoccupation with human "states of emergency" or sovereign states of exception, or market failure, to the emergence of material assemblages that amplify human events such that they reverberate throughout the assembly of things. Within the analytic frame of capitalism, the microbial devastation of the Americas as well as the subsequent annihilation of forms of life would be incidental to the project of expansion rather than at their core.

As a matter of historical development, the industrialization of war preceded the industrialization of civilian factories. The two most important components of industrialization—interchangeable parts and the assembly line—were developed because of the demands of larger and larger armies, not larger and larger civilian markets. Furthermore, the demand for American industrialization was not the result of an "invisible hand" but a directive of the War Department to create operations for arms production using interchangeable parts. Congressional funding for the directive resulted in a number of inventors competing for government contracts to mass-produce weapons that could be repaired more easily in the field. The winner, Thomas Blanchard, developed the technology for mass production that reverberated across the Atlantic, inspiring the images of cams featured in Jean le Rond d'Alembert and Denis Diderot's *Encyclopédie*. The *invention* was Blanchard's conceptual change that allowed cams to turn irregular forms. These early lathes developed by Blanchard inspired a wave of production mechanization in the United States and Europe. But Blanchard's contract for the new lathes was exclusively with the Springfield Armory and only later was made available on the market for private factories. According to David Hounshell, the two main currents of American industrialization flowed from the Springfield armory system, and "the idea of uniformity or interchangeable parts was combined with the notion that machines could make things as good and as fast as man's hands, or even better."[97] The aleatory milieu of war has its own efficacy independent of the logic of markets and modes of production and should be given its due.

Despite the temptation to use capitalism as an explanatory logic for violence, the ecology of warfare as a martial logic exceeds the instrumentality of the market and profit drive of the market. Furthermore, the overemphasis on capitalism and the provincial image of the European battlefield also complicate the more recent development of the global civil war theses found in Carlo Galli's *Political Spaces and Global War* and Tiqqun's *Introduction to Civil War* that, similar to Hardt and Negri, derive from Foucault's account of biopolitics.[98] The civil-martial divide prescripted in the move to say all war is now civil war requires that the civil peace that foregrounds classical war assumed by Foucault has ever existed in the first place. The accounts of McNeill, Fuller, Braudel, and Mumford as well as the historical record of martial development in the Americas suggest that the divide between the civil and the martial is a convenient fiction not unlike the state of nature. And insofar as the claim to an internal civic space (the zone of policing rather than war) can be made, it is in a very limited capacity and one entirely idiosyncratic of a very narrow temporal and geographic slice of Western Europe basically between the Treaty of Westphalia (1648) and the Napoleonic Wars (1803–15)—but again only if we ignore the globalization of warfare happening outside this narrow territorial limit from 1492 forward. From the larger historical and geographical vantage point, there is no classical period *before* the global civil war once the Europeanization of the globe began.

From the beginning of European expansion, war was neither interior nor exterior, as the territory was in some sense a smooth space in that it had not been nationalized, much less internationalized. War took place, but it lacked the political quality that makes Foucault's inversion possible. As such, the laws of war found in Hugo Grotius and early discussions of Christian traditions of just war were flagrantly disregarded, just as they had been in the European Crusades. And furthermore, decolonization does not return us to the romantic image of international politics either. In the aftermath of formal-legal colonization, everything is nationalized but virtually none of the *new* national spaces resembles what could be called a sovereign territory except for those spaces that cannot be fully decolonized because they are settler colonies. This is similar to the period in Europe before Westphalia in which sovereignty was claimed by the Catholic Church and some kings but had little bearing on the organization of war in regard to questions like territorial integrity or "balancing." The postcolonial period following World War II is one in which sovereignty is unequally distributed to the point of needing to describe the behavior of states, like the U.S. as supersovereign.

There is a further anachronism in Tiqqun and other Agamben-influenced extensions of Foucault's war thesis. In almost all cases of the global war theses, and more generally the revival of Carl Schmitt's work on liberal international order, the loss of the classical state system is a result of the exceptional politics following the American state of emergency in 2001 or a component of a lateness in capitalism generally starting after World War II. Others peg the creeping national security state to Cold War politics following the ideological struggle also after World War II. In the American case, the slippage between war, state making, and security begins long before September 11, 2001, and the rapid securitization of life that takes place during the Cold War. Such presentist accounts require a selective forgetting of how indebted Cold War and war on terrorism practices and politics follow traditions of counterinsurgency and nation building throughout the settlement and expansion of the American colonies, before and after the modern republic.

Put simply, the classical mode of warfare between isomorphic, mutually recognized sovereign states occurred at best provincially within the continent of Europe for a very short time. This period from which Foucault draws his logic describes neither the brutal and deterriorialized wars of religion that raged in Europe before the Treaty of Westphalia, nor the wars exported beyond the shores of Europe thereafter. Therefore, the attempt by Foucault and others to draw a distinction, even temporarily, between security's project of policing and the sovereign state's external application of war, while historically significant for a fleeting moment, is not helpful for understanding war before and after the eighteenth century. The simple distinction between internal war (security) and external war (war/international relations) is consumed in the empirical forge of the actual history of warfare. On this point, I concur with Antonio Negri conceptually despite our differences of origin and periodization: "War has always pretended to have an ordering finality in postmodern capitalism. . . . Ontologically, resistance immediately denies that claim. Modernity's order can no longer suppress postmodern disorder; Hobbes crumbles before Guernica and Fallujah."[99] Negri's formulation would be quite accurate if instead of postmodern capitalism it was the longue durée from 1492 to the present. The fight is not between a once great and constraining modern order and a new reckless "postmodern order" but modernity itself as a global project in the operationalization of the planet for war. This does not mean we should *forget* Foucault's inversion of the Clausewitzian formula between war and politics. It is usefully provocative, even if misleading. I think it is better instead to read Foucault as an amplification of rather than an inversion of Clausewitz. To say that politics is war by other means still requires the affirmation of Clausewitz's original formulation of "politics as war by other means," without which Foucault's war as politics would have no object. For Foucault, of course, modern war has the population—its survival, productivity, and health—as its object, both as a target and as a site of mobilization.[100] If there is a concept to be developed, it is one of war as a force that complicates, forms, and deforms the political landscape of the planet. Therefore, there ought not be an escalating number of concepts called war's *others.* Security is as intimately a part of colonial, postcolonial, and precolonial warfare as policing is a formative element of international politics. The difference, at best, is one of intensity. Instead, as I try to develop in the previous section, particularly in the Americas, state making and politics more generally in the settlement context is a martial art and inseparable from practices of violence.

So in this sense, despite a kind of Foucauldian or biopolitical turn in international relations scholarship, there is little if any historical evidence to support a contemporary "liberal way of war" or some more insidious humanitarian version of warfare that came about in the aftermath of 9/11, or as part of the civilizing mission of the Cold War, as those seduced by Carl Schmitt's romantic vision of "real war" or Agamben's post-9/11 state of exception have suggested.[101] War in the five-hundred-year epoch of the Eurocene has always thrived in the interregnum of definitions, national boundaries, racial classifications, humanizing missions, settlement practices, and economizing projects, or even entrepreneurial efforts at reordering the planet. War did not fall from grace. The Eurocene has never known grace, only war.

Therefore, while war is not timeless, it has axiomatic tendencies that run through at least the last five hundred years rather than the last twenty or the last one hundred, and war has a semiautonomous trajectory of change that proceeds and exceeds the rise and fall of ideologies and changes in modes of production—many of the most supposedly outmoded like primitive accumulation still thriving in the form of resource extraction, particularly in obvious states of war. And nowhere has this transideological consistency been more true than in the colonies and subsequent postcolonies in which the continuity of war overshadows the changes in justificatory discourse. The Eurocene as a martial geology has proceeded and exceeded mercantilism, various stages of capitalism, and communism, and will likely continue through the "postideological" throes of our new authoritarianism. Rather than pointing to either a mode of production or a particular mode of destruction, the bits and pieces of quantity theories of war, annihilation over exhaustion, capital drives, desires for order and settlement, and the technics of Europe as well as the innovations of the Americas and elsewhere have found new assemblages in new environments.

The Eurocene, like Sloterdijk's "extended operational space . . . of European earth-users," takes place in the creative and formative collision between the never fully exogenous refinement and expansion of war on the European continent and the radically brutal laboratory of free-fire innovation in the periphery.[102] Unlike Hannah Arendt's description of a boomerang where there is a point of origin and a return, the ecological history of the Eurocene is one of attraction and vitality, renewal and mutation, experiment and habituation.[103] The distance and difference between the continents, technics, enmities, and habitats were for the martial assemblage of the Eurocene the "free energy" of Schrödinger's voracious systems devouring negative entropy. What emerged from this habitat was then leveraged and refined to create what can now be called a global order. However, what is, I think, apparent is that such an order from an ecological perspective has amplified particular attributes and forms of life over others rather than equally consuming bodies and nations.

What it means to say that we live in the Eurocene is neither to say that we are now all European or that a Euro-American constituency has a proprietary claim to the planet. Instead, the Eurocene conceptually demarcates forms of martial life that emerged at the expense of other forms of life and other ecological orders often by means of annihilation such that this order, the Eurocene, is now apparent in the planet's geological and geopolitical history and its continuing present, at least until it is otherwise interrupted by something likely already incipient but not yet sufficiently catastrophic. Resistances persist and even flourish despite the best efforts of annihilation. But no other form of life has gained the virulence or capacity yet to displace the European geological and geopolitical epoch. The rising powers of Russia and China certainly do not represent alternatives to the Eurocene. At best, these states and their appetites will add insignificant phenotypic characteristics on par with stripes or the ornamental filigree of Charles Darwin's finches. But these shifts and changes are not inevitable, just so far very likely. There is no "cene," Euro or otherwise, in an instant, and reality is not path dependent. But conceptually, the Eurocene is visible in its artifacts, practices, habits, relations, and the forms of life tormented and exterminated by those routines. In the next section of the book I explore three particular "operational spaces" in war's ecological becoming. Bombs, blood, and brains each represent microterritories of the Eurocene's ecological expansion and homogenization as well as the Eurocene's limit as each resists explication and operationalization.

PART II. OPERATIONAL SPACES

4. BOMBS AN INSURGENCY OF THINGS

Our time is perhaps the time of an epidemic of things.
—TRISTAN GARCIA, *Form and Object*

What is sovereignty can only come from the arbitrary, from chance.
—GEORGES BATAILLE

Life is nothing more than a Frogger game with IEDS.
—MATT GALLAGHER, *Kaboom*

In a chapter titled "In a Little Plastic Bin," sergeant and blogger Matt Gallagher tries to capture a fleeting moment of sanity after a day as part of the armored cavalry. Gallagher describes in his weaponized beat poetry manifesto an attempt to "embrace the suck" in the comforting seclusion of a porta potty that he calls his "sanity box." Compulsively checking the lock on the door, he spews forth:

> ShootMoveAndCommunicateBOOMBOOM
> Scouts Out.
> ShootMoveAndCommunicateBOOMBOOM
> Scouts Out.
> Emotional burnouts. All of us. Life is nothing more than a Frogger game with IEDs . . .
> iWar?
> Yeah. iWar. iWar. Fitting, in that succinct, catchy pop-culture kind of way . . .
> I War. Subject. Verb. Where's the object? We're still looking for it, some five years later.
> I don't care about you, I don't care about me, and I certainly don't give things. Anythings. Everythings. Things . . . Life makes sense in this little plastic bin. (Gallagher, *Kaboom*, 125–26)

Gallagher is seeking refuge in his "little plastic bin" after a patrol that ended as a tense standoff in the dark with an overturned crate obscuring a suspicious plastic rectangle and a spool of reflective wire leading out from under the crate. In the second before radioing in the improvised explosive device (IED) specialists, hoping against hope that he will not be blown up, Gallagher experiences a gestalt shift. The rabbit is a duck. The suspicious object is not an explosive device; it is a cassette tape. In point of fact, the spooling tape mistaken for wire is coming from a copy of Bon Jovi's *Slippery When Wet*.[1] In a split second Gallagher has gone from staring into the possibility of meaningless death to the intense embarrassment of having almost called in the cavalry to disarm the greatest hits of a hair band from New Jersey. This is the iWar of the IED. The IED can be anything; anything can be an IED. The world can explode at any moment. Life only makes sense in a porta potty.

So if an IED is not a cassette tape, what is it? This chapter makes steps toward defining how to go about asking that question. The unsatisfying answer is that an IED is an assemblage of things. But what those things are is difficult to say. It could be fertilizer, palm oil, a wooden box, and homemade chemicals; a forgotten land mine mated with a cell phone, strung together bits of old copper wire, a nine-volt battery, and a rocket-propelled grenade (RPG) shell; a dead goat stuffed with artillery shells rigged to set off a daisy chain of other explosives buried in the road. The problem is that an IED is a real thing that has changed the course of two major wars but it is not one thing or any particular thing. An IED exemplifies what Timothy Morton calls "a strange irreductionist situation in which an object is reducible neither to its parts nor to its whole."[2]

Morton uses the example of a coral reef. What makes a coral reef possible is not the homogeneity of its parts but a heterogeneous multiplicity. A reef is made up of the calcified skeletons of coral, a particular salinity and turbidity, and a vibrant community of organisms from fish to bacteria. If the turbidity is too high, there will be insufficient sunlight for the microorganisms that form the trophic foundation for the rest of the species. If the salinity is too low, the coral polyps will not thrive and form communities. However, even coral that gives the coral reef its name is not the essence of a coral reef. Artificial reefs formed from concrete, crashed airplanes, or sunken ships can attract a swarming community of sharks, groupers, tang, shrimp, and eels. As these "artificial" structures become inhabited by communities, they can themselves become the foundation for coral to thrive. So the assemblage that makes a coral reef a coral reef has no essence or essential list of parts but, as Steven Shaviro writes, "has its own autonomous power."[3] Further, despite an irreducible set of possible additions or subtractions, we have no problem differentiating between a coral reef and a parking lot.[4]

Gallagher was, among other things, struggling with this very problem. In equal parts, his life and his ego depended on the ability to discern one assemblage of a banana box, wire, and a potential detonation device from another assemblage, banana box, Bon Jovi tape, trash. The stakes of this philosophical and sensory conundrum were his mortal fate. Yet, in the end, Gallagher *recognized* that this collection of objects was not an IED. The rabbit was a duck.

In 2006 the U.S. Army began to categorize IEDs by their method of detonation and delivery in an effort to fix this problem of irreductionism. Explosive devices that are part of an assemblage with a motor vehicle are classified as vehicle-borne improvised explosive devices (VBIEDs). Others are part of human delivery methods in which someone has himself joined into a relationship with an explosive devices such that he is now a bomb. These are classified as suicide vest improvised explosive devices (SVIEDs). Other IEDs are detonated by long wires so that an operator can time an explosion with an approaching target. These devices are called control wire improvised explosive devices (CWIEDs). In addition, IEDs exist in the electromagnetic tributaries that transmit and monitor contemporary life. Human-triggered devices that use cell phones, garage door openers, or old TV remotes are classified as radio-controlled improvised explosive devices (RCIEDs).

Finally, IEDs referred to as victim-operated improvised explosive devices (VOIEDs) are the most common devices found in Afghanistan. These autonomous IEDs do not have a human detonator and commonly respond to the pressure of feet or tires; some feel movement or heat. Like RCIEDs, VOIEDs also make use of the electromagnetic spectrum. However, they are not restricted to receiving human orders. Instead they listen and detonate themselves when they hear radio signals or even jamming devices. Despite the best efforts of identification and taxonomy, the vast majority of IEDs are classified as unknown (UNK). This is because their actualization as technical objects often results in the negation of their existence. Forty-four percent of IEDs do not leave enough behind to be categorized. It is also possible that the pieces that remain are not distinctive enough to be separable from the debris of cars, goats, humans, trash, buildings, and culverts. The IED enacts a zone of indistinguishability between bomb and its surroundings. Even a building or entire alleyway can be part of directing the concussive force of the explosion by either happenstance or planning.

Furthermore, these categories do not exhaust the rich machinic fauna of IEDs. However, this taxonomy is what was developed by the U.S. military for record keeping. This chapter draws from a database of 7,528 IED incident reports (see table 4.1 at the end of this chapter for a summary of the data). The reports were leaked by Private First Class Chelsea Manning in January 2010. After these reports were released by Wikileaks later that year, *The Guardian* began a project to organize the data in ways that would be accessible and meaningful to the public. The War Logs project created a number of interactive data formats that allow readers to explore the vast amount of detailed information visually. Some interfaces give a sense of scale through basics such as pie charts and bar graphs. Other interfaces make use of the geographic information contained in the files to show animated plotting of events such as IED attacks.

This chapter makes use of the full spreadsheet available for public download by *The Guardian.* The incident reports vary a great deal. However, every report contains a tracking number, time and date of filing, description of the event recorded, latitude and longitude, and casualty reports broken down by combatant type or civilian status. Although reports on IEDs in Afghanistan go back as far as 2004, the U.S. military did not begin consistently classifying IEDs by the typology described above until January 12, 2006. It is not clear if there was a specific directive or if it was just the spread of tacit knowledge. The coding process for all its shortcomings in trying to classify a deluge of heterogeneous garbage and burnt ends was accompanied by much more detailed reports. The effort to identify, often speculatively, the type of IED led to a very different attention to the bits and pieces left behind. As a result, the information about the actual designs and components of the IEDs improved consistently after January 2006. A month later, Department of Defense Directive 2000.19E formalized the Joint Improvised Explosive Device Defeat Task Force originally created in 2003 as a permanent program to be known as the Joint Improvised Explosive Device Defeat Organization (JIEDDO).

The program was proposed as a "Manhattan Project" to defeat the improvised explosive device. The mantra of JIEDDO is Attack the Network, Defeat the Device, Train the Force.[5] Record keeping certainly improved under the management of JIEDDO, and there is now a vast amount of data for scholars to sift through, but it is not clear that any of this research did much to defeat the IED. In six years, JIEDDO spent $23.26 billion trying to achieve its three goals. During that same period of time, IED attacks in Afghanistan went from 797 attacks in 2006 to 15,222 attacks in 2012. In that time, 53,997 IEDs and their human collaborators injured more than 11,416 U.S. soldiers and killed more than 1,298 soldiers in Afghanistan. If you include Iraq, IEDs account for almost two-thirds of all U.S. soldiers wounded and killed in both wars.

It is unclear how much IEDs cost, but estimates range from US$30 to US$267 per device. If there is a human detonator involved, often there is an additional fifteen dollars for the task of setting and detonating the device. However, even at the highest estimate, the financial cost to the insurgency for 53,997 IEDs was only $14.5 million compared to the $23.26 billion spent to stop them—or $1,613 spent by the United States for every dollar spent by the insurgency. Given the IED was the most common weapon used in both theaters, to really understand the financial asymmetry one would have to include the price tag for the tanks, fighter jets, battalions of women and men, and most importantly the U.S. version of a drone—the multibillion-dollar unmanned aerial vehicle (UAV) program—in the comparison to really do it justice.

In addition to the amount of violence that IEDs can create at such a low cost, they have also been decisive. The U.S. has not achieved victory against IEDs or their users in Iraq and Afghanistan for the $23.26 billion spent. Instead, the U.S. has pulled out of Iraq and has been trying to withdraw, unsuccessfully, from Afganistan since the end of 2014 with less than victorious standing. The "Manhattan Project" failed.

Very few articles and no scholarly books have been published about improvised explosive devices. Those articles that have been published focus either on technical details for using countermeasures developed for detecting or jamming IEDs or on the *human* social networks as a critical site for targeting.[6] The only piece to think about what IEDs do—rather than reducing them to a technical problem or obscure them as a social problem—is a short piece by Norman Jones and others researching the psychological effects of different IED types on British veterans.[7] This study found that there is a slight statistical variance in favor of command wire IEDs causing more intense post-traumatic stress disorder (PTSD). The researchers conclude that CWIEDs cause more terror because there is a human whose intent it is to kill you. The research infers that accidents are easier to cope with than malevolence.[8]

Given the tidal wave of ink spilled about UAVs or nuclear weapons, it is surprising that IEDs have gained so little attention. Even the AK-47, a similarly revolutionary device, has been the star of several books, and small arms more generally are taken quite seriously by security studies.[9] One can speculate on the paucity of research, but the significance of the IED for the future of warfare is undeniable. In part the lack of interest may be due to precisely what makes the IED so powerful. The IED is not a thing. The IED is a condition of possibility present in almost all contemporary life; IEDs are native inhabitants of a world of global relations and things that hover on the edge between tool and weapon. Unlike the AK-47, there is no Mikhail Kalashnikov or great inventor of the IED. There was no significant scientific breakthrough like Enrico Fermi's sustained nuclear reaction that gave birth to the IED—no proper names, no essential identity or even consistent components; IEDs are the weaponization of the throbbing refuse, commerce, surplus, violence, rage, instant communication, population density, and accelerating innovation of contemporary global life. In this regard, we can learn a great deal about the phase state of the human species by studying what makes something like an IED possible and lethal.

The First IEDs

In some sense, IEDs are not new. There is a nearly continuous use of traps resembling the mine throughout the recorded history of military conflict, and IEDs are only the most recent of "victim-operated devices" or booby traps used in conflicts over land. However, most of the history of mines falls under the category of defensive weapons, the most basic mine being the digging of trenches that were then filled with sharpened and hardened spears. In this case the only potential energy or trigger for the mine was the relationship between gravity (falling) and the ability to puncture the skin because of the reduced friction created by the spear point.

Even with the advent of gunpowder and other explosives combined with pressure triggers and trip wires, mines were primarily used for maintaining defensive perimeters. The idea that such weapons could be decisive in combat is nearly absent from the historical record of military defeats and victories. Before 2001 the only other record of a battle being won as a result of the use of a mine is in 52 BCE during Julius Caesar's conquest of Gaul. Vastly outnumbered by a loose confederation of Gallic tribes, Caesar commanded his troops to build traditional trenches to protect their camp in the open field facing the fortified town of Alesia. Facing a larger force and Gaul's defensive home-field advantage, Caesar also ordered his troops by cover of night to dig short and randomly distributed spiked trenches throughout the field of battle and then to cover the trenches with loose dirt and twigs. The result was chaos on the battlefield. Caesar reversed the telluric advantage of Gaul by deterritorializing the field of battle. Locals no longer felt at home on their own soil, and the resulting chaos broke the fragile confederation of tribes. So while there is more phenotypical similarity between the modern smart mine and the IED, the morphogenetic history suggests that the capability of the IED laid dormant for 2,064 years, as the IED has more in common with what

Caesar's troops called the "lila" or lily of the field than the modern land mine developed for segmentation and the building of perimeters.

To explain how a primarily defensive weapon with an ancient past became dominant in contemporary warfare requires that we break from the developmentalist account of technology. Such accounts start with a simple version of any given device that is then refined or innovated to become more sophisticated. These accounts proceed as if there is a straight line from the chariot to the automobile, each stage of the device passing from one technological breakthrough to the next. However, in the case of the IED, very few breakthroughs account for the technical device itself as even the main "types" of IEDs have a seemingly infinite number of variations in component and placement. In this way, IEDs are not *merely* tools or weapons. They are technical objects as named by Gilbert Simondon. For Simondon, "The technical object is a unit of becoming . . . just as in the case of phylogenetic sequences, any particular stage of evolution contains within itself dynamic structures and systems which are at the basis of any evolution of forms."[10] For example, an automobile is neither a mechanical chariot nor combustion engine with wheels. Instead, in the car is an "internal resonance" with an environment that emerges with it. After all, one cannot separate the explosion in road development and the automobile; each requires the other. According to Simondon, every technical object is "at once technical and geographical."[11] This is what distinguishes a technical object from a tool or utensil. A technical object makes concrete a virtual environment through its own internal consistency.[12]

Following Simondon, Félix Guattari inverts Martin Heidegger's thesis on technological thinking, arguing that it is the machine that demands of us the question of technology, not the other way around.[13] So rather than a general question of technology, "for each type of machine, we will propose a question, not about its vital economy—it's not an animal—but about its singular power of enunciation . . . [its] specific enunciative consistency."[14] This is the problem to be addressed. What is an IED if it can be seemingly anything? We are looking for its "enunciative consistency." Therefore, to account for the emergence of the IED as a kind of consistent effect, there has to be an explanation of the world that emerges with this technical object.

The IED and Its Milieu

The discussion of weapons and their diffusion generally follows the state and the armies that use them. Always starting with great powers and the most "sophisticated" weapons, studies of military diffusion chart the slow drift of devices to the periphery as the technical knowledge and need or prestige of a weapon interacts with the technological capability and in some cases cultural values of a given state-military apparatus. The structural features of an agent, generally a state, either inhibit or induce the adoption of a weapons or fighting technique.[15] However, the accounts of Emily Goldman and others break the technical object into software (content) and hardware (form) and ignore the internal consistency that makes a device "work." As is the case in particular of the IED, there is nothing singularly in its design that would count as a trade secret or invention.

Instead, we have to look elsewhere for the creativity and intensity of the IED. Rather than the laboratory or the inventor, the milieu or ecology of war is itself a source of mutation and innovation. No matter how much we revile the cruelty and devastation of war, it is a "metamorphosis machine."[16] War is generative, in the sense that it complicates the boundaries of the human, and it insists on distributive or assembled agency rather than the mythologies of great leaders or inventors. To this end it is necessary to think about war's creative materiality beyond the simple divides of bullets and bombs, or security studies, and bodies and rights, or human security. Those are all materialities in their own right, but their ways of understanding the world say little about either the broader ecology or the other entities and assemblages that populate war's ecosystem. After all, war has been creatively and formatively more than human since the first rock or bone was picked up to extend the striking power of the human arm or the first horse was mounted to shock the enemy. War did not "become" technological with the predator drone and the biopolitical intervention into the biome of the soldier's body. War has always been an assemblage of things of which any particular human played only a linkage or fulcrum of a larger, more heterogeneous order. So it is in this assembly that we can look for the IED not as an object but as a partner with the humans that reinvents and innovates their return to the field of battle.

So humans are not to be left out of this investigation. It is the humans who are confronted by a war of objects. The human body once immersed in the thick of combat resembles few of its enlightenment capabilities of reason or reflection; it is often the sinew between objects and forces from adrenaline to body memory to the attachment of a so-called suicide vest. War is a highly dispersed and entangled set of relations that are better described as a "continuous multiplicity" than through essentialist representation of discrete agents, things, or objects.[17]

So while it is necessary in this case to take an "object-oriented" approach, to see the world from the IED, we should resist transposing the subject onto the object. It is, at times, highly persuasive and therefore politically effective to anthropomorphize objects, but it is equally valuable in the milieu of war to objectify and then reassemble the operators—goats, mortar shells, wheel barrels—in the complex relay of change and mutation that makes the flow and tumult of war possible. We have to walk a fine line between loosening the grip of anthropocentrism while still holding on to the human element in the puzzle. To go too far into the world of the machine risks losing the rhythms and irruptions of war, an instrumental and mechanistic account of things.

Therefore we have to focus on how the milieu of war and its attendant objects are built into the very architecture and distributed in the economic flows that war inhabits. Economics here should be taken in its broadest sense of *oikos*—the same root for both economy and ecology. Roads, bridges, smuggling, dense urban zones, topographies of race, affect, sexuality, postcolonial legacies, sacred attachment, then overrun by surplus weapons, cell phones, garbage, and concrete all play their part. The oikos, once unrestricted, finds the economy of war written into the very ground beneath our feet. It is here that the *umwelt* or lifeworld of war's objects can be found.[18]

The IED is not merely the wires taken from any number of sources, or its explosive "package" ranging from original manufacture to the artifact of a forgotten war. Neither is the "device" reducible to its "improvisor," the artisan who crafted its new arrangement of receptors and detonation feedbacks. It is the strange attraction or "prehension," as Alfred North Whitehead would say, of all these things in their forces and arrangement. In Whitehead's words, "the whole world conspires to produce a new creation. It presents to the creative process its opportunities and its limitations."[19]

Like Darwin's finches, even within this narrow temporal and geographic corner of warfare there are varied attributes and different morphogenetic histories for each subspecies of IED. Some perceive light, heat, microwaves, sound waves, radio waves, and each an increasingly precise modulation of the electromagnetic spectrum. For detonation some require the interplay or sequence of many of these modes of communication. All require ways of being in the world, that is, ways of fighting war for which disrupting the flow of goods and services is strategically significant; IEDs would not stop the Ostrogoths of the sixth century or the Sioux Nation, nor would they be that effective against the castles and fortified towns of the Middle Ages. Despite the varied existence of caltrops and many other "victim-operated weapons" throughout military history, the effectiveness of the IED is intimate with the larger assemblage it is plugged into to disrupt.[20]

The depth of material conspiracy necessary for something like an IED to emerge as an entity that can devastate the world's most expensive military force compels those interested in the concept of war to look more closely at well-worn typologies of technological superiority as well as technological innovation, diffusion, and deployment.[21] The effect of such a rethinking is not merely instrumental; it does not provide us with a new toolbox to better predict or plan the prosecution of wars. Instead, to foreground the efficacy of things makes the global experience of military acid reflux sensible, as in perceptible.[22] This is how I read Stanley Hoffmann's prescient warning in 1965 that international politics would increasingly become "a global echo-chamber of swarm-life."[23] Partitioning the sensible in favor of the "things" of war shows the values of national security doctrines based on technological superiority and hegemony devaluing themselves as the objects of war fight their way back into politics.

The varied intensities or insistence of objects is not historical in the sense of development; however, there is something like a natural history that has come to pass such that particular technical or machinic objects make statements in more decisive ways. Gilbert Simondon posits the technical object as one that must create and inhabit its environment simultaneously. That is, the object cannot be disassociated from its ecosystem because each had to be virtually implied in the other for the actual object to emerge. For Simondon this keeps open the space of invention without the "new" being ex nihilo. As he puts it, the technical object "is caused by an environment which had merely virtual existence before the invention. The invention happens because a jump is made and is justified by the relationship which is instituted within the environment it creates."[24] The conditions of Hoffmann's echochamber or what William Connolly calls a global resonance machine require a processes of amplification and dispersal, intensification and territorialization.[25] Parts of an assemblage become more vibratory, singing with greater intensity, as in the case of American evangelical Christianity entering into an unexpected relationship with the crusaders of free market capitalism, or in the global feedback from both sides of the supposed civilizational divide, each providing sustenance to the other as in Connolly's example.[26] This raises the question: what is an amplifier? W. Ross Ashby, one of the founders of Cybernetics, describes an amplifier as "a device that, if given a little of something, will emit a lot of it. . . . Such devices work by having available a generous reservoir of what is to be emitted, and then using the input to act as controller to the flow from the reservoir. . . . It works by supplementation."[27]

What is the reservoir war draws upon? Surplus weapons, postcolonial injustice, nationalist affect, or just rage can all be drawn into amplification. As amplification is not just a linear increase, it can result in qualitative changes. Amplification of rage can transform discontent into a change in voting preference; what in chemistry is called a phase transition like the movement from solid to liquid or, more rarely and surprisingly, when matter skips states and jumps from solid to the excited molecules of a gas; from discontent directly to revolt. Manuel DeLanda explains that the extrapolation of physical phase properties to human endeavors allows more fluidity and creativity in the nonlinear history of human development than the "stages" often presumed by development economists or archaeologists. Humanity in its relationship to the environment has "solidified" and "liquefied" at different rates and times; "in other words human history did not follow a straight line, as if everything pointed toward civilized societies as humanity's ultimate goal."[28] This is how machinic statements proceed—amplification or incipient elements lured into actuality by any number of irritants or perturbations, that is, resonances or loose relations that draw upon the abundant resources at hand, intensifying or amplifying the otherwise imperceptible assemblage. And an ecosystem is the medium or milieu that this amplification reverberates through. An ecosystem and its study—ecology—represents a concept for understanding the profound entanglement of objects that make immanent creativity possible at every level of the system.

In the same moment that thought thinks, ecologically there must be some conceptual disaggregation of assemblages from environmental changes in intensity; differing folds or relations of entanglement become relativized such that disequilibrium, the engine for dynamism, is impossible—everything is everything. Guattari playfully refers to the danger of a pulsating undifferentiated substance as "cosmic pulp"—in fact at the cosmic scale it may be pulp.[29] But this is of little use to those of us slogging our way through space-time at the scale of human organic life. Local changes in equilibrium, intensifications, are vital to human experience and efficacy in the world. Therefore the anthropocentrism of some forms of connection, change, and event are necessary for limiting or territorializing zones of entangled matter such that they are perceptible to an embodied, mortal species such as humans. This is what, I think, Gilles Deleuze and Félix Guattari mean by a science: the positing of functives, constituting a molar perspective for the study of local orders and emergent ecologies such as war.[30] The territorialization of space-time into particular geographies of interest or study is the work of a function or, in the philosophical register, a concept. In this case the concepts of assemblage and ecosystem allow for the pulling apart of the world such that a difference in organization is perceptible and a milieu or medium can be at least temporarily diagrammed to show the movement or change in the organization of an IED without losing that this is something we can call an IED. So we need *consistency* for a technical object but we will not find a "condition of possibility" that we would want to call a cause. We are trying to describe the IED and milieu but not its cause.

Instead of looking for the IED's cause, my metaphysical wager is that orders, while not obeying laws of conformity, cluster or are attracted to particular lures or transient forms over periods of time. The important point here is the instance that abstract machines are created and transitory, this description resists the stasis of eternal forms or Platonic ideals. The vulnerability of abstract machines as well as their contingency upon their incorporation in concrete assemblages leave the universe open ended rather than finite and repetitive or merely combinatorial. Deleuze and Guattari refer to these as abstract machines, which they define as follows: "The abstract machine, or machines, is effectuated in forms and substances, varying states of freedom. But the abstract machine must first have composed itself, and have simultaneously composed a plane of consistency. Abstract, singular, and creative, here and now, real yet nonconcrete, actual yet noneffectuated."[31] This allows for there to be a "point" to naming something or identifying it without the necessity or permanency of identity. The space of possibility for the IED is a kind of strange attractor that traverses quantum, micro, meso, and cosmic scales, illustrating what Deleuze and Guattari call refrains, the lures of these strange abstract machines. Unlike theorists of identity or form, Deleuze and Guattari leave open the possibility that the creative advance at any level could also deterritorialize a given refrain, leading a new or differently creative thing to diverge from what came before it. So we are looking for consistency but not permanence. In Connolly's description of consistency, "each level and site of agency also contain traces and remnants from the levels from which it evolved, and these traces affect its operation."[32] This does not mean the world is all flux and therefore impenetrable to thought and investigation. It means we ought not be too disappointed if the concepts we generate have a limited or even fleeting shelf life. So orders? Yes. One order? No. Therefore, real creativity, extending well beyond the human estate, has room to breathe.

Why does a seemingly simple and technical device require so much theoretical or even metaphysical footwork? In part because IEDs do not continue at a single pace, experience, or rhythm of duration. This is why, to some extent, JIEDDO's attempts to settle on categorical and bounded definitions of IEDs obfuscate more than they reveal. Land mines, discarded bomblets, unexploded ordinances, victim-operated weapons, victim rage, exclusionary urbanization, permanent genetic damage passed from generation to generation, e-wasted dumping, cheap arms sales, racial formations, resource extraction—all exhibit different temporal flows and variably transversal connections that can *become* an IED. And the conflict or war that lured it into existence can dissipate and then reemerge days, months, years, decades, or even a century later. Five still-armed and dangerous American Civil War mines were unearthed and detonated in 1965 outside Mobile, Alabama.[33]

The milieu of war often lingers, mutates, and reemerges with differing levels of intensity long after "sustained combat" has ended. It is from the betwixt-between that something like an IED can emerge. It assembles in a maelstrom of decades of electronic garbage dumping, the accumulation of mines from past wars, new and old antagonisms, and stop/start development projects responsible for a morass of paved roads.

All of this raises the question of the relation between the part and the whole. If causality is not a sufficient category, then is an IED a subset of war? Is it part of the assemblage of war? Certainly the ecology of war is populated by heterogeneous assemblages made of people, affects, guns, hills, culverts, streets, mines . . . And those objects are transversally entangled in space-time. But what is the nature of such entanglements? How do we account for things as assemblages without losing either their distinctiveness as objects or their functions in larger machines?

I am looking for the answer in the entanglement between objects and the temporal rhythms of the IEDs, expressive statements that Guattari referred to as "enunciative consistency." Certainly an IED that explodes is part of its expressive statement. However, other bombs do that too. So the IED's capacity to explode bears the singularity of an IED because of its ability to wait, to listen, to become part of the road or city. The IED, like Caesar's lila, has the ability to produce a fundamental distrust of the familiar. The ground beneath one's feet is suspect. The interpenetration of these attributes, and more, is the expressive statement, the consistency, of an IED. The explosion is kinetic, thermal, affective, communicative, and implicated by the IED's ability to lie in wait while the former attributes were still virtual. Expressive statements replace what would otherwise be called an "effect" in "cause/effect" explanations but it is more than just an effect. The presumption of cause's independence from effect is not possible if we want to describe assemblages such as IEDs. The technical object and its milieu loses its consistency or "eventness" once we start trying to isolate or vivisect the components—six inches of copper wire, discarded cell phone, eighteen ounces of fertilizer—for positivistic description. The tension is between the permanence or identity of the object presumed by positivist modes of analysis and the object's relation to its movement or flow that constitutes that identity.

Alfred North Whitehead takes up this problem of objects in nature at the end of *An Enquiry Concerning the Principles of Natural Knowledge*. For Whitehead, objects can be described with relative ease using mathematical proofs and formal logic if an object is a particular function or thing that has uniform characteristics independent of space and time.[34] However, math and logic lose their footing for Whitehead once objects inhabit a world replete with processes of ingression. Not unlike the relationship between Euclidean and Riemannian geometry, the difference is between discrete objects in empty space and continuous space gradients that cannot be separated from the things that make it up. In the pithy words of Timothy Morton, ecological thought "permits no distance."[35] What Whitehead attempts to do in going beyond the limits of formal logic and mathematics is to give conceptual grip to what Henri Bergson calls *elan vital* or Jane Bennett calls the *vibrancy of matter*.[36] However, unlike Bergson and closer to Bennett, Whitehead keeps life or liveliness immanent to matter. The thin line to be walked is between a kind of panpsychism in which everything is equally agential and a mechanistic matter that dooms creativity or change.

It is worth following Whitehead through his attempt to work out a concept of rhythm that is not wholly reducible to matter but is still of *this* world. Whitehead begins by laying out the problem between the object and the event of liveliness that gives it meaning or identity:

The specific recognizable liveliness is the recognized character of the relation of the object to the event that is its situation. Thus, to say that the object is alive suppresses the necessary reference to the event; and to say that an event is alive suppresses the necessary reference to the object.[37]

The problem is one of the necessity of separation or discreteness in describing a thing. Any attempt to separate an object from its *process* or event so that each can be analyzed and described eviscerates the essence of the analysis. Whitehead identifies the problem and the solution in the question of the relationship of time to the object. However, this is not a matter of object + measure of time, that is, the traditional definition of change. This is insufficient to understand an object. The event is not merely the passage of time: T1 being the object before event, T2 being the object after event; the resulting slice now available to be metabolized for data. Instead, Whitehead gives the following explanation:

> Life (as known to us) involves the completion of rhythmic parts within the life-bearing event which exhibits that object. We can diminish the time-parts, and, if the rhythms be unbroken, still discover the same object of life in the curtailed event. But if the diminution of the duration be carried to the extent of breaking the rhythm, the life-bearing object is no longer to be found as a quality of the slice of the original event cut off within that duration. This is no special peculiarity of life. It is equally true of a molecule of iron or a musical phrase. Thus there is no such thing as life "at an instant"; life is too obstinately concrete to be located in an extensive element of instantaneous space.[38]

The ability to analyze cause and effect, object and event, separately such that one could establish "discreteness" requires a reliance on the instant, a moment without movement or life. No such moment exists and therefore in order to understand things in an ecological way, it becomes necessary to study them underway rather than embalmed in the freeze frame of efficient causality. The freeze frame is insufficient. Its limitation explains why Deleuze and Guattari say that science requires a slow motion of a particular slice of chaos rather than a snapshot. For Deleuze and Guattari, following Whitehead, multiple temporal flows give *life* to things. It is this process that must be captured by a concept, whether that be war or any other event from steam to cellular differentiation. This rhythm or consistency within the flow of becoming that has form acts as a lure—what Deleuze and Guattari call an "abstract machine." But what is lured in the case of the IED?

This lured creativity gives rise to what Major-General J. F. C. Fuller called the "constant tactical factor." According to Fuller's observations on the battlefield, there is never a moment of totalization in which technology equals dominance. Instead, Fuller argues that "the sole thing impossible in war is for it to stand still. That directly a weapon approaches or enters the master stage, the constant tactical factor comes into play."[39] The constant tactical factor, when extended beyond the ambiguous anthropocentrism of Fuller, is like the refrain or abstract machine; it is the intensifying force that amplifies those on the wrong side of a hegemonic mode of war. This rhythm of creativity organizes people, matter, technology, cities, spaces, geography, and flow toward disruptive innovation. Tinkerers, artisans, and inventors participate in a nomad science that is neck deep in the assembly of things. Reassembly and reorganization rather than ex nihilo invention is the mode of production for the constant tactical factor. The mutations and territorialization that occur in the wake of this abstract machine are more likely experimental or improvised rather than tested or budgeted for assembly line production. Nomadic warfare is, in Deleuze and Guattari's terms, a minor science.

It is clear that IEDs experience many such moments of mutation. In the summer of 2004, in response to the paving of roads in Afghanistan, insurgents stopped burying IEDs and started using explosively formed penetrators (EFPs) with explosive-shaped charges that created molten copper slugs traveling at twenty-six thousand feet per second and while activated near the road could be hidden as much as a hundred yards from the target. The millions of dollars spent paving and widening roads to make the burial of IEDs and remote detonation more difficult only benefited pressure-triggered and radio-controlled mines. These IEDs feel heat rather than listening to radio waves or feeling pressure. Even the $3 billion developing jammers was spent in vain.

Soldiers responded by creating decoys, hanging toasters and hair dryers on ten-foot booms in front of vehicles to set off EFPs. Within weeks, insurgents angled the IEDs so that the slugs shot beyond the decoy to strike the vehicle. By fall soldiers in communication with the Department of Defense (DOD) developed heat decoys that could be adjusted. These "Rhinos" were deployed along with jammers. Beginning in 2010 new EFPs were triggered by the jammers. Insurgents in Afghanistan developed a pressure-sensitive wooden box with no metal parts that could deploy ten or twenty EFPs at a time and was undetectable with any existing scanning or portable X-ray device. In the years between 2004 and 2006, the number of IEDs increased tenfold and the geographic distribution expanded from one or two provinces to the entire theater of operations. Soldiers poured into Afghanistan in an effort to quell the escalating violence. Between 2007 and 2011, the number of U.S. troops on the ground increased from 26,000 to 100,000. This fighting force was backed up by the JIEDDO's $4 billion budget and staff of more than two thousand experts. In 2012 sixteen thousand IEDs were deployed.

J. F. C. Fuller's account of the constant tactical factor assumed that only the artisans of war were the engines of difference and mutation. However, in the case of the IED it is the stubborn perdurance of hi-tech and manufactured waste dumping that provides the near limitless flow of materials and parts from place to place. The protocols of production, waste disposal, and consumption habits—which are never entirely human—generate the exteriorization of waste from the centers of cutting-edge commerce to the periphery. The Global South is a kind of eddy in the flow of global capital's need to hemorrhage the objects of planned obsolescence to make room for new demand

and new products. But these seemingly disposable things that break so easily never break completely. The dominion of human control is limited. The insistence of things exceeds the attempts to ship, bury, burn, and smash them. The bits reassemble in the hands of other artisans, or even simply in the new space-time in which the perdurance of a dormant machine from another time reemerges, exploding on a battlefield in ways unimagined by its inventors or manufacturers.

The constant tactical factor is a limit to war's totalization because the machinic capacity of things exceeds the enframing of humans. Anthropocentric views of technology (either pessimistic or optimistic) fail to capture the dance of distributive agency in the obstinacy of things. Creativity is a grand conspiracy. In the words of one army colonel, "The enemy found a seam. I don't think they knew it was a seam, but it just happened."[40]

The Milieu of the IED Is Saturated by War

> War is not simply the antithesis of civilization. Historians refer constantly to war without really knowing or seeking to know its true nature—or natures. We are as ignorant about war as the physicist is of the true nature of matter. . . . During the fifty years with which we are concerned, war punctuated the years with its rhythms, opening and closing the gates of time. Even when fighting was over, it exerted a hidden pressure, surviving underground. —FERNAND BRAUDEL, *The Mediterranean and the Mediterranean World*

If it appears that the mutational character of the world is gaining momentum, becoming more perceptible, it is because we have crossed the *pharmakon* of technics. The Bergsonian image of humans as the rise of toolmakers is overwhelmed by a world of surplus, a built environment reaching the escape velocity of its builders. To take just a few examples, fifty million tons of e-waste are generated each year. This is enough to fill a train long enough to encircle the entire planet. The e-waste comes from the 716 million new computers that will go into use this year, up from 183 million five years ago. Each will have an average lifespan of two years rather than the six-year average in 2005. Around 700 million cell phones will be sold this year, and their initial life span is about eighteen months.[41]

To put all this in the context of two current conflict zones, Iraq and Afghanistan have 10 million land mines each (the population of both countries is 30 million, so that is already enough land mines to kill one-third of the population). The top-ten most mined countries have an estimated 101 million mines. A global count is impossible to determine, particularly if we include the countless improvised mines built from surplus munitions and then mated with one of the millions of castaway cell phones.[42] According to the Small Arms Survey, there are 900 million firearms and that number is increasing by about 8 million per year. It is an estimated $6 billion market. Small arms account for the overwhelming majority of the 208,300 direct conflict deaths between 2004 and 2007. The number dying from armed attack inside and outside official "conflicts" is closer to 740,000 a year.[43]

Further, the overwhelming majority of that violence, whether by direct attack or the flood of waste and surplus objects of military and nonmilitary origins, finds its way into the expanding slums of the world. For the first time in human history, more than half of Earth's total population lives in cities.[44] With the rise in urban living, the number living in slums has also increased dramatically. The percentage of those urban dwellers who live in slums is over 1 billion and is expected to double by 2030.[45]

While I said Earth is saturated, the global diffusion does not resemble an equilibrium. Of the top-ten most mined countries, all are in Africa, Central Asia, Southeast Asia, or the Middle East, with the exception of Bosnia. All direct conflict deaths took place outside the United States and Western Europe. This is also the case with urban slums. In sub-Saharan Africa 72 percent of the urban population lives in slums, in South Central Asia 59 percent, in East Asia 36 percent, in Western Asia 33 percent, and in Latin America and the Caribbean 32 percent. The highly industrialized nations have a 6 percent slum rate by comparison.[46]

This description seems to suggest that the study of war's ecology ought to be a question of political economy, given the arms market, the growth of cities, and the burgeoning slums. However, it is not the arms that are sold and purchased that have altered the outcome of two major wars but the weapons left behind, the actual surplus from previous wars that then combined with the deluge of electronic waste shipped, dumped, and smuggled throughout the Global South. It is also not the overpopulated struggling cities of the Global South pumping out billion-dollar military platforms or invading nations halfway around the world.

Political economy thought in traditional terms is ill-equipped to deal with the far-from-equilibrium systems of global life. After all, economic explanations of urban growth failed to predict the dramatic increases in urban size and slum intensity. The prevailing wisdom of development economics was that the persistent and intensifying economic decline since the 1970s experienced throughout the developing world would slow the growth of cities.[47] The exact opposite has occurred. In fact, as jobs and economic opportunities have diminished substantially along with the buying power of wages in cities throughout the Global South, the number of people relocating to cities has spiked upward, seemingly undeterred by the economic conditions. Paul Virilio makes a compelling case that these newly emergent megaslums ought not be called cities or even really urban. They are makeshift settlements whose only apparent similarity to earlier settlements called cities is the density of population. The ultracity, as Virilio calls it, does not follow the rhythms or any organizational structure we would commonly call urban. In many cases the megaslums have a kind of persistent temporary status. This is seemingly a contradiction in terms but one that aptly describes the situation in which many of the slums began as refugee camps or emergency living facilities and, while outlasting their temporary or emergency expectations, still have extraordinarily transient populations as well as a nearly continuous rate of creation and destruction as the result of attempts by the state to raze settlements and the often fragile structural integrity of shacks and shanties. Virilio uses the phrase *revolution de l'empart massif* to describe the conflicting movements of internalization and exclusion.[48] Mike Davis argues that austerity measures, wars, and the general failure of cash crops and resources in export-based development models have created a kind of perfect storm for the displacement and relocation of rural populations to urban centers.[49] Some of these factors are within the traditional jurisdiction of economics; however, none represents a smooth relation between economic indicator and outcome. Therefore it is useful to restore oikos to its larger purview that supersedes economy, to return to ecology all of its political, material, and historical inflections.

The presence of traditional military resources such as tanks and fighter jets that would be used to judge the material power of opponents in conflicts is no more indicative of outcomes than economic growth is predictive of urbanization. In these calculations, tanks and planes would be quantified to make judgments about military strength that small arms and abandoned mines would not. For instance, by traditional judgments of material power, Saddam Hussein was better armed and prepared for war in 1991 and therefore stronger than in 2003. However, the first Gulf War was a resounding success by military standards because Saddam Hussein fought the coalition forces using a large stockpile of weapons and training received from the licit and illicit Western and Soviet arms markets.

Technological assistance referred to in the U.S. budget as "security assistance" involves a very different flow of goods and services than the emergent objects of waste and surplus that form the milieu of the IED and the seemingly insignificant trade in small and light arms. When states play by the norms and procedures of modernist warfare, a *quantity theory of war* is operative and exhibits relatively predictable outcomes. J. F. C. Fuller coined this phrase to describe the mobilization and execution of nationalist war under the power of steam. For Fuller, a new episteme of military thinking emerges from an assemblage of mass nationalism, gunpowder, steam engines, communication lines, and capitalist economies, such that "it is war that shapes peace, and armament that shapes war."[50] To have more of something, a quantitative advantage, the units being measured have to be interchangeable or roughly isomorphic. If the application of force by opponents is similar in kind, then the quantity of force applied in the form of bullets, tanks, and bombs may determine the outcome of war. In the case of the first Gulf War, quantity was on the side of the coalition forces even though the Iraqi military was well armed.

The current conflict in the Persian Gulf and Afghanistan, however, is intensively and extensively asymmetrical—not in the sense that it is uneven, as is often the understanding of this term, but in the sense that there is a mismatch or incommensurable difference between the resources and tactics of each side. James Der Derian refers to this mismatch of contemporary global politics as heteropolarity rather than multipolarity as the poles are compositionally nonidentical.[51] The forces being mobilized are not comparable at the level of tactics, organization, or agenda. Therefore, quantity is not predictive of the outcome. In fact, quantity is not quantifiable. Clausewitz calls this the impossibility of polarity in war that results because there is seldom anything approaching equilibrium.[52] So to say that the U.S. military possesses more tanks than the Taliban or Pashtun fighters is true, as neither possess any armored vehicles. But that accounting of relative strength would only be relevant if both sides were fighting a tank war. For example, IEDs could be counted and compared, say, to their kissing cousins the "smart mines." However, that would tell us little about the possible outcome of the conflict. The United States could possess twice as many mines as its competitors and the uncertainty would still persist.

In part this is because of the differential flows and organizations of the opposing forces. The United States relies on major roads that can accommodate large caravans of trucks and vehicles. Without a nearly constant flow of goods and soldiers, the U.S. military would starve. Therefore, IEDs consistently do damage because convoys, whether for supply or patrols to ensure the passage of supplies, are highly susceptible to disruption. And that vulnerability is not reversible. Understanding this asymmetry requires understanding the assemblages of things that organize the differing lifeworlds spanning the theater of operations.

While it may be impossible to exhaust the census of "things" or give a causal accounting of all the objects that make up a milieu, the multiplication of objects that take part can provide a foothold for navigating the emergence and recurrence of conflicts. The improvised explosive device is exceptional for this pursuit because its recurrence, mutation, and advance have ravaged the roadways and urban corridors of the present U.S. occupations of Iraq and Afghanistan despite the best efforts of the DOD to target the supplies used to construct IEDs and the initiatives to track down and eliminate the humans who build them. According to a 2010 report in *The Guardian*, IEDs have accounted for nearly half of all combat deaths and half of all casualties in Iraq and 30 percent of deaths and 50 percent of casualties in Afghanistan.[53] Since the $30 billion counter-IED effort began in Afghanistan, the number of lethal IEDs has tripled. It should be mentioned that the numbers of civilians killed by these machines is difficult to measure, but estimates suggest that numbers exceed U.S. combat deaths of all kinds.[54]

The sluggish response of U.S. and NATO forces to detect and counter these weapons suggests that while the U.S. prepares daily for multiple scenarios of nuclear conflict, its preparation for these weapons was nil. Ironically, in 1996 the United States opposed Lloyd Axworthy's attempts to organize support to ratify an anti–land mine treaty because of the defensive advantage the DOD believed land mines represented in multifront wars against nonstate enemies. The U.S. position on land mines shifted 180 degrees after the Bill Clinton administration. While Clinton did not sign the treaty to ban land mines, he did stop the U.S. production of mines that would have been in violation of the treaty. The George W. Bush administration began production of devices that would be in violation of the treaty. The Barack Obama administration has also cited military necessity, arguing that signing the land mine ban would make it impossible to meet "national defense needs" and "security commitments."[55] The use of mines was thought to represent a national security asset that far outweighed the cost to civilian life throughout the Global South.

The U.S. military was blinded by an anthropocentrism incapable of recognizing the machinic character of the land mine, its ability to evolve in unpredictable ways. The human security position that advocated the treaty was equally misguided in its presumption that something defined as a land mine could be "cleared" or banned via the restriction of production and trade of objects defined as "land mines." There was a technological essentialism that presupposed land mines to be a particular, discrete, whole object differentiated from an entire ecosystem of other current and surplus objects. It failed to appreciate the fecund zone of indiscernability between military and nonmilitary *things.* When evaluated under the rubric of military necessity, the calculations of risk and reward presumed a technological hubris that assumed the United States would only ever encounter mines as a tactical device for defensive perimeters. The deployment of mines against the U.S., the most powerful and technologically advanced military on the planet, was thought to be little more than an inconvenience. Hi-tech armor, meandering winding magnetometers, satellite surveillance, and the increasing preference for airpower in the projection of force abroad underwrote the confidence that land mines were a device to be used by the U.S. rather than being reassembled from First World debris to be used against the United States.

The attraction or singularity of the mine is—more than can be deduced from its disaggregated parts—demonstrated by the heterogeneity of seemingly interchangeable parts that form the whole of any particular IED. To describe the components of the IED as highly varied only begins to characterize the shifts in assemblages that make an IED possible; IEDs are ambient, integrated, and distributed by methods that make it difficult to detect and combat. Unlike precision weapons, IEDs are neither smart nor dumb. *They are aware.*

The IED Is Ecological

As was discussed in chapter 2, war is an ecological system of deep relations and nonscalar mutations. Change happens all at once without a defined origin point or mappable process one could call initial conditions. For the IED, the urban battle space, entered by foreign invasion, created and then intensified new connections between a disconnected or disinterested group of objects, including the detritus of globalization such as broken garage door openers, old artillery shells, bits of wire, and half-dead batteries. The point of detailing such an ecological history, in the words of DeLanda, is to "specify the structure of spaces of possibilities, spaces which, in turn, explain the regularities exhibited by a morphogenetic processes."[56] These are the "slices" of slow chaos from which we can learn something about the function of a particular organization of the world.

Encountering the improvised explosive device as a species, in DeLanda's terms, is valuable for exhibiting the egregious arrogance of humanism at war; a morphogenetic account foregrounds the efficacy of extraordinarily inhuman actants. The "ontological theater," to borrow Andrew Pickering's term for quixotic technological devices, or the dance of human and inhuman agency, as Jane Bennett would put it, of the IED resists the instrumental view of war as a tool that is ready-to-hand.[57] It charts the practices of war that convert and condition open-ended bodies recognizably human and otherwise-than-human into alliances that can all too easily be organized and deployed for violent ends.

The very plasticity of people and technics that goes into the planning, mobilization, and deployment of the "things" of war belies the hubris of believing in the monopoly of human agency. The dilation of causality, efficacy, creation, and destruction in a veritable menagerie of actants makes more legible the surprising self-organizing actions of those "things" that are too often believed to be under the dominion of the humans that developed and deployed them.[58] Whether that be the supposed docile bodies of Iraqi civilians that *become* an insurgency or the innovative introduction of obsolete weaponry or more precisely the emergent assemblage of the two that *become* an IED, we must attend to multiple types of agency in war ecologies. To borrow again from Pickering, "things are unenframable."[59] Jane Bennett calls this "thing power," "the curious ability of inanimate things to animate, to act, to produce effects dramatic and subtle."[60] Often, much to our dismay, agency is promiscuous and gregarious and finds its efficacy where it finds attraction and connection.

So an ecology of the IED involves not just the objects; it involves connections that resonate through their chancy efficacy. However, this network of connections that underlies the concept of ecology is not just the recent advent of a vast series of tubes known as the internet or the satellites that make global telecommunication possible. Further, the network is much more than the distributed information systems of John Arquilla's netwar or the system of total battlefield awareness dreamed of by the American Defense Advanced Research Project Agency. From these presentist and instrumental perspectives, networks are little more than operational means rather than ontological processes, as the instrumental network—following its Cold War roots—is only relevant and desirable because its dispersed and redundant organization impedes targetability and decapitation. However, the humanism of such an interpretation misses the productive and directive capacity of nonhuman forces in networks better called assemblages. An assemblage is neither a metaphor nor a human-designed architecture but an actual plurality of relays, resonances, and physical interfaces that emerge as an assemblage. An assemblage is a real order of consistency. Therefore, to take the ecology of war seriously, to be humbled by its challenge to a humanist will-to-control, is to resist operationalizing its complexity so that thought can remain open to be provoked by the IED's aleatory and often horrifying creativity.

The information regarding the materials used in the assembly of IEDs as well as their tactical placement and success or failure to detonate in Afghanistan is culled from the 7,526 U.S. military incident reports filed between 2004 and 2009 and made available by Wikileaks. These data have been compiled and organized by *The Guardian* and are publicly available as Declan Walsh, Simon Rogers, and Paul Scruton, "Wikileaks Afghanistan Files: Every IED Attack, with Co-ordinates," *The Guardian*, July 26, 2010. Information of Iraqi IEDs comes primarily from the public audit by the General Accounting Office (GAO) of the Joint Improvised Explosive Device Defeat Organization. The GAO report was released as "Defense Management: More Transparency Needed over the Financial and Human Capital Operations of the Joint Improvised Explosive Device Defeat Organization," March 2008, http://www.gao.gov/highlights/d08342high.pdf. Specific references will be cited throughout. However, the landscape perspective I have developed regarding IEDs comes from sifting through all the data made available by Wikileaks, which is more specific and detailed than is helpful for the argument made in this chapter.

Additionally a number of NGOs, media outlets, and think tanks were used to fill in the holes in the casualty and IED data:

- Tom Vanden Brook, "IED Casualties Dropped 50% in Afghanistan in 2012," *USA Today*, January 18, 2013.
- "Afghanistan Civilian Casualties," *The Guardian*
- United Nations Assistance Mission in Afghanistan, "Afghanistan: Annual Report 2013; Protection of Civilians in Armed Conflict," February 2014, http://info.publicintelligence.net/UNAMA-CivilianDeaths2013.pdf.
- Rick Atkinson, "Left of Boom: 'The IED Problem Is Getting Out of Control. We've Got to Stop the Bleeding,'" *Washington Post*, September 30, 2007.
- Luis Martinez, "Last of 33,000 Surge Troops Leave Afghanistan," *ABC News*, September 20, 2012.
- "How Many U.S. Troops Are Still in Afghanistan?," *CBS News*, January 9, 2014.
- Civil-Military Fusion Centre, "Executive Summary: A Global Review (2012–2013) of IEDs and ERW in Afghanistan," *Afghanistan in Transition*, September 2013.

TABLE 4.1. IED statistics, Afghanistan, 2001–2012

Year	U.S. soldier deaths	Civilian deaths	Civilian wounded	U.S. soldiers wounded	IED attacks	JIEDDO budget in millions	Number of U.S. troops
2001	0	0	0	0	0	0	0
2002	4				22	0	4,100
2003	3				83	0	4,100
2004	12	122	96	62	191	0	20,300
2005	20	47	126	135	366	0	20,300
2006	41	347	770	279	797	3,700	20,300
2007	78	360	993	405	1,147	4,400	26,000
2008	152	518	1,257	790	1,632	4,300	35,600
2009	275	793	1,569	1,215	3,420	3,100	68,000
2010	368	881	874	3,441	15,225	1,900	98,000
2011	252	949	941	3,542	16,554	3,460	100,000
2012	132	868	964	1,744	15,222	2,400	68,000

5. BLOOD VITAL LOGISTICS

Human blood may seem like a peculiar choice for a book about trying to simultaneously decenter the human from international relations and politicize the making of current global political order, but as we will see with the brain in the next chapter, even those *parts* that we take for granted as being subsumed by our sovereignty as individuals challenge the continuity and authority of the human as well as the mechanistic cosmology that underwrites a particular form of life emerging as dominant among others. The idea of blood is not foreign to our thinking about nationalism and belonging. The geopolitical tradition of international relations marched through the Rhineland of the nineteenth century, tracking blood and soil through the circuitous pathways of geographers like Alexander von Humboldt and nationalist historians like Leopold von Ranke, and blood was certainly still present in the triumphant unification triad of land, state, and people by Carl Schmitt. Blood as the nomenclature of national or tribal continuity is readily used to this day to draw lines of enmity that constitute the political of international politics. So blood is a major player in the nation-state and world of nation-states and has been for at least as long as something like geopolitics has existed.

Blood politics is not restricted to Europe's heartland or the international; it was a defining feature of the gigantic emerging federal power across the ocean as well. There is no object of American jurisprudence and legislative history more infamous than the single drop of black African blood. The still commonly used reference to the one-drop rule refers to Virginia's antebellum hypodescent laws, which codified a long-standing mythology of blood and blood difference that hopscotched from biblical interpretation to phrenology to Nazi anthropology and back to American eugenics. The 1924 Racial Integrity Act and the subsequent eugenic policies restricting miscegenation and institutionalizing compulsory sterilization that continued late into the 1970s demonstrate the formative and pervasive horror of sanguine logic. As Alexander Weheliye argues, "juridico-political territorialization of racial hybridity frequently serves to solidify the ordering of humans along racial lines rather than heralding the suspension of racializing assemblages."[1]

As this story unfolds, it is clear that this metaphoric blood was not the blood of circulation, oxygenation, and coagulation but an imaginary blood—sacred and profane—for the provincial world of human affairs. The real blood of platelets, hemoglobin, and lymphocytes has no use for this sordid history and in fact resists its own signification through its insistent indifference to racial difference. The metaphorical droplet of blood is no match for the pipeline of plasma that ran through Hawai'i and then Europe and Northern Africa and then returned to the Pacific during World War II and continues to support global military operations. This actual blood alters the course of conflicts, undermines long-held beliefs about racial evolution, and continually countermands the exuberant will-to-control of twentieth-century science.

Blood finds its material footing at precisely the moment biopolitics emerges as an organizing principle of global total war. The ecology of the global U.S. alliance system from World War II through the Cold War and beyond is not solely based on the civic republican ideals that were said to bind the Allies against the Axis. More than ideology holds the United States and its European and Asian allies together. In addition to treaties of mutual defense are the practical treaties of blood exchange.[2] The symbolic pacts of loyalty are sealed with very real blood oaths that call upon vast infrastructures for the movement, preservation, and acquisition of blood products and even whole live blood for the so-called walking blood banks.[3]

Despite being the most overrun metaphoric dumping ground for intensive human drama and divisiveness, blood is affirmed in this chapter as a real thing that has a place and a role in the formation and creative advance of the international-cum-global. Human blood is both fugitive and indifferent as well as formative and insistent. The materiality of blood resists both the provincialism of human-manufactured racial difference and the hubris of a scientific mastery that believes itself capable of control via the breaking down of heterogeneous assemblages into their fundamental or component parts. The former greatly disadvantaged the Nazis during World War II and the latter requires that in many cases the U.S. military is only able to use 1.5 percent of its forward-deployed blood products before they rot on the shelf.[4] The great breakthroughs of blood pressure supports such as the protein albumin and other blood products demonstrate the ubiquity and indifference of blood. However, the failure of such methods also performs the limits and failures to master blood and shows just how insufficient parts are for the sustaining of life. Blood is differentially generic and insistently univocal in its heterogeneity. To put it another way, blood is an assemblage that defies essence or formula while being predictable and consistent.

The complexity of blood is to be found in it being an assemblage of objects such as proteins, lipids, cells, water, and minerals. Further complicating matters, the object blood is also refracted through a series of technical and somatic connections that range from proximate to global. A natural history of the adoption of abstractable blood and blood products as a key component of modern warfare and its extraordinary waste under the guise of military readiness is meant to be more than a set of hot facts about the excesses and failures of global empire. I find blood and its peculiar resistance to mastery and signification quietly heroic and worthy of its own exploration. Rather than playing some passive role in the ascendency of science as the lingua franca of the biopolitical state of affairs, blood sluggishly nags at the very grounds of war.

Blood nobly refuses to submit entirely to humanity's petty squabbles. To try to capture blood's virtue, I will sketch out the emergence of the U.S. Armed Services Blood Program. Further, I will detail the role militarized blood procurement played in steering blood from a medium of medical intervention to a national strategic resource with all its attendant and troubled global networks of acquisition, flow, and policing, such that blood and blood products became a global commodity. On the way to the state of current practice, we will take a few necessary detours through Nazi Germany's failed blood program and the demands for and success of the American-led Blood for Britain program despite its resonances with the racial logic of the Third Reich.

What I have in mind by object and assemblage is not much more than the commonsense definition, but there are a few attributes of these concepts that are not as common to sense. In casual conversation, an object is a thing. So far so good. However, commonly things are grammatically and causally subordinated to subjects or in some cases first causes like gods. So there are two necessary subtractions necessary before proceeding. First, everything is an object. And when I say object, I do not mean to imply the opposite of a subject. Instead I simply mean those actual things that perdure or hold together against the grain of a universe that is winding down. Objects, as science fiction writer Stanisław Lem calls them, are "islets of decreasing entropy" in a sea of noise.[5] Jane Bennett has similarly championed objects by channeling Baruch Spinoza's concept of *conatus* to describe the will or tendency of all things to hold on to existence.[6]

The second subtraction is somehow more invisible to our language and description. Objects are not passive receptacles or mechanical pieces in a Rube Goldberg device of subjective agency and causality. Objects have powers, capacities, and attributes that make them formative and collaborative. Nothing gets done without a crowded room of things working in relation with each other. Objects are real and continue to be even when we are not looking. Trees fall in the forest and make a sound even when humans do not hear them or more importantly do not have the idea of hearing them. Objects are real and formative and continue to be so independent of human perception and cognition of them. So there is not a divided world of formative conscious things (humans) and inert usable things (objects). Humans—like all other things in this story of blood, race, and conflict—equally have a role but not a lead role. All things are constrained and enabled by capacities and relations and so are ontologically equal. Put succinctly by computer programmer and alien object advocate Ian Bogost, "anything is thing enough to party."[7]

Assemblages are heterogeneous collections of objects whose relationships are differentially intense. As the intensity and organization of the collection or herd of objects changes, so does the expressive effect of the assemblage even if the population of things remains the same. Things do not dissolve into an assemblage but neither can one have an atomistic explanation of an assemblage: "a strange irreductionist situation in which an object is reducible neither to its parts nor to its whole."[8]

Similarly to Morton's coral reef example, discussed in the previous chapter, we can discern the difference between whole blood, plasma, red blood cells, hemoglobin, lipids, white blood cells, and so forth. Following Morton, this discernment is neither merely epistemological nor a subordinate relationship between part and whole. Despite the innovation of a technique called "blood fractionation" to discern and describe the parts of blood's distinct functions, the reassembly does not neatly add up to the collective we call blood. In fact, the body responds much better to fresh whole blood as compared to the defrosted cocktails of the reassembled parts. In the case of blood, the body responds differently to whole blood versus plasma versus the volume-creating protein albumin versus the oxygen-carrying red blood cell or the protein hemoglobin that bonds with oxygen so that the red blood cell can hold on to oxygen.

Also, blood does not quite seem to not be blood simply because of a low occurrence of one of these components. We do not say immunosuppressed individuals do not bleed simply because the fluid in their veins is missing substantial numbers of white blood cells. As is seen in the fits and starts of fractionation research, the responses of the body to concentrated forms of blood components are whole responses; they are not part of a response that adds up to the whole assemblage, blood. Thus the difference between blood and plasma or plasma and albumin is a real difference that is experienced by our body as much as it is known or captured by our concepts, even though blood contains both plasma and plasma contains albumin among many other proteins.

I am saying that things can be different things depending on their relationship and that such differences are neither predictable nor fully knowable even retrospectively. However, before dismissing such an outlandish claim for its logical contradiction, consider for a moment that it may not be the description that is contradictory or illogical but reality itself. In this case, why should we expect the world to live up to our standards of what we wish that it was, such that our logic could be neatly operative? What we are confronting is the difficult and irresolvable tension between atomism, form, and movement. Unfortunately, none of it exists at the instant or as we would like it. We cannot catch the becoming of objects or their collaboration as assemblages in the act, so to speak. Instead, we fumble around as provocatively as possible, in hopes of learning something from the world. This is what I think is meant by an object-oriented thinking.

Raced Matters

It shall hereafter be unlawful for any white person in this State to marry any save a white person, or a person with no other admixture of blood than white and American Indian. —Racial Integrity Act of 1924

At the turn of the twentieth century, blood was suffuse with meaning. European and American humans in particular invested in their crimson fluid the legacy and essence of their civilizational difference. It is this concept of blood that is meant by phrases that equate tribalism or race wars, such as blood feuds, blood rivalries, bloodlines, blood is thicker than water, and differences are in the blood. Blood was and often is a synecdoche for race, and so differences of blood are not merely differences; they provide the distinctions for superior and inferior inheritance that is meant to justify or at least give grounds for war, colonialism, and slavery, and the subsequent radioactive fallout of racism.

The persistence of this racial story of blood is closely related to the sacred logic of blood. That something in blood was constitutive of one's essence is an affectively powerful belief. Blood sacrifice, blood ritual, blood oath, blood brothers—the list goes on. To quote Buffy the Vampire Slayer, "it is never a lymph ritual"; it is always about blood. The fact that the symbolic tradition of blood has lost its sway on public policy and practice is due more—in my estimation—to the insistence and relative indifference of actual blood rather than the attempts at demystifying race. In the case of World War II, the demand for blood—its biopolitical need—began to overshadow blood's resonance with earlier mythological technologies of sorting populations. The use of transfusions by some daring field medics in World War I had demonstrated the salubrious effect of additional blood in trauma cases. Despite limited knowledge of how to extract or store blood, it became immediately apparent to those treating the wounded that the ability to replace lost fluid substantially improved the chances of survival.[9] However, the use of blood was very limited because of a lack of knowledge and technology to carry out transfusions, and the homogenous fighting populations of World War I provided few encounters for dispelling the myths of pure and impure bloodlines.

Blood, for many reasons, is very difficult to utilize even under perfect conditions for exactly the same reason it is such an asset. Blood's almost immediate inclination when removed from the body is to clot. Blood is also not entirely indifferent. Rather, it is quite attentive and observant of its surroundings and cohabitants whether revival antigens or the presence of gasses or injury.[10] While the human concept of race is meaningless to blood, foreign cells and microorganisms provoke an almost immediate and often violent response. In some sense, blood is racist. The antigens present on the red blood cells that give the fluid assemblage its color are in some cases very specific as to what other kinds of blood they will party with. Type A blood is accepting of other type As and also type Os but everyone likes type Os. Type B antigens play well with other type Bs and of course like Os. Type AB is the Andrew W. K. of blood and will party with anyone. Sadly, while type O is infinitely generous to others, its blood serum cannot tolerate anyone's antigens. So from the standpoint of blood, there are more or less four races of humans, but that racial difference exists between parents and children, aunts and nieces, cousins, and so on. However, in cases in which this difference cannot be tolerated, the consequences are lethal. Blood without antigen compatibility results in agglutination, a sudden and destructive clumping together of the unlike blood.

Two major breakthroughs emancipated blood from the path dependency of one body's vessels, allowing for a serious application of transfusions and subsequently blood's global adventure. Sodium citrate was found to block the clotting agent in blood so that usable viscosity could be maintained long enough to collect and then administer blood to a patient in need.[11] Before this discovery, arteries from the donor had to be directly sutured onto the recipient, often resulting in permanent damage to the donor. The second discovery was the ability to test for blood antigens so that type could be determined before a transfusion. These two innovations made blood portable and abstractable, and a new kind of war made portability and abstraction necessary.

The intense aerial bombing of England by the Germans in World War II, in particular the bombing of urban areas, led to serious injuries requiring transfusions but also made the ability to collect blood and store it extremely difficult. Even though sodium citrate could prevent clotting, blood quickly degenerates unless refrigerated. Frequent power outages and the need for medical resources to be on the move was in direct conflict with the demands of blood storage.[12] Across the Atlantic, blood research had been progressing in the U.S. since World War I. The standardization of blood typing and the ability to perform transfusions without permanent injury to the donor resulted in an immediate for-profit market in the U.S. Under the watchful eye of the Blood Transfusion Betterment Association, professional donors were issued books to record and approve all donations. To become a donor, one had to have a clean bill of health, refrain from drugs and alcohol, and most importantly have a telephone. As storage was nearly impossible, donors had to be reachable at all times.[13]

In 1937 Doctor Bernard Fantus coined the term *blood bank* and along with many others around the world began to investigate means for increasing the ability to accumulate and store blood. In Russia, cadaver blood initially appeared to be the solution. A massive hospital in the middle of Moscow known as the Sklif had thousands of beds and nearly constant trauma and emergency traffic filling it every day.[14] The centralization of care and the sheer number of bodies entering the door made it possible to acquire sufficient blood from the recently deceased, discarded placentas, and other parts brought to the hospital.

News of the success in the USSR inspired Fantus to begin collecting and storing blood at Chicago's Cook County Hospital.[15] Although the hospital was not willing to harvest cadavers, the idea and method of storing blood was a conceptual breakthrough that inspired the possibility of blood's serious commodification, as it meant that donors were no longer needed to be present at the time of transfusion. Shortly thereafter, Charles Drew, a doctor from Howard University who had recently completed an advanced medical degree at Columbia University based on his blood research, attempted to push the capability of storage a little bit further. Drew made an exhaustive study of all available knowledge on blood and blood characteristics and ascertained that blood could be effectively broken down into its major components if allowed to settle.[16] He further refined the process of using the spinning of blood to increase the effectiveness and speed of its separation. One of the components, a syrupy yellow substance called plasma, remained stable for much longer periods of time without refrigeration. Although plasma lacked the oxygenating or immunity properties of red and white blood cells, Drew's research showed that it had the ability to raise and stabilize blood pressure that was found to substantially lower the death rate in victims in which severe shock would normally cause a rapid decline.

Again plasma was only a temporary solution for hemorrhaging patients, as it could not provide oxygen. But the temporary solution was all that was often necessary to stabilize patients so that more complex procedures could be completed. The breakthrough of plasma was not just in its storage life. In fact, what makes plasma so interesting is that it has no antigens. The degree to which blood distinguishes between other types of blood is not present. So in plasma Drew had found a thing that was completely generic and indifferent to human difference while still having the effect of stabilizing patients.

The finding was of immediate interest to the U.S. Department of Defense (DOD), which was trying desperately to supply England with blood. Plasma represented something that if acquired in sufficient quantities could be shipped to England. As can be expected, there were immediately concerns over purity by both the DOD and the British government. Although desperate for blood, their fear of killing people with tainted blood was quite high as was the general reticence to have a substance from another body pumped into one's own, as transfusions were still quite novel. The process of producing plasma in significant quantities while assuring quality control also required extraordinary and unprecedented technical expertise, and so concerns were reasonable. In pursuit of this goal, the Blood for Britain program was created and given substantial resources to create the infrastructure necessary to provide England with safe and pure plasma. The DOD decided, against medical advice, that part of that high standard for purity included collecting blood only from healthy white donors for plasma production.[17]

After substantial consultation, and two national meetings of all available experts, there was a consensus that only one man possessed the knowledge and capability to create and execute such a program: Charles Drew.[18] The only problem was that Drew was black. Despite the racial policy regarding donors, necessity overwhelmed the DOD's racism in choosing a program di-

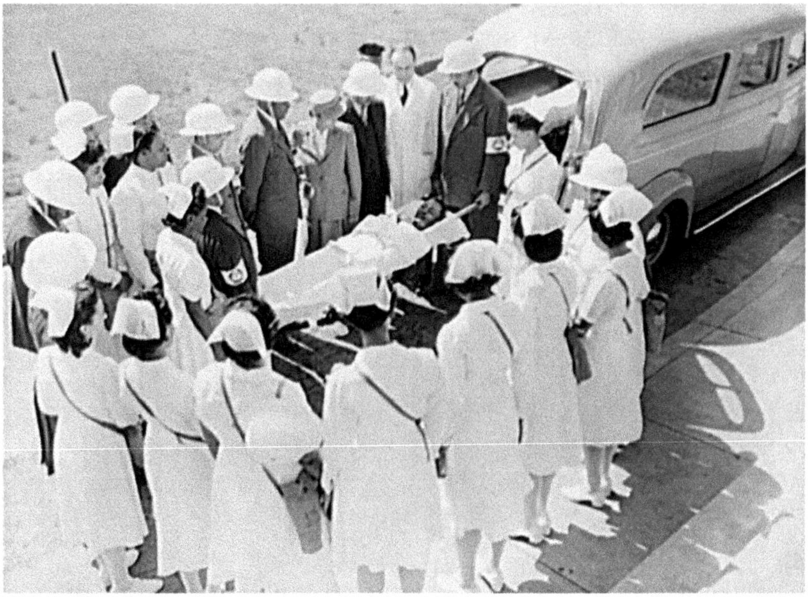

FIGURE 5.1. Charles Drew demonstrating treatment of air-raid victim. Photograph by Roger Smith, 1943. Library of Congress Prints and Photographs Division Washington, D.C. 20540. Farm Security Administration—Office of War Information Photograph Collection.

rector. Drew was offered the job and despite the race policy accepted the position and honorably executed his duties. The program was successful despite the incredibly laborious task of collecting sufficient blood and producing tested and controlled plasma. Drew saved countless lives through his efforts but could not donate his own blood to the cause.

The irony was that Drew's innovation had made it possible to produce a substance at an industrial scale that could be used in any body regardless of blood type, much less race. In 1942 the DOD under pressure from black newspapers and activists allowed the American Red Cross to begin collecting black blood for black soldiers.[19] Previous to this, black soldiers were told that they could receive white blood as it did not contain the impurities that made black blood threatening to white bloodstreams. It is difficult to confirm how many African Americans lost their lives as a result of blood prioritization; however, there were definitively shortages in both the European and Pacific theater that could have been offset by willing and vocal potential African American donors.[20] It is a matter of historical contingency and good luck that the population of African Americans in the U.S. armed forces and

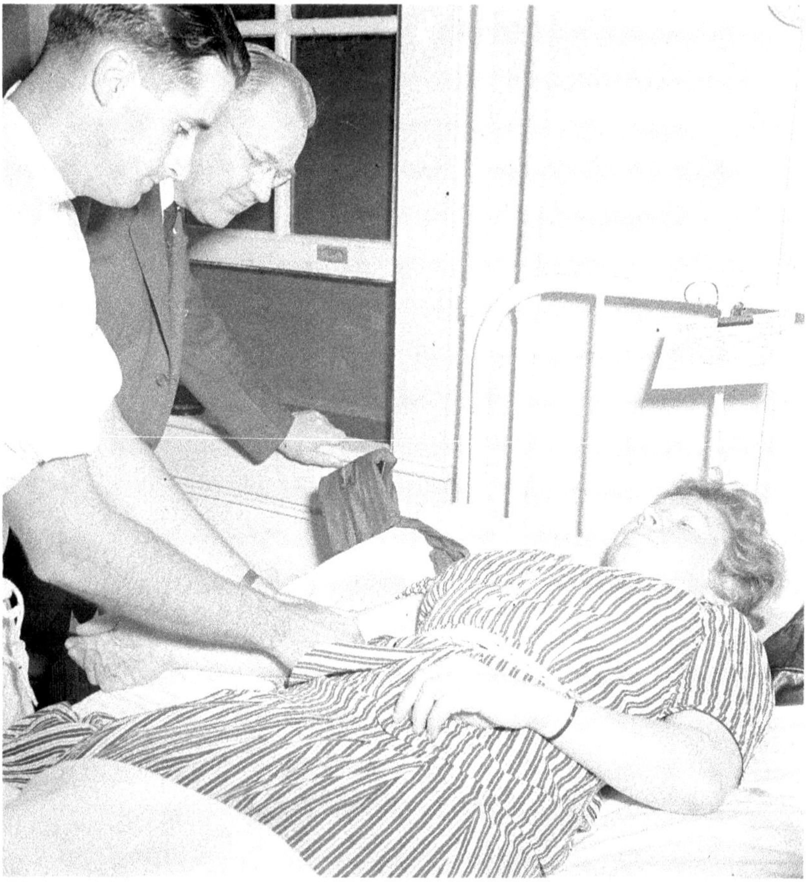

FIGURE 5.2. Civilian volunteer blood donor. Photograph by Marjory Collins, 1942. Library of Congress Prints and Photographs Division, Washington, D.C. 20540. Farm Security Administration—Office of War Information Photograph Collection.

in the donor population was relatively small. Otherwise many more U.S. lives would have been lost as a result of the DOD's refusal to listen to the racial indifference of blood.

The soldiers of Nazi Germany were not as lucky. Unlike the United States, which relied on the rumors of racial mythology and left the policing of that mythology more often to personal prejudice than legislated penalty, Germany developed a robust scientific literature to support the empirical basis of racial difference. After substantial research, a less than 2 percent variance in the frequency of occurrence of type A blood among so-called Aryan populations as opposed to type B blood in Slavic and Jewish populations came to form the basis of the 1935 Nuremburg Blood Protection law. The first man punished by the law was a Jewish doctor who valiantly saved a patient by transfusing his own blood while performing surgery. Unfortunately, the patient was not Jewish and Dr. Hans Serelman was sent to a concentration camp for "polluting Aryan blood."[21]

An overwhelming majority of the best doctors and research scientists in Germany were sent to camps and murdered because of their so-called blood race. This had devastating consequences for the German military as the Nuremberg laws were enforced on the front lines as well. Legal blood transfusions were almost impossible because of the fear of tainting blood. The metaphoric "one drop" could, after all, be hiding in anyone. Further, research into transfusions and other lifesaving medical procedures was hobbled by the imprisonment and massacre of the German medical class. After the capture of Dr. Paul Schultze in 1942, it was revealed that battlefield injuries for the Germans were resulting in dramatically higher death rates than Allied forces and that the only method developed for coping with hemorrhagic trauma and severe wounds was a derivative of vinyl that was being injected into soldiers in an attempt to increase clotting, as additional blood was just not feasible to acquire.[22] Comparatively, in 1942 the U.S. successfully exported 31,250 gallons of blood and plasma.[23]

The U.S. made further improvements because of Dr. Edwin J. Cohn's development of a method called fractionation that allowed for the isolation of the protein albumin identified as responsible for plasma's ability to expand and support blood pressure. The stabilizing aspect of plasma could now be isolated and importantly dried into an easily transportable high-density powder. The breakthrough was immediately classified as a permanent Allied edge. Albumin, like plasma, was indifferent to blood type but unlike plasma could be easily transported in smaller quantities and reconstituted with minimal risk of contamination and could transmit almost no diseases, substantially lessening the demands of testing and quality control.

After the classification of albumin, blood and blood product were a permanent part of the American arsenal. Blood was, like other resources for war, a strategic reserve that had to be maintained and sufficiently stockpiled to guarantee military readiness, just like rubber, iron, and petroleum. By March 1945 more than 2,000 units of blood were being shipped per day to forces in the Pacific and Europe, totaling more than 500,000 units, or 62,500 gallons of blood, in thirteen months.[24]

By 1950 the extensive research on blood and the horror and embarrassment of Nazi blood laws led to a standoff between the American Red Cross

FIGURE 5.3. Plasma inspection. Photograph by Arthur S. Siegel, 1943. Library of Congress Prints and Photographs Division Washington, D.C. 20540. Farm Security Administration—Office of War Information Photograph Collection.

and the Department of Defense.[25] The decision was made to end the segregation of blood in part because of the political mobilization to end the practice, but that mobilization was significantly aided by blood's properties, which had insisted, empirically, that race was superstitious animus, not reality.[26] The making biological and sanguine of civilizational difference represented by nineteenth-century concepts of species based on the coordination of phenotype and geography and then further indexed by lingering and conflicting mythologies of multiple descent were in policy, at least, coming to an end. In its place another sense of the term *race* began to emerge as described by Foucault.[27] Described as the petty normative, culture and class differences that characterize the internal war of politics and are externalized by modern war as a productive national body politic slowly displaced earlier notions of blood difference with a concept of difference that would pit the race of Americans against the race of Soviets. Foucault terms this "state racism."[28] Race became a squabble over forms of life rather than strains of blood. Blood's material necessity for the nation eclipsed the value of archaic notions of blood race.

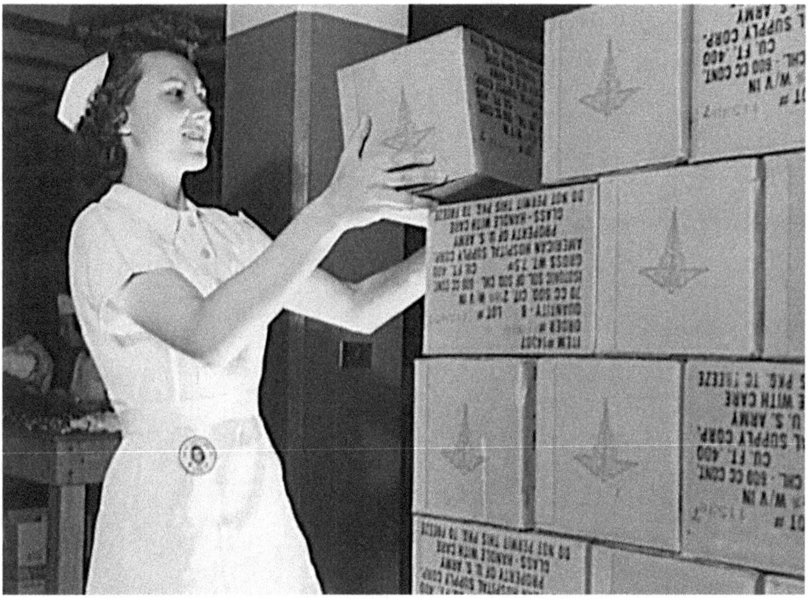

FIGURE 5.4. Transfusion bottles ready for shipment. Photograph by Howard Hollem, 1942. Library of Congress Prints and Photographs Division Washington, D.C. 20540. Farm Security Administration—Office of War Information Photograph Collection.

However, biopolitics is not sufficient to explain this process of deracialization. Rather, the capacity of blood and blood type to continually defy racial logics plays a significant part in the possibility of national security winning the upper hand against archaic racial blood. Blood's indifference to the human superstitious investments in difference undermined attempts at reinventing racial mythology. Even in the case of Nazi Germany, in which the resources of a whole nation were at the disposal of race thinking, manufacturing scientific racial knowledge was both ultimately impossible and catastrophic. The plasticity of meaning could not in the final instance hold up to the recalcitrance of blood. Further, blood's ubiquity and mysterious capacities inspired and brought together a global community of scholars. The research necessary to execute Dr. Drew's program for Britain, and its extrapolation into the industrial-scale production of blood for the Allied war effort, required that the circulation of information generated by blood's peculiar capacities and attributes crisscross the planet from Paris to Chicago to Moscow and even more unlikely across the American color line.

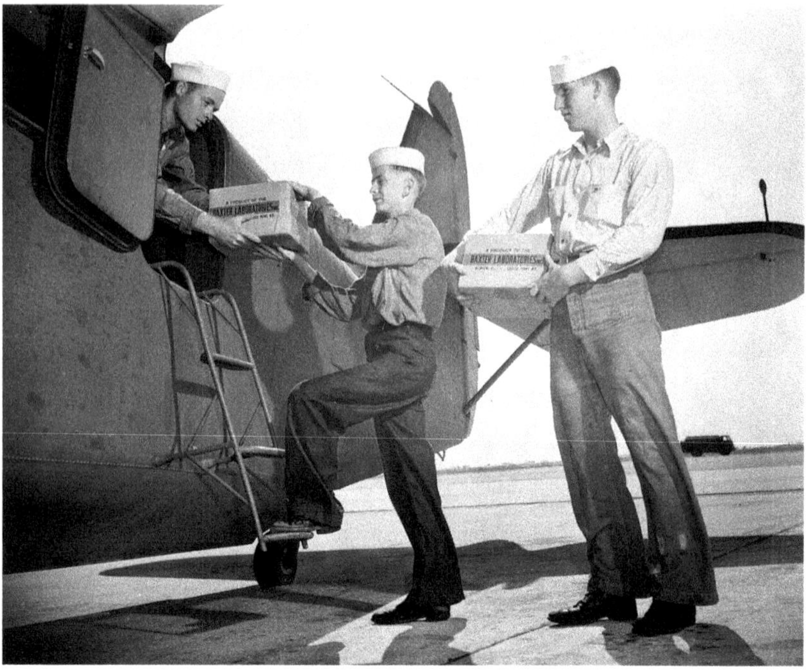

FIGURE 5.5. Blood shipment to the war front. Library of Congress Prints and Photographs Division Washington, D.C. 20540. Farm Security Administration —Office of War Information Photograph Collection.

The Assemblage Strikes Back: Fractionation and the Limits of Atomistic Science

If blood provided a challenge to racial thinking, it also continually asserted itself in the face of scientific control. Although in many respects the initial results of plasma and albumin were promising and ultimately made a huge difference in the war effort, it became clear well before D-Day that neither was a substitute for whole blood. Observations from field medics and doctors on the front lines were reporting cases of soldiers who initially were stabilized by plasma or albumin but would begin gasping for air and ultimately die.[29] The soldiers were suffering from a lack of oxygen because they did not have sufficient red blood cells to bond with the oxygen being drawn into their lungs. These soldiers were suffocating from the inside out. In the North African theater, American doctors had noticed that French and British doctors less enamored with the technological breakthroughs of U.S. blood products were producing lower morbidity rates by using whole blood transfusions. It is worth noting that the French were capable of this because they wholly embraced African and Arab

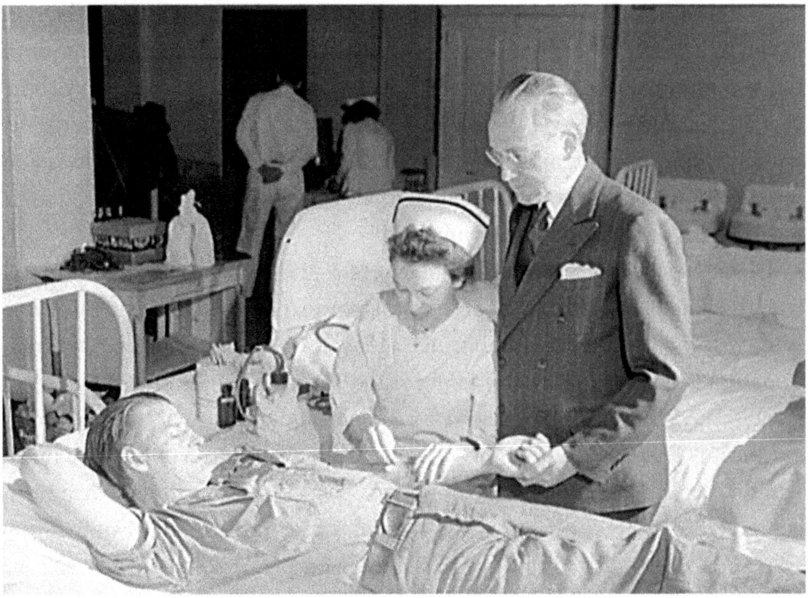

FIGURE 5.6. Prisoners at San Quentin prison giving blood. Photograph by Ann Rosener, 1943. Library of Congress Prints and Photographs Division Washington, D.C. 20540. Farm Security Administration—Office of War Information Photograph Collection.

donors as patriots fighting for the French cause. The French Republic in exile did not recognize phenotypic racial difference at the level of blood, and blood in turn affirmed the Republic's egalitarian practices of transfusion.[30]

Subsequently, the presumed causal object of pressure stability was increasingly under scrutiny. Albumin and plasma would continue to be important tools for trauma doctors, but the full assemblage of blood was necessary to ensure higher survival rates among soldiers. The challenge was to similarly commodify and industrialize whole blood so that it could be brought to bear on the battlefield. This was not an easy task. In fact, to this day the ability to collect and store whole blood is severely limited.

Portable refrigeration and other techniques would be deployed, but it became common knowledge that the window for useable whole blood was only about two weeks. This temporal fragility of blood is expressive in two ways. First, blood expresses the failure of its own disaggregation. Second, it is now clear to anyone paying attention that the stockpiling of blood means imminent invasion. Unlike the stockpiling of bullets or tanks, the window for successful use of blood is very short, and as whole blood must be collected in a form that cannot be concentrated, the number of donors requires the last-minute, vast recruitment of the civilian population. As a result, the closely guarded secret of when D-Day would commence was inadvertently announced by the scramble to gather sufficient blood.[31] The limitations and insistence of blood's integrity as an assemblage to produce the desired salutary effect made secrecy nearly impossible.

Despite even more significant advances in the storage of blood, it is still true that changes in the flow of blood are now expressive of coming war and thus closely monitored by intelligence agencies.[32] In part this is because fractionation is not capable of atomizing blood so that it can be stored and reassembled. Fractionation has also not been capable of manufacturing a blood substitute. There is no True Blood, despite the fact that extensive chemical and physical analysis of each component part of the assemblage has been completed repeatedly. The parts rendered and concentrated have measurable and important effects, such as stabilizing pressure or oxygenating or improving clotting. However, it is not possible to somehow reassemble or synthesize these in a way that fulfills the body's demand for blood.

These limitations on blood's shelf life and the failure of a viable substitute have produced a vast global network for the U.S. armed forces. The Armed Services Blood Program (ASBP) was institutionalized after the failure to restart the World War II ad hoc blood process during the Korean War. It became clear that blood had to be continuously on demand wherever U.S. forces might be. Further compounded by the post–World War II reconstruction combined with the extraordinarily high casualty estimates and diverse geographic scenarios for potential war during the Cold War, the ASBP, like the U.S. military, is deployed worldwide. This places demands on the program that require not only a vast internal infrastructure for collection, screening, storage, and transport but also a number of treaties and procedures for the local acquisition of blood when the ability to acquire and deliver sufficient supply from home is not possible.[33] The result of the treaties and regularization of blood flow creates a kind of Allied blood supply. Organized and codified through NATO, blood's fragility combined with the demands of U.S. empire have created a new racial-geographic bloodline. Safe and secure blood comes from an assemblage of bonded countries and populations held together by mutual defense, a vast interoperable medical infrastructure and surveillance network, and the exchange of sacred fluids. Despite blood's resistance to archaic racism, its circulation as a commodity has been institutionalized such that blood's collection, storage, and transport is organized

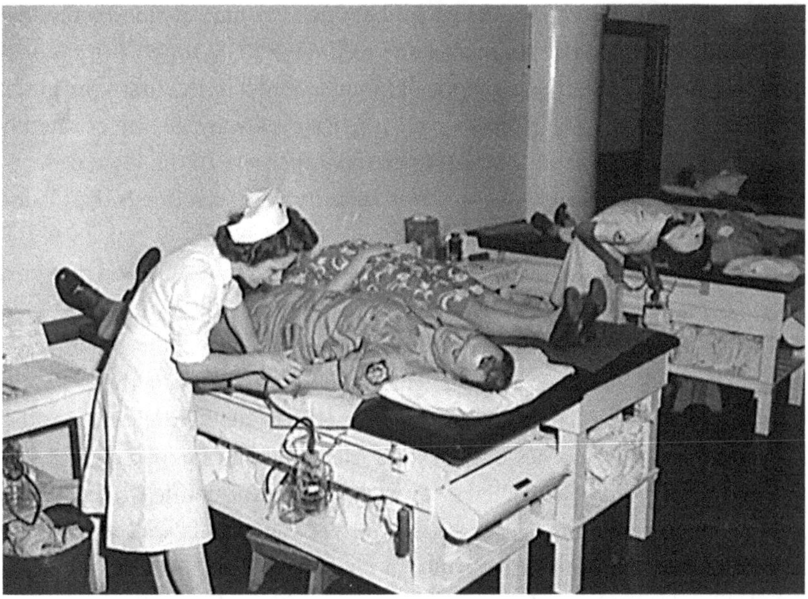

FIGURE 5.7. D-Day blood donors. Photograph by Howard Hollem, Edward Meyer, or Mac Laugharie, 1944. Library of Congress Prints and Photographs Division Washington, D.C. 20540. Farm Security Administration—Office of War Information Photograph Collection.

around lines of enmity not so distinct from the racial lines of the previous century.

So in part the creation of the ASBP was driven by the failure to master blood and the vulnerability created by that lack of mastery. There was a strategic and mutative interaction with this medical institutional failure and the material constructions of new identities being shaped by the terrain of violent geopolitics. Further amplified by months of very high casualties in Korea, the ASBP became a permanent organization whose job it is to oversee and coordinate blood collection and the assessment of blood needs for each of the military branches. Further, the practical necessities of blood governmentality developed in tandem with new investments in authority and security as a form of governance animated by future threats rather than existential needs. The ASBP empowers the Surgeon General of the United States to set the level of the strategic blood supply based on the assessment of the domestic and global threat level.

The shift from fulfilling practical demand to the logic and organization of security is most visible in the immense waste produced to overcome the short shelf life of blood. The ASBP mandates that five days of blood estimated on theater conditions be stockpiled and resupplied at all times.[34] In the second Iraq War, of the 11,250 gallons of blood shipped in the first year of the conflict, fewer than 212 gallons were used. That is a wastage rate of roughly 98 percent.[35] Blood is an interesting material measure of the logic of security, demonstrating the degree to which security's construction is built using more than words and discourse.

During the Kosovo conflict, for every one hundred units of blood provided to the battlefield, fewer than two units were able to be used before the blood had to be disposed of.[36] With comparatively few casualties, this rate of waste was sustainable. However, at the height of the second Iraq War, the U.S. military was consuming ten thousand people's worth of blood every six months. In many cases this was not only beyond what could be shipped over; it was beyond what could be collected or even purchased within the United States, despite significant advances in storage for red blood cells and other blood products that are used successfully in civilian settings. As a result, fresh whole blood (FWB) transfusions were common during Operation Iraqi Freedom despite the high risks of disease transmission. This practice runs contrary to every basic public health requirement for transfusions dating back to the 1920s and yet is demonstrative of the power of blood as an assemblage. The FWB transfusions have been demonstrated to interrupt cycles of coagulopathy common when high amounts of frozen red blood cells or other blood products are used. The transfusion of whole blood also reduces cases of hypothermia and rebleeding that occur in trauma patients receiving large volumes of frozen blood or blood products.[37] However, FWBs can only be performed with proximate donors like those of the earliest blood banks. In the 1920s, on-demand donors were called "blood on the hoof," but in the military context such individuals are now referred to as "walking blood banks."[38]

Security's unquenchable demand for blood in its encounter with the limitations for controlling and sustaining blood stockpiles has created unique problems for the ASBP. The result is the necessity to institutionalize policy for the collection of blood in theaters beyond the circle of friends created through NATO. Blood and security's peculiar future-oriented assemblage has further deterritorialized the imposed boundaries of enmity. As stated in the army's Field Manual 8-55, frozen blood is only meant to support soldiers while self-sustaining blood programs are set up. The field manual is explicit that in mature theaters, blood supply is based on fresh liquid red blood cells and fresh frozen plasma from the donor base.[39]

To achieve this often-impossible goal, host countries are assessed for their level of cooperation and supply, which in the case of Iraq and Afghanistan includes a population that ranges from financially desperate, to friend, to enemy. Civilian and World Health Organization studies of blood supply and infrastructure, to be reviewed ostensibly for development goals and humanitarian aid, are funded by the U.S. Armed Services Blood Program to gather sufficient intelligence about disease and infrastructure. For example, an article in *Transfusion*, one of the leading blood journals, titled "A National Mapping Assessment of Blood Collection and Transfusion Service Facilities in Afghanistan," details the location, quality, and supply liabilities for the entire nation of Afghanistan, province by province, not unlike a geological survey for oil, listing statistical variance of infectious disease and capacity limitations. The last section of the article is titled "Conflict of Interest." Under the heading it reads, "The authors declare no conflicts of interest," despite the fact that the article's research was fully funded by the Military Infectious Disease Research Program and Armed Services Blood Program and further reviewed by the Walter Reed Army Institute of Research for any classified or objectionable material.[40]

In addition to the assessment of supply, the field manual goes further to clarify the legal status of enemy blood and its use. According to the manual's interpretation of the Geneva Convention, there is nothing restricting enemy prisoners of war from "donating blood." As such, the field manual recommends that the blood type of prisoners ought to be collected and this information ought to be included with other intelligence regarding available resources. Further, blood captured from the enemy ought to be turned over to blood bank platoons to be used in medical treatment facilities.

The imposed lines of friend and enemy or the biopolitical redefinition of state race produced around the securing and threatening of whole populations is institutionally defined through blood treaties, but in conflict zones, blood even exceeds those boundaries. The need for blood and its generic quality supersedes the vital difference of the enemy even as it is often racialized or signified as blood. The U.S. military cannot help but make a policy that demonstrates that enmity is arbitrary in the face of actual blood that is the real substance of the enemy.

Enmity can of course be said to be merely political and therefore a different realm from biological considerations, but the political is often a decision at a scale of war that can only be found in the essence of a mass enemy. Without essence and identity, the contemporary techniques of war cannot take extreme risks with whole populations. At some level "they" must be somehow fundamentally like the enemy that hides among them. At least "they" are sufficiently different from "us" because if they were not, how could we take such risks? The archaic blood cannot help but sneak back on to the battlefield as war requires not objectification of the enemy but racialization of the enemy in the making of a population. Such racialization may take the form of treating those that we fight as objects, but that only raises the more fundamental question: why do we treat objects so badly? If one takes for a minute the position of blood-becomes-object rather than objectifying, it is impossible not to see the irony in practice of modern biopolitical warfare that is waged in the name of a population difference (race) by human bodies that seemingly have no problem sharing blood before they spill it.

6. BRAINS WE ARE NOT WHO WE ARE

What Shall Be Done with Our Brains, Or, the Problem of Plasticity

The only laws of matter are those which our minds must fabricate, and the only laws of mind are fabricated for it by matter.
—JAMES CLERK MAXWELL, February 1856

But suppose that there is a mode of tragedy in which what we witness is the subjection of the human being to states of violation, a perception that not merely human law but human nature itself can be abrogated. —STANLEY CAVELL, *The Claim of Reason*

Don't you see, Peter? I'm not safe. It's my mind. Ever since the pieces of my brain were reimplanted, it's been changing me back to the man I was before. Bit by bit, I'm losing the man that you helped me become. —WALTER BISHOP, character in *Fringe*

In J. J. Abrams's genre-bending masterpiece *Fringe*, the lead character, Walter Bishop, is confronted with his double from a parallel universe. In this twisted version of the twin paradox, it is not the speed of light that marks the difference between the Walters but a single decision decided differently by each Walter. The Walter quoted here experienced a moment of shock when he discovered that he almost destroyed the entire universe by traveling between dimensions. In this brief moment of humility, reminiscent of J. Robert Oppenheimer's response to the atom bomb test, Walter decides to make himself dumber.

Walter compels his best friend to give him a selective lobotomy in hopes that his hubris will be restrained and his ability to achieve such scientific feats will be safely limited. Evil Walter (Walternate) from the parallel universe is genetically identical, and his history is also identical in almost every way. However, a slight difference causes Walternate to embrace his intellect and confidence. The result is an unmatched brilliance but one that no longer has any regard for human life, as every cost can be rationalized, thought to its final conclusion. The result of unfettered brilliance is a Walter who is sociopathic, maniacal, and almost unstoppable.

In order to counter the threat of evil Walter, the sweet and quirky lobotomized Walter must reimplant the lobotomized tissue. The caring and kind man who has grown attached to this world and its inhabitants is struck by the horror of plasticity. Despite his *choice* to cut out pieces of his brain and his *choice* to reimplant the lobotomized pieces, he cannot control his return to hubris and destruction. The man who cares about the world enough to risk becoming something he hates is becoming a man who will no longer care about the world. That is, the indexical value "care for the world," which distinguishes the two Walters, is lost in the moment of reimplantation. There is no going back; the man Walter will become *cannot* care that he is what he has become. The brain is a kind of degree zero for the Eurocene as it presents the horizon for what makes us human but also the entry point for a politics armed with a mechanistic metaphysics bent on making and unmaking the human. In Peter Sloterdijk's words, "the good old possessive pronouns sound like presentations of grammatical folklore."[1] To begin unraveling the brain, to open the field of thought to the field of engineering or design, is to usher in an apocalypse simultaneously philosophical and practical.

This is the paradox of action and ethics for modern humans. Catherine Malabou's question, "What should we do with our brain?" is a real question, but the choices before us are not under our control. They are leaps and bounds, and where we land erases the point from which we leapt. This problem of a disappearing measure of change is a limit to plasticity not as a process but as a thinkable question. So the investigation of one's own plasticity is always a kind of thought experiment. When the process actually takes place, whether subtle or dramatic, it is often imperceptible. Like Walter, we have flashes of transition. Similarly, stroke victims and those suffering from Alzheimer's become frustrated when they feel capabilities they once had but now seem cut off from. Schizophrenics have moments of pharmaceutical and nonpharmaceutical clarity in which they fear a return to the other cognitive order or sometimes regret the things they have done. In these more dramatic cases, it is clear that something like psychoanalysis is well out of its league, as are any number of chemical treatments and mechanical interventions. Plasticity is a frustrating causal conundrum and is the real condition of our daily neuronal existence: "the self is synaptic" and vice versa.[2] And despite that maelstrom of operators in our daily neuronal life, most people still cling to the idea of their own determination, of being masters of their domain.

In presenting plasticity as both empirical fact and philosophical question, Malabou compels us to think something thought to be unthinkable and something that Malabou reminds us we do not know. There is an epistemological question about how the embodied brains we call scientists study other embodied brains we call objects of study. The question goes something like this: how can we make knowledge out of the condition of possibility of knowledge without reaching some kind of logical paradox? The knowledge of a thought in itself would either be merely a representation (say, an MRI image) or the idea of a thought in itself (the idea of that MRI image). Malabou's position has gotten support from many prominent neuroscientists championing neuroplasticity, as well as from political theorists like William Connolly and Brian Massumi, who see in neuroscience the terrain of contemporary political engagement. Further, the backlash against such positions has reached a screeching pitch pursued in the name of defending the dignity and uniqueness of the human. To address these concerns for the fate of man, I take Immanuel Kant's resistance to neurophilosophy as emblematic of the reactionary humanist position—the argument has changed little in the intervening 220 years—the hope being to lay out the philosophical landscape in which contemporary neuropolitics unfolds before thinking through the neuropolitical practices that have proceeded with little attention to the lack of philosophical consensus on its possibility.

In the preface to *Anthropology from a Pragmatic Point of View*, Kant declares the investigation of "cranial nerves and fibers . . . a pure waste of time."[3] For Kant, the brain as mind is noumenon, in the sense that it cannot be appreciated by the senses, and the brain as brain is phenomenon, to the extent that it can be appreciated by the senses but not known in itself. On this point I almost agree. However, there is no reason that Malabou's materialist adventure with the brain requires a naïve realism in which things are self-evident and waiting only for sufficient scientific explanation. Malabou is not making an argument for material transparency. Instead, what is truly discomforting for Kant and provocative for Malabou is the insistence of the brain as matter, that which gives and receives form, to be thought.

Further, Kant's rejection of this kind of speculation goes much further than the philosophical proof of the unreachable thing in itself. For Kant, the problem of the brain is a threat to the very image of "Man" required by his thinking. Kant cannot deny the natural and physical nature of the human, but such a nature requires a difference in kind from the "irrational animals," of which he says one can do with "as one likes."[4] Kant accomplishes this task through a developmentalist account of human maturation.[5] The transition of early childhood to the possibility of a free human is for Kant the difference between "feeling one's self" and "thinking one's self," marked by the linguistic shift of speaking in the third person to the use of the "I" to describe one's actions.[6] The internal coherence of thinking rather than feeling is at the heart of Kant's disdain for brain science. According to Kant, the material problems of the brain's work must be fully under the dominion of the thinking subject, or man is merely an animal. The sense data of sensation must then also be a purely passive stream, of which thought is in command. Kant lays out the stakes of this view of cognition in a section he titles "Apology for Sensibility."

> Sensibility . . . monopolizes conversation and is like an *autocrat*, stubborn and hard to restrain, when it should be merely a *servant* of the understanding. . . . The inner perfection of the human being consists in having in his power the use of all of his faculties, in order to subject them to his *free choice*. For this is required that *understanding* should rule without weakening sensibility (which in itself is like a mob, because it does not think), for without sensibility there would be no material that could be processed for the use of legislative understanding.[7]

So Kant does not deny the materiality of consciousness; in fact, he insists upon it. However, the brain of the mind must remain a substrate under the command of the faculties. The substance of these faculties is elided by Kant's command-and-control model of consciousness by sleight of hand, not explicit argument.

In light of what is lost for Kant, that the brain is of serious importance to the question of thought leads me to think that the rejection of brain research as a "pure waste of time" is not the traditional Kantian epistemological problem. I want to read his rejection as an overreaction caused by the fear that the brain would unravel the *free choice* of human action. The faculties then are a kind of artifice to bridge a gap Kant would rather not think between the brain's ability to give and receive form. The thing that changes itself confronts Kant with the horror that the human is an animal among animals. We are an organism with moods, instincts, bodies, brains, nerves, perceptions, affects, encounters, lesions, tumors, headaches, hallucinations, parasites, ideas, interfaces, implants, desires, frailties, glasses, memory, recordings . . . So rather than an epistemological question we have an ontological question: "A brain that changes itself. That is exactly what 'I' am."[8] It is in this weird space of Kant's "mob," the melee of forming and being formed, that Malabou's questioning takes place. However, like Kant, many others find such questions not only a "pure waste of time" but a threat to what it is to be human.

Three Images of the Brain as Prelude to Neuropolitics

It is typical of our time that it turns to men from contemplation to action, and the sciences from theory to experimentation. Astronomy and epistemology have taken on an increasingly experimental aspect in our age of spaceships and of machines steered by automatic light or heat receptors. Something similar has been happening in the study of society. Where in the past the formation and rise of nations were merely observed by scholars, today statesmen and voters increasingly want to *do* something about the process.
—KARL DEUTSCH, "The Study of Nation-Building"

Despite Kant's judgment that neurobiology was a waste of time, at least three images of the brain, rather than mind, have come to displace the Kantian model. I will outline three alternative conceptions of human intellect that insist we are brains, not minds, and then consider the philosophical and political fallout from the insight of all three. In all three cases, I want to speculate about a future in which, thanks to advances in neuro and cognitive sciences, the "intangible" character of the human may cease to exist. Although in no way meant to be exhaustive, the folding of intellectual life back into the gray matter of our brains has taken at least three broad forms. The neuroelectric brain, the neurochemical brain, and the informatic brain all figure heavily in research during the twentieth century. There is substantial overlap between all three models being discussed. However, it is worth partitioning each to see how each concept provokes a different image of thought and a different conception of the human as an instrumental resource.

In the early days of cybernetics research, an image of the brain as something other than a storehouse for representations and ideas began to emerge in the often informal experiments of scientists such as W. Ross Ashby and W. Grey Walter. The experiments ranged from shock therapy on institutionalized humans to the designing and building of simple electric brains, as in the case of Walter's tortoises. In the artificial and human case studies, what was being proposed was an image of the brain as a governor or mechanism that had the dual capacity to remember (stability) and react differently from or even opposed to memory (adaptation). This duality between the storage capacity of the brain and the adaptive attributes of the brain was undergoing reconciliation in an effort by Ashby, Walter, and others to banish the superstition of the mind.

The idea of the mind as something distinct from empirical reality was for these scientists a holdover of the soul or spirit. The mind was a kind of secular placeholder for earlier iterations of human exceptionalism. According to Ashby, this residual spiritualism was unnecessary to understanding the workings of human capability. Although the human brain was physiologically unique and advanced as compared to nonhuman apes, the brain was not magical. In fact, brain function is, as Ashby writes, a "logic of pure mechanism, rigorous as geometry."[9] Inspired by what Ashby saw in electroshock therapy as the ability to "reset" or alter the brainwaves of patients, he attempted to set out relatively simple attributes of the nervous system that allowed for humans to execute learned knowledge as well as develop and acquire new knowledge. According to Andrew Pickering, Ashby joined others in displacing Kant's image of a representational brain in favor of a brain akin to a device. What was intangible could be made tangible.[10] In fact, according to Ashby, "the making of a synthetic brain requires now little more than time and labour. . . . Such a machine might be used in the distant future . . . to explore regions of intellectual subtlety and complexity at present beyond the human powers. . . . How will it end? I suggest that the simplest way to find out is to make the thing and see."[11] Science has not yet proven Ashby right. However, as a practical matter of capability, advances in brain imaging and machine–brain interfaces suggest Ashby was headed in the right direction. The European Union and the United States are in fact betting more than $4 billion on this model of human consciousness in the form of Henry Markram's Blue Brain project. Rather than designing a mind in a bottle as representational theories of artificial intelligence attempted to in the twentieth century, Markram is building a brain from the ground up, one simulated neuron at a time. The Blue Brain project has already successfully constructed a rat brain using this method and demonstrated the artificial rat brain as proof of concept.[12]

That Markram's project would bear fruit would not have surprised Ashby. At the end of *Design for a Brain*, Ashby writes, "Life on earth must thus not be seen as something remarkable."[13] For Ashby, like life, the brain is merely another adaptive system nested in the larger adaptive matrix of organisms, populations, species, and ecosystems. For life to exist at all is sufficient for the full complexity of human intelligence and creativity. What makes the brain distinct from something like genetic information is its ability to adapt or rewrite in relationship to the environment. While genetic information provides basic parameters and capacities, Ashby saw the brain as a kind of interface with the outside world such that learning could alter response and even, in the case of many animals but humans in particular, alter the environment such that organisms could better achieve their ends. Unlike Kant and others who relied on the nether world of the intellect to explain these advances, Ashby saw the external world as a storehouse for ideas that could "amplify" adaptation and learning. According to Ashby, "The environment acts as dictionary."[14]

In addition to undermining the concept of mind as something taking place in an incorporeal realm, Ashby further undermines the idea of the individual or author as the "cause" for seemingly original ideas. The process of "invention" is dynamic and mostly external to those organisms that come to standardize or "remember" those new behaviors, techniques, or technics that emerge from individual interaction with the broader environment. As will be discussed later in this chapter, Ashby is already foreshadowing the more robust theories of extended mind put forth by Lambros Malafouris, Alva Noë, and others.[15]

The mechanical or neuroelectric brain proposed by Ashby profoundly undermines legal understanding of uniqueness, authorship, and creation. As research moves forward in the capacity to map, read, interpret, and even build Ashby's "brains," we will likely face unprecedented changes in how one seeks to pursue the already fraught dialectic between promoting innovation while protecting what is innovated. If we are to fully accept the death of mind as the last vestige of the theological in our contemporary image of the human, then it is no longer clear what even constitutes the intellect much less the process by which the products of that intellect are created and identified or distinguished as "ours" from other brains with contending claims and connections to what Ashby would call adaptations rather than inventions or ideas. In some sense, following Ashby, every act of creation must be a form of neurocognitive mimicry as the "environment" is our "dictionary." As technological means continue to catch up with Ashby's theories, the "intangible" and "incorporeal" will be shown to have been technical, not metaphysical, problems.

The neurochemical brain ventures even further into the problem of first cause in relationship to the intellect as well as the basic foundations of sensory experience. Whereas Ashby was attracted to the regularity of electrosensitive brainwaves, John Lilly and others like Timothy Leary developed techniques for intervening in the functioning of the brain. Lilly began his research a decade after Ashby under the auspices of national security funding created to develop knowledge about the mind that could be used to counter brainwashing techniques used by the Chinese and Russians against U.S. soldiers. Lilly's research first focused on the ways the brain could be altered by depriving the body of sensory experience. Lilly's famous sensory deprivation tanks were capable of inducing vivid hallucinations as well as changes to the perceptive capabilities of the human body. Experimental subjects reported that inanimate objects moved or "undulated" as if they had come alive. Lilly's early research provides further substantiation of Ashby's incipient extended mind thesis. Not only is the brain not a generator of "representations" in the Kantian sense, but brain function itself is structured by experience and sensation rather than the other way around. Kant's category of "a priori knowledge" necessary for structuring the intelligibility of experience was turned on its head by a few simple deprivation experiments. Extended deprivation, according to Lilly, could return individuals to fetal states in which brain activity flattened out and the body twitched rhythmically. It was as if the experimental subjects were losing their capacity to be subjects without the external stimulus of light, sound, and touch.

During the 1960s, Lilly and others began to experiment with lysergic acid diethylamide (LSD-25) to induce and alter the perceptive capabilities via chemical alteration of brain function, mirroring many of the effects of extreme isolation. Apropos of this chapter, the expansion of LSD research coincided with the expiration of the patents on LSD held by Sandoz labs. What was evident to Lilly in both the deprivation experiments and the chemical interventions into human consciousness was that all humans are "programmed biocomputers." In fact, Lilly went so far as to argue that "no one of us can escape our own nature as programmable entities. Literally, each of us may be our programs, nothing more, nothing less."[16] Although Lilly's research ventured off into the mystical realm of the 1960s counterculture, his insights were well received by his funders at the U.S. Department of Defense. Furthermore, psychedelics research has expanded dramatically in the last ten years.[17] Parallel to the "mind-expanding" drugs of the 1960s, the same decade saw that adaptation of first- and second-generation antidepressants. In addition to altering or programing of perceptual capabilities, pharmacology was developing more and more sophisticated ways to alter mood. Furthermore, the adaptation of stimulants such as methylphenidate, more commonly known as Ritalin, to treat behavioral and learning disorders established the neurochemical brain as a significant market and area of research and development.

Perception, mood, and behavior are now all seen as capacities that can be fundamentally altered by chemical intervention. From the perspective of pharmaceutical companies, Kant's enlightened self has been replaced by a chemical self. There is nothing fundamental about one's cognitive capacities that cannot be altered by Prozac or Wellbutrin. To put it somewhat differently, no amount of willpower or intellectual maturity could stop the effects of these drugs. Psychoactive drugs are not something that the body experiences and the mind makes sense of. The very capacities of sense making are altered by the introduction of these chemicals to the brain and nervous system. In a disturbing line of research, high doses of beta blockers have been used to inhibit the production of adrenaline and other hormones associated with the formation of particularly traumatic memories. Clinical tests show the ability to change the intensity and in some cases even erase memories through the application of these drugs.[18]

Psychoactive drugs further extend the idea of the brain as mechanism into the popular imagination as well as the practical realm of research and development. The human consciousness is increasingly seen as a programmable system. We are our brains and brains are media. Brains can be written and rewritten through an increasing number of interventions. The demand for so-called smart drugs has created a significant black market for stimulants and even off-the-shelf creativity-enhancing supplements. Whether existing nootropics enhance creativity, the neurochemical image of the brain as an extension of the neuroelectric brain suggests that stimulating or even creating creativity is again a technical problem rather than a problem of inspiration. The once mythical muses may soon be swallowed or inhaled as the particular brain states of creative thought may one day be modeled and sold via machine–brain interfaces. As an individual or even collectivity of brains come to be seen as a chemoelectric media platform rather than as unique and singular individuals, one wonders how long before the makers of pharmacological or electric stimulants for creation come to claim partial ownership for what gets created. At what point do those who intervene in the process of creativity become collaborators or partners rather than tools?

The third variation on the brain image substantially overlaps with the previous two but is likely more significant than both as it represents the possibility of converting either electrical or chemical process into digital information. In some sense, both of the previously discussed images of the brain are analog images. That is, the neuroelectric and neurochemical brains are dependent on organic processes that can be altered by environment or intervention but are still dependent on the confines of a single individual's gray matter. The informatic brain sets those analogue processes free. Following Alan Turing's proposal of a universal computing machine, John von Neumann proposed that the brain was a computer.[19] According to von Neumann, no particular computation within the brain was sophisticated. Instead, what made the human brain or any brain capable of complex phenomena was the sheer magnitude of different and parallel computations happening at the same time. The difference between a simple calculation and something more closely resembling creativity was not a difference of kind but a difference of degree as measured by quantity. Therefore, like Ashby, von Neumann saw no reason why the workings of the human "mind" could not be programmed into a computer. The question was one of whether the process being programmed had been sufficiently broken down into its simplest constitutive components. If this could be done, then the "calculations" of the human brain could be represented by the digital language of ones and zeros. No "idea" was immune from becoming digital. In this sense, according to von Neumann, everything was information. In fact, von Neumann went so far to say that at the scale of the single neuron, the human is "*prima facie* digital."[20]

The brain as computer thesis advanced by von Neumann and others has been vociferously challenged by many weary of its reductionism.[21] The failure of artificial intelligence (AI) research to produce something like a working brain is often leveraged against von Neumann. Despite skepticism, the informatic brain is a powerful image of the brain even if it turns out to be closer to a metaphor. Furthermore, von Neumann's more basic insight that extraordinary complexity can emerge in relatively simple structures has been further developed and demonstrated persuasively by Stephen Wolfram via digital automata. Wolfram calls this the "principle of computational equivalence."[22] To put it simply, all phenomena, natural or artificial, are processes and have to follow some kind of rules. Therefore, the obstacle to computational description is merely sufficiently describing the rules and components of a process. Other theorists have taken this idea further to see all systems, from organism to the family, states, and corporations, as fundamentally informatic. For political scientist Karl Deutsch, this meant that all complex systems had the capacity to develop something like a mind whether it was as state or an individual. The constraints for thinking were not biological or spiritual but technical in that what made a mind possible was the ability to, as Deutsch described, store information and act on stored information while possessing the capability to learn by adding new information without disrupting the stability necessary for order and functionality.[23] The informatic image of the brain demonstrates that thinking and creation can exceed the boundaries of any particular human body, suggesting that collectivities can create and even suggesting that nonhuman singularities and collectivities have the capacity to invent. The seat of creativity is a set of processes rather than an attribute of a particular human being.

Coming full circle, insights from all three brain images have made possible the development of digital-neural interfaces, suggesting unprecedented assemblages of people and things for the creation, sharing, and remixing of ideas. For example, MRIs have the capability of visualizing images and even ideas, such as who a person is thinking of, which can be determined from scans of brain activity.[24] Furthermore, research at the University of Washington has demonstrated the ability to send information directly from one brain to another.[25] One can already imagine the intellectual property (IP) disputes between the next generation of Steve Jobs and Bill Gates when the ideas in question were created via linked brain communities or research being done on a wireless neural net. How ideas, concepts, processes, and techniques of all kinds will be recorded and tracked or altered in such a world boggles the mind but will no doubt play a significant role in politics, military affairs, and the economy. As philosophically specious as the creator or inventor myth is, innovation in brain research will materially alter debates around freedom and personhood in ways that mere argument simply could not. I am not a neuroscientist and I have no idea which model of the brain will come to dominate the applied sciences of brain intervention. However, all three images suggest that the problem of plasticity must be thought if we are to consider what we can become as a species.

The Crisis of Plasticity

Despite the increasingly dominant position that privileges brains over minds, there is a kind of philosophical panic among, in particular, ethical philosophers over the loss of the mind. Following Kant, somatic fundamentalists like Jürgen Habermas, some neo-Arendtians, self-described intentionalists, and humanists of various stripes see the recourse to the brain as the death of man. For these thinkers, neuroscience is just the next normalizing discourse to follow psychoanalysis in a long line of expert power moves that seek to subjugate people through the naturalization of pathology and control.[26] Furthermore and somewhat contradictorily, they fear that the instrumentalization of the brain will lead to an objectification of the human, destroying the dignity and intrinsic value on which human rights are premised.[27] The image conjured is of a mechanistic world where freedom is abolished via genetic, surgical, educational, or pharmaceutical intervention. As Slavoj Žižek has rightly pointed out, the irony of holding both positions is that one affirms both the ideational view of the mind's autonomy and that the mind, dignity, or what you will is, while independent of material existence, directly threatened by its modification. According to Žižek, "It's not so much that we are losing our dignity and freedom with the advance of biogenetics but that we realize we never had them in the first place."[28] This is the affective horror that animates the reactionary position of Kant and other enfeebled humanists. Those who cling to intentionalism or a mind-dependent existence feel plasticity gnawing away at the certainty of their values.

At this moment, while I am writing about the problem of plasticity, twenty-five milligrams of sumatriptan are dissolving into my bloodstream. As the drug circulates through my body and finally passes into my brain, it will stimulate serotonin receptors. The purpose of the drug is to interrupt my pounding migraine headache. The theory is that migraines are caused by the swelling of blood vessels in the brain. Stimulating the receptors causes contractions that compress the swelling in hope of ending the headache. However, this very mechanical reaction is paralleled by the effect on my serotonin. As a serotonin agonist, the medication also blocks the signal of the pain caused by the migraine. Further, serotonin is a critical neurotransmitter for my sense of well-being. Well-being is something we code as a psychological state in humans, but in both humans and even simple invertebrates, serotonin signals both need and satisfaction regarding food, sex, and other basics.

I am feeling better already, and I have less pain. Is my change in mood because of the alteration of my internal brain chemistry? Is my internal brain chemistry changing because the drug has changed my sense of "well-being" or because I "feel better" from the pain relief? My mood has definitely improved, but I do not *know* why it has. And most importantly, while I "chose" to pop a pill, it is difficult to say I "chose" to feel better. The epistemological morass does not, however, foreclose the ontological provocation. Instead, the pill, my brain, my neurons, my serotonin, and my general mood are agents provocateurs of thought, each alien yet intimate.[29]

Plasticity takes a darker turn away from the everyday when the cause of such a change is not a pill "I" popped but the intervention of the alea.[30] The relationship between plasticity and chance for Malabou takes the form of traumatic brain injuries, lobotomies, Alzheimer's disease, and sudden shifts in hormones or other regulating neurological chemicals. The aleatory world of the brain puts the "I" in the potent grip of forces well beyond our control. However, unlike the more general confrontation with mortality, what Malabou calls "destructive plasticity" forces us to confront the fact that our identity, the "I," is not essential. In fact, the horror of plasticity is that our life can go on without us. The "us" is changeable and can be lost without the cessation of life. Not only are our minds not autonomous or independent of our bodies; they can undergo metamorphoses that leave no trace of what we once were, erasing the consciousness of what we have become. This, for

Malabou, is a critical element of plasticity. Unlike elasticity, plasticity has no promise of return. Both concepts suggest a limit point at which the system breaks. However, the change in each is different, as something elastic returns or can return to form after its change. Plasticity names an unredeemable metamorphosis. So the "I" and its attendant and mysterious "faculties," which are essential to moral freedom for Kant and others, always has the potential to explode and return to the mob of sensation only to reemerge as something altogether different.

This is the terror that keeps Jürgen Habermas up at night. What will we do if we discover our freedom is contingent, that our very nature can change?[31] The plan of the rest of this chapter is to follow Malabou's provocation through to the heart of that terror. However, unlike Habermas and others, I hope to do so without the sentimental attachment to a humanity that never existed. Rather, I want to push Malabou's concepts further. I want to consider what happens to plasticity and destructive plasticity when they are let loose in the wilds of politics. As knowledge of our formable and forming nature becomes not just known but made practical, the nightmare of humanists becomes real. Populations of human bodies without essential identities can be altered or, in the language of cybernetics, steered. Unlike understandings of power, even subtle forms of disciplinary power, plasticity on the scale of the individual and the polis represents the possibility of change without subjection, that is, without resistance. Rather than the relations of power that make subjects in the Foucauldian image, we have the possibility of designing or steering subjects that have no index of what they were before, such that something could resist. Instead, plasticity represents the possibility of a frictionless change—in the sense that one can imagine (and has imagined) the alteration of the brain or the assemblage of bodies-brains-semiotics-technics that is often called the social—that would leave no trace of what could be called an alternative. This is not the failure of resistance to produce an outcome, as in the case of the noncompliant prisoner who nonetheless remains imprisoned. Instead, plasticity raises the question of techniques that produce bodies that do not know that the "they" that they once were wanted to resist or even are imprisoned.

The incorporation of plasticity into politics raises the specter of Gilles Deleuze's "societies of control," in which individuals become "dividuals"—humans as counters in a flexible and constantly "modulating" economy.[32] Control in this context is often read systemically, as if only the "society" views humans as counters but the "dividuals" themselves, like Robert Duvall's character in *THX 1138*, yearn to be free, to be unique. Confronting plasticity and its explosive potential to obliterate precursors poses a different dividual, a real dividual: bodies stripped not just of their identity but of the desire to have an identity or to have an identity nonidentical to the identity that preceded it. Control represents the real possibility of order without the leverage or friction of ordering.

Normatively I do not disagree with Habermas and others that this reduction of freedom to an engineering problem is horrifying. The point of disagreement is that arguing against the existence of such a possibility will have an effect on the probability of this nightmare.[33] The attempt to safeguard humanity through the scapegoating of materialist thinking is self-defeating, as it insists that human freedom and dignity are independent of the brain while also decrying the possibility of each "becoming material." In its cruelest form, this line of argument amounts to trying to cure someone with Alzheimer's by scolding her about the intrinsic dignity and rationality of humans. Lesions beat argument every time. So rather than taking recourse to moralize the horror of control, it is necessary to take seriously the possibility of control as a material configuration enabled by the inessential nature of humans: their plasticity. Furthermore, I ask the reader to affirm the horror of destructive plasticity rather than look away or flee into the arms of humanist sentimentality.

Dreaming of Control, or the First Age of Neuroplasticity

The generation of cyberneticists that came of age in the 1940s was inspired by the brain, in particular brain pathology. The now commonplace understandings of networks, feedbacks, chaos, self-organization, and complexity that organize the functioning and understanding of everything from the internet to the global climate system come from this extravagant period of intellectual innovation. Nearly all the influential thinkers of the period, save Norbert Wiener, started with the brain as inspiration and model for their new ontology of nature and science.[34] The brain was the black box par excellence.[35] Furthermore, it is telling that schizophrenia was the first object of neuroplastic research. The seemingly far-fetched conundrum of Dr. Walter Bishop, how to purposefully alter the functional structure of the brain, was the holy grail of the first era of neuroplasticity. Unlike Kant, these thinkers were drawn to the seemingly impenetrable object of the brain. But unlike the early brain scientists that Kant mocks, who started with the cranial nerves and fibers that make up the brain, the cyberneticists tried to understand and model the brain's function rather than its structure. The early success of this modeling led almost immediately to its application. Rudimentary brains, the homeostats, were put to use regulating machinery, creating temperature-controlled homes, regulating the firing of naval artillery, and being put to use on actual human brains.[36] The goal in almost all cases was designing systems or altering them so that they could "strive to hold back nature's tendency toward disorder by adjusting its parts to various purposive ends."[37]

Imagined in the first generation of cyberneticists was a kind of secret functionalism that could be leveraged against the otherwise entropic tendencies of the cosmos. So a brain, for cyberneticists, was a machine that could receive information, store information, abstract that information, recombine information, and communicate or output new information that might otherwise be called action, all governed by a life principle to persist in completing in these tasks. The earliest of these works, W. Ross Ashby's 1952 *Design for a Brain*, captures the goal of such research in the first sentence of the text: "How does the brain produce adaptive behavior?"[38] The subsequent works—W. Grey Walter's *The Living Brain* (1953), Pierre de Latil's *La pensee artificielle* (1956), John von Neumann's *The Computer and the Brain* (1958), Stafford Beer's *Decision and Control* (1966) and *Brain of the Firm* (1972), and Gregory Bateson's *Steps to an Ecology of Mind* (1972)—all focus on the same question: what is it that allows a brain, even if not a human brain, to self-steer and adapt in complex ways despite being mechanical at the level of each operation? That is, the basic chemistry of a single neuron is relatively simple. Importantly, in each iteration of this question, every one of these thinkers, like Malabou, begins from the premise that a soul or mind independent of the brain is an insufficient answer.

Why return to the beginning, so to speak? In part, because each of these texts is philosophically sophisticated and challenging in its interplay between the experimental material world and the questions provoked by the various machines and even brain experiments—and in part because much can be learned from cybernetics as a trial run of neuroplasticity. Cybernetics is all the more demanding of Malabou's question because of Ashby's use of his theories to support and perform shock therapy, not in spite of them.

While Ashby's questionable use of electroshock therapy undermined the place of many of these thinkers in philosophical debates and traditions, cybernetics failure in theory did not follow in practice.[39] Environmental study, artificial intelligence, systems biology, climatology, robotics, complexity theory, anything digital, informatics, techno music, and video games—all are explicitly indebted to cybernetics. We live in a cybernetic age even if we do not know it. And if there is a unifying theme in all these strands of cybernetic thought, it is plasticity, the ability of systems to sustain integrity while changing. To be formed and give form while surviving is not a philosophy to come, as Malabou thinks it, but one at the heart of the technological revolutions of the twentieth century.

In light of this convergence, it is worth considering the almost immediate hope and proposal to apply the insights of plasticity, particularly brain plasticity, to human systems.[40] If the brain can be modeled, maybe it can also be hacked, to borrow an anachronism from cybernetics' digital future. Can political systems and the brains that constitute them be steered or completely reengineered to produce a new functionality? Neuropolitics in this iteration is a hope for self-sustaining, systemic, and dynamic homeostasis—that is, control.

This chapter examines two neuropoliticians. The first is the political scientist Karl Deutsch, author of *The Nerves of Government: Models of Political Communication and Control*; the second is the founder of the Yale University neuroscience program, José Delgado, whose book is *Physical Control of the Mind: Toward a Psychocivilized Society*. In each of their political programs, I see a species-scale revolutionary potential for plasticity and the horror of a designed world. The hope is that in these two early attempts to apply insights of neuroscience, we see the political possibilities in the concept of plasticity.

Deutsch begins his book with a question redolent of Malabou and the cyberneticists: "What is the capacity of this political system for self-transformation with significant preservation of its own identity and continuity?"[41] Deutsch, from the beginning of his exploration of the "essential connection between control and communication," is concerned with Malabou's problem of continuity, the delicate balance between plasticity and a destructive plasticity in which self-transformation obliterates the self that undertook transformation.[42] The stakes of the question for Deutsch are apocalyptic. In 1963, at the time of the book's publication, the dark shadow of the Cuban missile crisis loomed large. As a seasoned international relations theorist, Deutsch saw the world as an anarchic system of militarily competing states. That competition creates a security dilemma in which each state is compelled by the danger of anarchy to improve its defensive capability. Tragically, as goes the theory, the indistinguishability of defensive and offensive military capability leaves every other state feeling more threatened by the first state's improvement in military defense. The result is escalating arms races, until crisis and a lack of transparency produce war. As a result the international system, for Deutsch and many others, was a cycle of conflicts and wars created by the opacity of intention and threat. Furthermore, following this view of the international order, as no military innovation has ever been withdrawn, global nuclear war was not a question of *if* but *when*.[43]

The Nerves of Government shows that Deutsch understands the international system and its repetition of the security dilemma as a problem of communication and control. The international system was a defunct brain. Imperfect information and the inability to learn from the past pain of war (i.e., memory) led to repetition compulsion. Therefore, unless an international system could be designed that overcame what he called the "pathology of power" in favor of plasticity—the ability to learn—war would be dominant until life was no longer dominant. Without a combination of memory and plasticity, Deutsch argues, a "society becomes an automaton, a walking corpse."[44] So beginning with a cybernetic theory of mind, Deutsch attempts to develop various theories of "collective personality," "group mind," and "group learning."[45] Society is, according to Deutsch, a "plural membership of minds," whose autonomy and self-steering is contingent upon the degree of *control* or the "pattern of information flow" rather than the amount of power a state can wield.[46]

In an extreme form of American exceptionalism, Deutsch also equates the U.S. model of democracy with a functional mind and takes the Soviet system as the archetype of a pathological system; the critical difference was the flow of information. For Deutsch, the ability of a system to learn and adapt toward increasingly effective survival was dependent on the flow of information. In a fabulous example of Michel Foucault's repressive hypothesis, Deutsch insists that the truth will in fact set us free.[47]

Unfortunately for Deutsch, he underestimated the warning of his best friend and inspiration, Norbert Wiener, regarding the integrity of information. Wiener argues in *The Human Use of Human Beings*—an apt name for this kind of designed politics—that the value of information is highly dependent upon its integrity, not just the quantity of information or the freedom of its flow.[48] In agreement with Deutsch, Wiener argues that the political system is first and foremost a communication network, with multiple levels of feedback that disseminate and habituate value. However, Wiener sees in the U.S. not a model for democracy but precisely the opposite. Wiener describes an increasing incentive for corporate and national interests to use bluff and sabotage in the information networks such that what is received and proliferated are the competitive and violently instrumental values of market ideology and state militarism.[49] The national echo chamber is a collective mind adapting, according to Wiener, but it has no transcendent liberal value or telos. Instead, the cybernetic nature of politics and control amplifies the dominant logics we now call neoliberalism.

What is significant for our story is that Deutsch and Wiener are in agreement about the stability of control as a self-organizing and self-amplifying tendency of a complex system of brains. Plasticity is present. However, the difference is that Deutsch's naïve faith in transparency and his romantic view of existing American values blind him to the effect of that self-amplifying system. For Wiener, what is necessary is the production of new values and techniques to overcome what closely resembles Antonio Gramsci's hegemony or Alexander Galloway and Eugene Thacker's exploit.[50] For all four of these thinkers, tendencies within systems gain dominance even if the distribution of the system is not seemingly hierarchical. Rather than hierarchy or topology generally, what makes an order dominant is precisely the ability to remake the individuals in that system through the various feedbacks of information and plasticity.[51] In Wiener's view, ideology is not false consciousness, not because market ideology and state militarism are good or natural but because there is not true consciousness. Rather, market ideology and state militarism become the native operating system of the U.S. According to Wiener, there is, as in an Apple computer, no command prompt for individual users in a national system. Therefore, those who hack or steer the system determine its outcome. According to Wiener, "a block of human beings to increase their control over the rest of the human race . . . may attempt to control their populations by means not of machines themselves but through political techniques as narrow and indifferent to human possibility as if they had, in fact, been conceived mechanically."[52]

I am not sure Deutsch ever came to agree with Wiener. However, by the second edition of *The Nerves of Government* (1966), Deutsch expresses a deep disappointment that the techniques of cybernetics will not catch up with his hopes for the practical application of control to replace the pathology of power.[53] There is for Deutsch an inexplicable lag in the academy and the political system, a failure to wake up to the functionalism of an adaptation based on the free flow of information and learning.

It is precisely Deutsch's lag that José Delgado responds to in *Physical Control of the Mind.* Also inspired by Ashby's *Design for a Brain,* Delgado lays out a manifesto based in part on his applied research on animals and humans utilizing electrical current to alter brain activity. If it seems we are returning to the fictional world of J. J. Abrams, it is important to remember that Delgado was an MD/PhD in Yale's physiology department faculty and later founded and organized the medical school at the Autonomous University of Madrid in Spain.

Physical Control of the Mind was published in Harper's World Perspectives series, whose mission statement expressed the view that "man is in the process of developing a new consciousness which . . . can eventually lift the human race above and beyond the fear, ignorance, and isolation which beset it today. It is to this nascent consciousness, to this concept of man born out of a universe perceived through a fresh vision of reality, that World Perspectives is dedicated."[54] *Pace* Malabou, it is also worth noting that the series was dedicated to the reunification of the humanities and the sciences in the pursuit of a materialist philosophical thought that could overcome "the false separation of man and nature, of time and space, of freedom and security" in the hope of escaping the "present apocalyptic period."[55] As further evidence to this strange convergence, Delgado's volume in the series was published alongside those of theologian Paul Tillich, psychoanalyst Erich Fromm, Nobel Prize–winning physicist Werner Heisenberg, and Marxist literary theorist Georg Lukács. I rehearse all this to make the point that what follows is not marginal or out-of-hand rejected by many as lunacy. Instead, Delgado was and is considered by many to be a public intellectual and scientific genius who contributed directly to the future of humankind.

Wasting no time, Delgado begins *Physical Control of the Mind* with a grand and apocalyptic tone consonant with the series mission. In the opening chapter, "Natural Fate versus Human Control," Delgado lays out an argument for overt human intervention into a stalled human evolution in order to save the species from the autogenocide of nuclear war and industrial excess. To set the tableau, Delgado writes: "Manifestations of life depend on a continuous interplay of natural forces. Worms and elephants, mosquitoes and eagles, plankton and whales display a variety of activities . . . which escape human understanding, obeying sets of laws which antedate the appearance of human intelligence."[56]

Although Delgado shows reverence for the complex behavior of the natural world, the appearance of man is a break with evolution thus far. Delgado, close to contemporaries who hope to name our current geological epoch the Anthropocene, argues that humans differentiated themselves from other organisms through their "ecological liberation."[57] By this, Delgado means the use of technics, a built environment that alters not just the chances of a species' survival but the system in which that species survives.

Like many contemporary systems biologists, Delgado insists upon the epigenetic character of humans. What "liberates" humans is the ability to learn techniques for environmental modification, whether the building of shelter or eradication of disease, and pass that information down through the ages. So an archive of distributed knowledge forms a kind of exogenetic code—hence *epigenetic* legacy—that supplants or changes the determinism of internal genetic traits.[58] Importantly for Delgado, the epigenetic heritage enables what he calls "freedom of choice." However, this freedom is not of the traditional liberal variety; what is important is not freedom at the level of the individual but the species' ability to steer its development with or against genetic possibilities. Thus, argues Delgado, "our activities are less determined by adaptation to nature than by the ingenuity and foresight of the human mind, which recently has added another dimension to its spectrum of choices—the possibility of investigating its own physical and chemical substratum."[59]

Despite this newfound avenue for control, humanity, according to Delgado, has disproportionately focused on the development and design of technics rather than alterations to the meat sacks of human being. The result is the accumulation of extraordinary transformative power but in exchange for a "servitude dominated by levers, engines, currency, and computers."[60] Power for Delgado, as for Deutsch, is pathological and regulated by an economy of zero-sum competition that has invested in the destructive technics of "atomic overkill" rather than in human betterment. Therefore, responsibility to the species demands a new awareness that can overcome "behavior . . . composed of automatic responses to sensory inputs."[61] The goal of such an awareness would be to counteract the automatism of individual humans and fundamentally alter behavioral patterns. Unlike Deutsch and Wiener, Delgado does not think this is achievable at the level of information transparency and flow or that the creation of new values will be sufficient—although all of this is necessary. Instead, we should "re-examine the universal goals of mankind and pay more attention to the primary objective, which should not be the development of machines, but of man himself" and thus as a species consider intervening physically in the "intercerebral mechanisms."[62]

So in hopes of overcoming the crisis of power and the dead end of politics, Delgado declares the discovery of human plasticity as the next phase of epigenetic evolution.[63] The genetic—or, as he reads it, neural—determinism and cultural construction have converged in the discovery and understanding of neuroscience and genetics.[64] What Delgado describes as control is not different in kind or degree from Malabou's concept of plasticity. The difference is in the center of gravity or fulcrum of formation and formability. This difference is not an empirical difference, however. Instead, the weird material and discursive similarity between the two thinkers belies a severe metaphysical difference. For Malabou, it is not defensible "to advocate an absolute transparency of the neuronal in the mental."[65] Instead, what she calls "a reasonable materialism . . . would posit that the natural contradicts itself and that thought is the fruit of this contradiction."[66] The wager is that "the brain does not obey itself" because the alea is in reality.[67]

For Delgado, it is precisely the opposite. The stability of the substrate of a plastic brain can be brought into harmony with thought and value. In some sense, Delgado represents the specter of Kant. Nature is a kind of mechanistic matter to be brought under the control of the faculties.[68] In the case of Delgado, those faculties can be engineered. Ultimately, though, the difference in concepts of plasticity as process is minimal and may in fact be resolved someday empirically, which could leave Malabou's speculative metaphysics inert. Leaving that aside, it is important to see how much differently Delgado's neurorevolution proceeds from Malabou's precisely because of the bifurcation point between their thinkings' respective commitment to an ordered versus aleatory nature.

Again like Malabou, Delgado sees a promise in neuroplasticity. We are called, as Malabou agrees, by the question, "What should we do with our brains?" But Delgado's answer comes from extensive experimentation with brain electrodes rather than as an open-ended philosophical conundrum. Delgado claims to have developed sufficient knowledge of the brain's terrain and function to target aggression instincts that, he argues, are at the heart of human violence. Made famous by dramatic video footage, the Córdoba bull experiment demonstrated Delgado's ability to alter the mood and behavior of complex vertebrate animals.[69] In the experiment, Delgado steps, as a matador, into a bullring with a bull infamous for goring bullfighters. Delgado is armed only with a large remote control. The bull, he explains in the voice-over, has been fitted with an intracranial electrode calibrated to alter its aggression response. Another matador in the ring uses his cape to get the bull's attention; the bull charges. When Delgado engages the electrode, the bull stops charging and grows increasingly uninterested in the fluttering cape. Delgado then approaches the bull, at which time rather than charge, the bull retreats and cowers in fear.

On the basis of this experimental work, Delgado proposes that the perfection of such electrodes represents the possibility of ending the threat of nuclear war and saving humanity from its current fate. Aware that many would find his methods objectionable, Delgado lays out a strong argument for preferring the freedom of the species to the freedom of any one individual. But

Delgado is not a communitarian in the sense that he would argue that there are competing collective and individual values the greater good ought to trump. Provocatively, Delgado makes a different argument, one based on the instability and inauthenticity of individual identity. In some sense, Delgado presents us with an affirmative case for Malabou's destructive plasticity.

Like all good technophiles, Delgado begins by making the case that altering the human brain is inevitable, as no state regulation will be able to stop the progress of "scientific advance."[70] According to Delgado, the physical control of the mind is, like a knife, neither good nor bad. In a pithy phrase, Delgado states that "science should be neutral, but scientists should take sides."[71] By this he means that morality lies in how we put the inevitable technological development to use. His argument begins again to converge with Malabou's when Delgado switches from the tack of inevitability to an argument against the presupposition that there is something lost when minds are altered: "The mind is not a static, inborn entity owned by the individual and self-sufficient but the dynamic organization of sensory perceptions of the external world, correlated and reshaped through the internal and anatomical and functional structure of the brain. Personality is not an intangible and immutable way of reacting but a flexible process in continuous evolution, affected by its medium."[72]

From this, Delgado extrapolates an equivalency between a kind of constructivist position that argues for the cultural and social production of subjectivity and the intentional alteration of neurochemical processes. Similarly, as Malabou rightfully points out, this difference is indeed thin, as discriminating between culture and nature is as arbitrary as it is circular.[73] Furthermore, Malabou writes that in the concept of plasticity, "the entire identity of the individual is in play: her past, her surroundings, her encounters, her activities; in a word, the ability that our brain—that every brain—has to adapt itself, to include modifications, to receive shocks, and to create anew on the basis of this very reception."[74]

Not unlike the minimal difference in argument between Jeremy Bentham's normative account of utilitarianism and Michel Foucault's account of discipline, Delgado and Malabou share a view on the ineluctable absence of either a constitutive inside or outside to the formation of the subject. However, unlike Malabou, Delgado has not left to the imagination how such an insight should be put to use.

Delgado further extends his analogy into the accepted realm of liberal control by comparing physical interventions to the broadly accepted role of education. In a broad appeal to common sense, Delgado writes, "Culture and education are meant to shape patterns of reaction which are not innate in the human organism; they are meant to impose limits on freedom of choice."[75] Here Delgado aligns his proposal with earlier progressive education advocates, from Immanuel Kant to John Dewey. In an extension of his famous editorial "What Is Enlightenment," Kant argues in *On Education*, "Man can only become man by education. He is merely what education makes of him, for with education is involved the great secret of the perfection of human nature."[76] Like Delgado, Kant argues that education is necessary because "discipline must be brought early; for when this has not been done, . . . undisciplined men are apt to follow every caprice."[77] Despite the differences between Kant's transcendental idealism and Delgado's naïve materialism—a difference substantially impacting the technique of intervention—the narrative of paternal development is nearly identical. Malabou's relative silence on technique places her in strange company given her deconstruction of the binary of nature and culture.

Delgado ends his section "Ethical Considerations" with the promise that electrical stimulation can never be total. Instead, electrical stimulation is in a "dynamic equilibrium" with the other forces, "a new factor in the constellation of behavioral determinants."[78] Delgado concludes then that the real threat to human freedom is not brain alteration but remaining "slaves of millenniums of biological history."[79] On the final page of *Physical Control of the Mind*, Delgado implores: "Shape your mind, train your thinking power, and direct your emotions more rationally; liberate your behavior from the ancestral burden of reptiles and monkeys—be a man and use your intelligence to orient the reactions of your mind."[80] One can hear in Delgado's manifesto resonances with those, like Ray Kurzweil and other accelerationists and posthumanists, who drive to take charge of our brains.[81] Although more humble and open-ended, Malabou also implores us to overcome ideological critique and take plasticity as a fact as well as a philosophical provocation; she warns, "so long as we do not grasp the political, economic, social, and cultural implications of the knowledge of cerebral plasticity available today, we cannot do anything with it."[82] I do not think for a moment that Malabou wants to do with our brains what Delgado has in mind. However, what is apparent in both Malabou's call to take the brain seriously and the underlying principles of Deutsch's and Delgado's thinking is that the future of neuroplasticity is politically fraught and no longer science fiction. To put it another way, the challenge of neuroplasticity is necessary but insufficient to formulate a politics or an ethics. Instead, liberal goals of human maturity (Kant) and perpetual peace (Kant again), when armed with techniques of neuroplastic intervention, animate the desire to obliterate the aleatory in favor of a designed order.

The Soft Machine: The Mutation of Control

> Cut word lines—Cut music lines—Smash the control images—Smash the control machine—Burn the books—Kill the priests—Kill! Kill! Kill! Inexorably as the machine had controlled thought feeling and sensory impressions of the workers, the machine now gave the order to dismantle itself and kill the priests.—WILLIAM BURROUGHS, "The Limits of Control"

William Burroughs's essay "The Limits of Control," which inspired both Gilles Deleuze and later Foucault, is written in direct response to Delgado's proposal of universal mind control.[83] Burroughs, in his signature paranoid style appropriate to the American moment in which he was writing, speculates that techniques of "mind control" have already been deployed in the United States, in particular in the assassination of Robert F. Kennedy. Burroughs speculates that Sirhan Sirhan was under "posthypnotic suggestion."[84] For Burroughs, the brain, at least the physical intervention into the brain, is not ultimately what is threatening. It is language that is the milieu of control: "Suggestions are words. Persuasions are words. Orders are words. No control machine so far devised can operate without words, and any control machine which attempts to do so relying entirely on external force or entirely on physical control of the mind will soon encounter the limits of control."[85] According to Burroughs, control requires "opposition or acquiescence. . . . If I establish complete control somehow, by implanting electrodes in the brain then my subject is little more than a tape recorder, a camera, a robot. You don't *control* a tape recorder you *use* it."[86] For Burroughs, it is not possible to imagine control without a controller or, at least, the exteriority of control. His subject/object split is misleading and obfuscates a vital distinction between oppositional power, a kind of kinetic microphysics of resistance, and the frictionless character of control.

In this section, I argue that Malabou's conception of neuroplasticity pushes the horizon of control beyond Burroughs's prudish humanism and, further, that plasticity, with its obliteration of an outside to the material seat of consciousness, contributes to a more provocative understanding of control in Deleuze and Foucault.

Instead of entertaining the possibility of total or destructive control, Burroughs extends his sentimental attachment to a residual human nature—an other of control—by elevating resistance to a life principle, not in the sense of what a good life is but as a condition of living at all: "When there is no more opposition, control becomes a meaningless proposition. It is highly questionable whether a human organism could survive complete control. There would be nothing there. No person there. *Life is will!*"[87]

Burroughs's concept of life as resistance is very close to Spinoza's *conatus*, in which a thing is characterized by trajectory, momentum, or force to perdure in its thingness, on its own terms. However, what happens when control is understood in the tune of the steering or homeostasis of cybernetics rather than in opposition? If exteriority—the forming rather than formed—is ephemeral, then what is life or *conatus* opposing? Instead, we would have to consider the possibility that *conatus* can follow control. I see this as the critical insight of destructive plasticity.

Therefore, Burroughs is at odds with Delgado and in some sense Malabou, who sees in the interplay of brain and alea "neither an inside or outside world."[88] For Burroughs, control requires a controller and the time and distance to enact control on an object of control.[89] In Burroughs's lingering humanism, there is somehow a remainder of the human that exceeds the brain and its ordering functions. So for Burroughs there is a kind of species-wide fail-safe. Following Malabou, exteriority, as well as the distinction between the physical and mental, does not survive the explosive concept of plasticity or its empirical demonstration. So to inject plasticity into the term "control" pushes us to hear Deleuze's Spinozist refrain of "we know not yet what a body can do" in the key of horror rather than hope. We have to consider a body with no essential limit and therefore a concept of control without a humanist horizon. This understanding of control draws on the earlier period of neuroplasticity found among the cyberneticists rather than the connection to Burroughs. Among the cyberneticist thinkers and inventors, control is architectural or emergent, that is, imminent to a system or machine rather than imposed.[90]

However, pushing Deleuze via Malabou beyond Burroughs does not necessarily land us in the necessity of total control. Instead, it subtracts the human or human essence, as well as interiority and exteriority, from the presence or absence of control. In their study of network behavior, Galloway and Thacker persuasively make the argument that horizontal, even endogenous, network architecture is not necessarily more democratic or less controlled than more hierarchical or externally imposed structures, as has often been advanced by those who celebrate the supposed democracy of the internet revolution.[91] I might say the same of consciousness and the brain. In fact, this is what marks the transition from disciplinary institutions to control societies—the shift from power or coercion to control: a multivalent, graduated continuum of modulations. In a society of control, interventions occur at the level of populations, whether human or neuronal. What Galloway and Thacker refer to as protocols can alter the arrangement and formation of bodies without anyone being at the wheel (in the anthropocentric sense of the term). A protocol according to Galloway is

> a set of rules that defines a technical standard. But from a formal perspective, protocol is a type of object. It is a very special kind of object. Protocol is a universal description language for objects.
>
> *Protocol is a language that regulates flow, directs netspace, codes relationships, and connects life-forms.* Protocol does not produce or causally effect objects, but rather is a structuring agent that appears as the result of a set of object dispositions. . . . Protocol is always a second-order process; it governs the architecture of the architecture of objects. Protocol is how control exists after distribution achieves hegemony as a formal diagram. It is etiquette for autonomous agents. It is the chivalry of the object.[92]

It is important that an ahuman or inhuman etiquette or chivalry can emerge without having been designed and then become recursive. The neural Darwinist explanation for the embedding of culture in physiology and genetics demonstrates the absurdity of genetic determinism and illustrates the possibility of emergent control in a plastic brain.[93] There has been some controversy about the so-called maternal instincts as an effect in both men and women of a hormone called oxytocin. The mechanistic view that explains humans as a series of brain states is happy to discover such a hormone. However, the existence of oxytocin is insufficient to explain anything other than the existence of a process. For the process to become active in the deep human past, an ecology of inducements and constraints selected individuals with increased social attachment and receptors for hormones that promote a sense of attachment, as well as more complex brain structures for expanded social existence.

However, genes are not rules. They are incipient expressive objects influenced by material and cultural networks—ecology. According to neuroscientists, when humans are born, they have some hundreds of trillions of synapses, but only those that are used, that is, activated by experiences, survive. The neurons that remain dormant start to disappear around the age of eight and diminish throughout the course of human life. The synaptic sequence that responds to oxytocin, the so-called parenting hormone, must be used before too long into human maturation, or the individual will not have been primed for the hormone.

Without the responsive synaptic structure, you could fill a person artificially with oxytocin, and it would have no effect. That is to say, there must be a biocultural relationship of care that causes the release of the hormone, fires the synapses, and feeds back into the expression of genes needed to make more oxytocin. All of which again presupposes what can be called a behavioral or cultural characteristic of care.

What emerges is a strong tendency of people to love infants. What makes this process interesting is that there is an ecological history of this hormone and the care of offspring that is not reducible to genetics, physiology, or culture but nonetheless has produced a consistent "etiquette" characterizing hominid behavior since long before *Homo sapiens.* The complex but regularized relationship between all these agents—DNA, culture, environmental pressures, and brain plasticity—is coordinated by an etiquette or chivalry that, while not a law, has consistency of interaction necessary to produce a million years or so with few Medeas and Zeuses. It is a protocol, neither solely learned nor hardwired. Control can emerge, assembled as it were, on the fly, but once established, exploits exist for intervention—hence the ability to tactically steer hormone ecologies, even weaponize them. In the case of oxytocin, the U.S. military has deployed oxytocin in training and high-stress battle situations. As with parent and infant, the artificial levels of oxytocin promote troop bonding. However, in the space of war, the artificially induced bonding is not universal; oxytocin is not love. Instead, those hormone-tightened bonds also provoke extreme hostility toward enemies seen to threaten the in-group. Oxytocin fuels militarily useful rage and violence where once it intensified care.[94]

Confronted with *modified* soldiers, the risk is that that protocol, once grafted onto the sociopolitical world, devolves into a kind of brute structuralism, a return to Delgado's dream of total control as an engineered order of the faculties. Protocols are so emergent, so immanent to the system, that resistance, even in its most descriptive sense—the microphysics of power redoubled by the friction or refraction of that power relation by the subject on which power was directed—ceases to have much application. The mobile and plastic nature of the modulation—control protocols within a network architecture—is part of a dynamic equilibrium, a range or average of control with an acceptable and even useful margin of error, that lacks the traction to push back. It is like trying to fight underwater without sand to stand on.[95] So the problem is not the totality of control but the particular organization of a control protocol. The control society, in this reading of Deleuze, optimizes rather than either represses (sovereign/juridical) or manages (specific) bodies. The control system is a difference machine with a refrain or protocol of control, not the rigid state of Burroughs's control system.[96]

The cutting edge of Malabou's destructive plasticity is consonant with this concept of control. Importantly, Malabou's reading of the brain gives control a conceptual reach distinct from concepts such as power and discipline on which Burroughs's oppositional notion of control relies. So if we take Deleuze at his word, that control societies are of a different order than disciplinary societies or carceral logics, we must take leave of Burroughs for an inhuman terra incognito waiting in the virtuality of the human brain and push it further into the polis, the species, and the planet. Destructive metamorphosis—a plasticity, not an elasticity—marks this epoch increasingly known as the Anthropocene. Malabou's distinction between elasticity and plasticity is precisely that metamorphosis is a cascade of transversal changes from which we cannot return. Significantly, at each level of mutation and formation, plasticity may confront us philosophically and practically with the paradox of a subject, even a species, that will not recognizably or self-consciously survive the crisis of philosophy it is called to answer. Destructive plasticity revolts and shudders against the dreams of sovereign steering and control. Rather, the sovereign is the alea, the emergent.[97]

Conclusion: The Horror of Being Things

This is reality, and we must accept it and adapt to it. . . . The concept of individuals as self-sufficient and independent entities is based on false premises.
—JOSÉ DELGADO, *Physical Control of the Mind*

If those arrangements were to disappear as they appeared . . . without knowing either what its form will be or what it promises . . . then one can certainly wager that man would be erased, like a face drawn in sand at the edge of the sea.
—MICHEL FOUCAULT, *The Order of Things*

Every species can smell its own extinction. The last ones left won't have a pretty time with it. In ten years, maybe less, the human race will just be a bedtime story for their children. A myth, nothing more.
—JOHN TRENT, character in John Carpenter's *In the Mouth of Madness*

What has to be navigated in the confrontation between Malabou and those who would cash in on plasticity is the tension between "the plasticity of the brain [as] the real image of the world" and "a vision of the brain [as] political."[98] Well-meaning neo-Spinozists of the Deleuzian variety see in the possibility of the brain a hope and the potential for newness. On the latter, we can agree without hesitation. However, what Malabou shows us is that the collapse of nature and culture is a beginning, not a sufficient ending. To push this point further, I am attempting to think the horror of those who would put neuropolitics to use in the existing political order and therefore take newness to be as capable of destruction as of hope. The fragility of things requires that we take the possibility of destruction seriously.[99] In particular, both Deutsch and Delgado are within the mainstream of liberal thinking and therefore ought not to be seen as outliers in the world of possible tactics of neuropolitics. After all, each focuses on procedural democracy as improved by accountability and the improvement of the populace through the diffusion of norms of open communication and order over conflict and violence.

Importantly, I see in Delgado and Deutsch an easy conquest for what Fred Moten and Stefano Harney call *government prospectors*.[100] Governmentality in search of innovation may in fact find neuropolitical proposals quite reasonable. Certainly the U.S. military already pursues Delgado's agenda, albeit in reverse. Soldiers' brains and bodies are continuously altered for performance, including but not limited to the manipulation of memory intensity, sleep needs, depression, and guilt; oxytocin is used to create stronger bonds between troops and a more violent territorial character.[101] These interventions are something quite distinct from the "ideology" of neuroplasticity, of which Malabou makes short shrift.

Techniques of control based on plasticity at the neural, polis, and species level are not discourses that make use of metaphors of networks, of self-organizing and self-healing systems. These techniques are real practices to be put to use in the engineering of an order.[102] Therefore, the pharmakon of neural plasticity is not exhausted by the opposition between Malabou's gesture toward a Foucauldian care of the self and a discourse of cognitive capitalism. Instead, the danger is the competitive struggle between attempts to reinvent and innovate the self in the face of the sedimentation of control practices gaining an increasingly global scope. As seen in the work of Edward Bernays, stretching back to the beginning of the twentieth century, the fight to steer the polis of brains is not new. It is merely the case that "manufacturing consent" has escalated from the cottage industry of early advertising firms to full-scale industrialized design.

This is why it is necessary to read plasticity in the context of control. Control here is not a synonym for power or oppression; it is the name we give to an emergent and surviving order often indifferent to those it controls. This is why Malabou insists on "the possibility of destructive plasticity, which refuses the promise, belief, symbolic constitution of all resource to come[.] It is not true that the structure of the promise is undeconstructable. The philosophy to come must explore the space of this collapse of messianic structures."[103] Beyond what is to come, what preceded an order is not always apparent either; *it is possible to lose things.* In this sense, the horror of plasticity and control is not the confrontation with the ineffable but the ineffability of what once was and can never be again. The fragility of things is real; freedom as we currently cultivate it can be broken.[104]

For Burroughs, the limits of control rely on a separation of form and humanity—as if "form could be left hanging like a garment on the chair of being or essence."[105] To move beyond Burroughs toward the bleeding edge of Malabou's and Deleuze's thinking is to accept, even affirm, the virtuality of control: a freedom no longer diminished or infringed upon or even lost but a freedom that never was, the victim of an "annihilating metamorphosis."[106] Such a vestigial appendage of freedom, if it persists at all, may itch from time to time or gnaw at an atrophied consciousness, but it will be relieved by the proverbial butt scratch. This is what Malabou calls "a power of change without redemption," but she also adds a "power without teleology" and "without meaning."[107] What has to be considered when plasticity—formative and destructive—enters into the orbit of control is the possibility of a change with the telos of meaning of an other that cannot be indexed as change by the body or polis changed.

In the brain, like the IED, and blood, we find limits to the Eurocene. To harken back to Lázaro Luis Iranzo's blessing of Captain Bernardo Vargas Machuca, the problem of nonbeing, nonhuman being, the nonhuman in human being collides and pushes back against the recurrent belief that "nature was made better."[108]

This is the horror of the lobotomy, political plasticity, and the subjective catastrophes of strokes and Alzheimer's disease; it is not the screams of the damned but the blank stare and a mouth hanging open. Horror is the body that need not know that it should scream. The plasticity of control and the accident possess the capability of "forgetting the loss of symbolic reference points."[109] In our current predicament, can we not already observe the body that cannot help but shop; the body of irresistible consent, thoughtless bland calculations; bodies consuming pink slime or worse yet consuming metabolized and sterilized politics lacking almost entirely any relation to the political? Contra Burroughs, there are worse things than death. There is surviving manufactured control. Can we not imagine what Eugene Thacker calls a blasphemous life, a "life that is living but that should not be living," rather than the dignity kill-switch imagined by Burroughs?[110] In this image of the end without ending, "life is weaponized against life itself."[111] In a world without an essential self, without a soul, we are left to ponder that kind of ending.

This is why destructive plasticity is better understood through horror than tragedy. Lesions, decay, dementia, shock therapy, brain manipulation—all demonstrate life's indifference to a particular form of life, not the mourning and melancholia of tragic loss. Tragedy is more about attachment than loss. Horror is the confrontation with a world that does not care or even know what is lost. In Malabou's words, the accident or the aleatory compels us to "think a mutation that engages both a form and being, a new form that is literally a form of being."[112] This form of being, insofar as it is one, is characterized by both fragility and perdurance. Such a thought leaves us with little grounding. It is mere life, a life without qualities—what Deleuze calls a life.[113] This is not reason to flee into the fantasy of a soul or the moralization of intentionalists. The horror of matter is real and cannot be assuaged by persistent argument to the contrary. We live in a world in which decay and catastrophe, not the human subject, are sovereign. That there is no transcendental index for the self that loses its self is without alternative. There is no ideology to attack; it merely is. That so many find this predicament unacceptable changes nothing.

Can we affirm such an inhuman position, or is this all fruitless and cynical nihilism? The affirmation is vital. To confront the unhumanity of the world demonstrates exactly what is at stake in fragility. The brain, the polis, freedom, the world, and Earth are breakable, even explosive, rather than necessary. It is from the impossible position of extinction or oblivion that providence is eviscerated in favor of the freedom of the aleatory. E. M. Cioran's account of decay is resonant here. Cioran writes, "We cannot elude existence by explanations. We can only endure it, love or hate it, adore or dread it, in that alternation of happiness and horror which expresses the very rhythm of being, its oscillations, its dissonance."[114] This is the task and the freedom of philosophy at this moment. The violence of destructive plasticity is the true limit of control. Reality, according to Malabou, always contains "a vital hitch, a threatening detour . . . one that is unexpected, unpredictable, dark."[115] However, the radical possibility of otherwise comes at an astronomical cost. This is the lesson of both formative and destructive plasticity at every scale. The thing that changes itself dwells at the precipice of nonbeing.

7. THREE IMAGES OF TRANSFORMATION AS HOMOGENIZATION

Despite the obstacles posed by *things*—from trash to our own blood to the plasticity of our brains—in the effort to systematically manage a martial form of life, new visions of how things could be are no less committed to the further operationalization of the luxurious metabolic demands of modern life. Global-scale *solutions* once thought fantastical are quickly gaining the pragmatic traction of reasonableness. Rather than consider the terms of failure, new coalitions committed to transforming the planet are redoubling their efforts. The technocratic managerialism of neoliberal states—a carbon tax here, a regulation on toxic pollutants there—is showing its insufficiency. A return oto "big ideas" has a renewed cache. It would seem that only a TED Talk can save us now.[1]

In part, the scale of the current crisis inspires these kinds of "revolutionary" responses, but because the scale is indexed by questions of survival and specifically the survival of moderns, a privileging of what Bonnie Honig calls *mere life* rather than *more life* takes place. As a result, more often than not these revolutionary visions for transformation are intensifications rather than reversals or detours from the global project of homogenization.[2] The narrow echo chamber of the Eurocene presents us with alternatives that bear a stronger fidelity to the same than they do to the possibility of other worlds. These visions of transformation on a global scale follow in the deep groove of Peter Sloterdijk's "earth-users" rather than considering how the very conditions of the crisis may intimately owe a debt to the very making of planetary projects in the first place. In this chapter, I discuss three of these transformation projects: recent efforts to double down on modernity, the upgrade of the global Marxist project, and an American military vision of a containable world.

Vision I: Liberal Hopes for Transformation, Transcendence, or More of the Same

Rather than inspire responses adequate to the crisis at hand, Euro-American humans are in large part fighting to continue exactly as we are. For many scientists and modernists, the critical question is how to maintain extraordinarily high levels of economic growth and consumption, because they see these practices as necessary conditions for human advancement and innovation. These inheritors of the cybernetics movement of the 1950s, called systems thinkers, see management and governance of the entire Earth system as the next logical step.[3] For many systems thinkers or more recently ecomodernists, what it is to be human is to alter our environment. Therefore, each step forward means taking greater control of the world around us. The most emblematic of these thinkers was R. Buckminster Fuller, whose cult classic *An Operating Manual for Spaceship Earth* advocated that the whole planet be seen as a ship that humans must learn to steer, just as they learned to navigate the oceans or grow crops. The next generation of systems theorist, after Fuller, came of age at the beginning of the computer revolution in the 1970s when the capability for truly planetary systems management seemed to be on the horizon. Stewart Brand was one of the leaders of that movement and founded a journal called *The Whole Earth Catalog*, which sought to be a forum for planetwide guided evolution.[4] It is not surprising that given the interest in engineering and global development, news of the limits of growth would be met with calls for transcending those limits rather than trying to live within them.[5] For Brand and other more explicitly industrial geoengineers like David Keith, who has become an outspoken advocate for industrial intervention into Earth systems, the solution to the Anthropocene can only be found in its cause, that is, human engineering. To quote the title of a recent essay by Keith, the hope is located in the question, "Can science succeed where politics has failed?"[6] For Keith and Brand, the hope is building what Brand and his collaborators call "a good Anthropocene" in their "Ecomodernist Manifesto."[7] Following proposals made by biologist E. O. Wilson and others, Brand and his team suggest setting aside half of the planet to be rewilded. To do so, the management of both human and nonhuman populations must be greatly intensified to accomplish what ecomodernists call decoupling.

According to the team of mostly Harvard and Oxford scientists working with Brand, modernity has greatly lessened the impact of humans on natural systems. Industrial-style agriculture, nuclear energy, and most of all urbanization reduce the dependence of humans on the environment at large.

Following their argument, if humans can be herded entirely into supercities, and human access to the rest of the planet can be restricted, then a renewal of the natural world can take place. Food, energy, and other natural resources will then have to be produced by synthetic means made possible by advances in materials engineering such as nanotechnology. Extreme forms of recycling would also reprocess and disaggregate all waste produced by humans into base materials to be used again. According to ecomodernists, if we can let go of the idea that nonmodern civilizations "lived more lightly on the land,"[8] then we can concentrate and accelerate the *advantages* of highly industrialized societies for the rest of the planet's human population. According to the manifesto, even "developing countries [can] achieve modern living standards."[9]

The externalities of these lifestyles, such as carbon emissions, which cannot be contained within cities, will have to be ameliorated through vast planetwide interventions. Solar shielding to regulate sunlight, albedo modification to alter the color and reflectivity of clouds, mechanical carbon capture and storage, and even planet shields of old compact discs are proposed as permanent features of future human existence and governance. Here humans and the rest of nature can share a planet while no longer sharing a world. The optimistic version of this program is that these more extreme means of planetary regulation can be replaced in a few hundred years after nuclear power and perhaps other alternatives can create "tens of terawatts" of electrical output.[10] In a weird resonance with critiques by Donna Haraway, Bruno Latour, Philippe Descola, Jane Bennett, William Connolly, and others who have effectively demonstrated that there is no "nature" external to humans, ecomodernists have conceded the point, declaring the need to invent precisely the human/nonhuman or culture/nature binary of modernist social theory via an unprecedented scale of global governance.[11]

The political or governance side of the vision is significantly less developed than the technical components of the manifesto. With regard to politics, the manifesto merely ends with a kind of disclaimer: "We value the liberal principles of democracy, tolerance, and pluralism in themselves, even as we affirm them as keys to achieving a great Anthropocene."[12] So whereas the "Ecomodernist Manifesto" sees a great deal of promise in invention and change for the technical future of the species as a question of governance, history for them is over, and a thin rendering of global liberal democracy is its endpoint. For these advocates, technical issues are not understood as politically charged, nor is this approach understood as a specific sociotechnical response to the crisis of the Anthropocene designed to buttress a specific form of consumer capitalism. The existing limitations of *real existing liberal democracy* are not addressed, nor are there experts on politics or governance included in the who's who of scientists committed to the ecomodernist project. The challenges to a "great Anthropocene," as Brand and his team call it in the conclusion of the manifesto, are technical rather than political. They are future-oriented rather than historical, and universal-planetary-species rather than particular. Returning to Keith, the goal of ecomodernism is for science to fix what politics cannot.

Jedidiah Purdy, in his book *After Nature: A Politics for the Anthropocene*, attempts to fill in where the ecomodernists leave off. Purdy similarly sees in the "end of nature" the possibility for a new, more equitable nature. While Purdy diverges from some ecomodernists on how and to what extent rewilding and urbanization play a key component in a good Anthropocene, he shares the commitment that modernism and a break from nature are necessary for achieving survival, and any egalitarian future. In a less technocentric vision of geoengineering and managed wild zones, Purdy proposes something like a Roosevelt revolution for resources and global democratic governance. In geoengineering as well as in other technological forms of intervention into the means by which humans survive, Purdy sees the possibility of minimizing scarcity and competition that otherwise undermine democratic principles, while supplementing the naïve technological optimism of ecomodernists with more robust forms of democratic governance.[13]

Thus, like Ecomodernists, Purdy sees the human character of the Anthropocene as an opportunity rather than just a catastrophe. If humans are capable of destroying the planet and altering the very material conditions of life, then they must be capable, via collective action, of altering the planet for good. As Purdy writes, "Most important, the Anthropocene is a call to take responsibility for what we make, as well as for what we destroy. It is the starting place for a new politics of nature, a politics more encompassing and imaginative than what we have come to know as environmentalism."[14]

Also like the ecomodernists, the main obstacle to pursuing a new ecological agenda is a naïve and nostalgic view of nature as "untouched" or "untouchable." Once we accept that there is no nature as such, we can better "spare nature" and live more abundant lives on the planet. In some sense, the scale of catastrophe is a kind of windfall, as it creates a set of challenges worthy of and inciting a true democratic cosmopolitanism that can overcome the binary of purely instrumentalist extractive views of nature and what Purdy sees as a "misanthropy" of contemporary environmentalism.[15] For Purdy, political interests in the Anthropocene are not entirely synonymous with human interests; however, when it comes to planetary survival, the problems of nature and other ways and forms of life will ultimately be mediated by the legal and civic orders of Euro-American moderns.[16]

While global in consequence, the crisis we confront is not singularly global in origin. It periodizes in very different ways from those suggested by the Ecomodernists and Purdy's Anthropocene. The attempts by ecomodernists and Purdy alike to "move on" from history is significantly dangerous for the prospects and success of their future visions for global abundance. The human-induced apocalypse, when viewed with an emphasis on the effects of the 1492 landing in the Americas, suggests that before the "great acceleration" after industrialization, there was a particular geographic and political, rather than species-wide, character to the geological era. Advocates for renaming our "cene" the Anthropocene focus too narrowly on climate change, and too expansively when attributing to the whole species the catastrophic transformation of Earth's atmosphere.[17] It is telling that so many of those who warn of the dangers of the Anthropocene so quickly use the same scientific evidence to make a case for the "great powers" of the world to take control of the planet through geoengineering. Ecomodernist proposals, for both our new geological epoch and the solution that follows closely with it, fail to consider the importance of the geopolitics that have brought our planet to this moment—where nature itself appears as a question for not just human but also Euro-American governance. If we are to meaningfully take on the challenges and even enemies of our current apocalypse, we must consider carefully the origins of the apocalypse we now face. To do so is to take seriously that there has never been an Anthropos for which we can now discuss the geological consequences of something like a single human species. Rather, the Euro of the Eurocene designates a vanguard among the European people who developed a distinctively mechanistic view of matter, an oppositional relationship to nature, and a successive series of economic systems indebted to geographical expansion.[18] The resulting political orders measured success by how much wealth could be generated in the exploitation of peoples and resources. The Euro assemblage of hierarchies, racial superiorities, economies, peoples, animals, diseases, and global resettlement is reflected in the geological record. What McKenzie Wark has called the Carbon Liberation Front was not a global phenomenon but a way of doing technopolitics originating within a narrow geographic region of the world made global by brute force.[19]

For those like Purdy and the Ecomodernists who wish to democratize the Anthropocene, the problem of power is still at the heart of the matter. Purdy writes, "In the Anthropocene, environmental justice might also mean an equal role in shaping the future of the planet."[20] Contrary to Purdy, I want to suggest that anything resembling environmental justice, or any future at all, will require an unequal role in shaping the planet. In fact, the transformation of the global political system necessary for environmental justice will require significantly constraining and even repressing those powers that continue the political-geological project of the Eurocene. Environmental justice will require the kind of struggle taken up by W. E. B. Du Bois and Frantz Fanon in the name of self-determination and anticolonialism but at a global rather than national scale.

On the eve of the creation of the United Nations at the Dumbarton Oaks conference, Du Bois saw the failure of a dream before it had even been fully formed as a thought. He saw that the vast new international body was little more than the institutionalization of what he called the global color line.[21] The great powers had insisted on a Security Council with veto powers and a General Assembly subordinate to its nuclear authority. The simple suggestion that the planet could be governed equally ignores that challenging vast systems of injustice will not be welcomed by those who benefit from exploitation. Continued settler colonialism, continued primitive accumulation, and continued violent power politics all reap enormous rewards for many in developed countries. To bring those interests to task will require a confrontation with a state system guarded by great powers that still use fifteen thousand nuclear weapons to deter change, and now deploy swarms of drones to hunt down those too small for the nuclear option. Therefore, Purdy is wrong to say, "There is no political agent, community, or even movement on the scale of humanity's world-making decisions."[22] Earth has global governance. We share a world governed by a few states with the capability of ending all life on the planet. Those states are, at the international scale, authoritarian, in that they rule by economic violence and warfare. That some of those states are liberal democracies domestically is of little consequence to the rest of the world.

The importance of the Eurocene is, as the Anthropocene was intended, to more accurately describe the world we live in and raise awareness of just how unequal that world is and by what means the great homogenization is being carried out. No global environmental justice movement can address justice or the environment in a significant way if it does not take as its starting point our current global environmental crisis as a *geopolitical* crisis.

Purdy and, to a lesser extent, the Ecomodernists, are in a position intellectually and philosophically to take these dangers seriously, even if their policy proposals offer little opportunity for things to turn out otherwise. However, there are fellow travelers within the future-oriented ecomodernist project who welcome this homogenization as if it were a kind of preordained convergence to truly live up to the promise of the human.

For the outer rim of the geoengineering movement, hypermodernization is meant to buy time until the species can escape the planet, or escape the organic character of the species. For these more overtly Promethean strains of geoengineering, the "downside" I identify in the trajectory of the Eurocene—total homogenization—is seen by them as an opportunity. One such movement even unironically calls its transcendent apocalypse *the singularity*, whose adherents follow the dream of Ray Kurzweil to become "spiritual machines."[23] This view of the future finds its salvation in a free market *doxa* that sees human transformation in what is believed to be the limitless innovation of profit-driven market competition. Silicon Valley is now home to a kind of scientific think tank of capital investors and cutting-edge researchers eponymously named Singularity University that is dedicated to such a future without limits. The bar napkin description from the website of this pop-up university reads: "Exponential Thinking, Abundance and How to Create a Mind."[24] In a perfect Singulatarian future, the current ecological crisis will become obsolete because humans will no longer need anything to live other than the energy necessary to power the vast computer servers that will host our digital consciousness. Earthly concerns are treated as quaint and trivial compared with the vast potential of the human mind freed of death and decay: all of life reduced to data, all of experience reduced to knowledge.

It is important to take seriously that other technological lines of development have been marginalized, even driven extinct, not because they are less functional or less innovative but because a particularly abstracted and mechanistic view of technology grounded in Western Enlightenment got lucky.[25] To mistake the fortuitous contagion of 1492 with the superiority of a European techne is as much a failure of facts as it is a failure of ethics. Instead, the contemporary line of technological innovation more closely resembles the trilobite explosion just before the Great Dying, in which changes in the environment benefited only one species. The vast homogenization and resulting monoculture resulted in Earth's third major extinction event some 252 million years ago. The cult of innovation incentivized by the market follows a very limited set of ideas that have varied little since John von Neumann laid out his model of the computer in the 1940s. The computer explosion has followed von Neumann to his logical end, but no further.[26] The "breakthroughs" of the last quarter century have basically been technical solutions to the contest between heat and processing speed. In almost eighty years, we have not produced *another* computer, and we are already running up against the limits of physics in the advance of the diagram in which we are currently stuck.[27] At this particular moment, moderns are like viruses that have plugged a very small number of technical diagrams into the engines of replication and production. As a result, we are drowning in a deluge of obsolete cell phones and many other larger and smaller interfaces for computers that do not fundamentally differ from their replacements. Each device, tablet, smartphone, smartwatch, laptop, desktop, smart car, smart refrigerator, and drone is the same device with a different interface for the world. We live in an age of one technical device, the digital computer, and it has a thousand faces. This is a sign of stagnation, not exponential innovation, and a foreclosure by the embrace of the liberal capitalist singularity in a modern world order.

Vision II: Kill Capitalism or Die Trying?

Wait, wait a minute there, yes it's Karl Marx, that sly old racist skipping away with his teeth together and his eyebrows up trying to make believe it's nothing but Cheap Labor and Overseas Markets. . . . Oh, no. Colonies are much, much more. Colonies are the outhouses of the European soul, where a fellow can let his pants down and relax, enjoy the smell of his own shit.—THOMAS PYNCHON, *Gravity's Rainbow*

Contemporary Marxist approaches to the environmental crisis fall broadly into two camps. There are those theorists who have returned to the logic of contradiction to overcome the melancholy of ascendant neoliberalism. There are also those who are tired of waiting around for history, and hope to forge a new vanguard to transcend the limits of ecological systems and even the terrestrial container itself. The new vanguardists see in the vast revolutions in logistics and materials engineering, as well as in the advances of climatology, a kind of engineered communism whereby the Malthusian limits of capitalism and communism can be broken through with the unleashed potential of labor after alienation, or even after labor altogether. It is worth considering both images of Marxist praxis together as they converge on a forgetfulness. The question of liberation for whom and at what cost still hides in the dark corners of both of these strains of contemporary thinking. Although more attentive to marginalization, the historical patterns of marginalization and the uneven distribution of annihilation are sidelined by these demands for a global concern. The recent explosion in contemporary Marxist theory is too vast to address in specific here. Instead, I will discuss just a few theorists emblematic of the two diverging paths between contradiction and vanguard transformation.

Jason Moore's *capitalocene* has greatly intensified the Marxist interest in ecology beyond the quiet work of longtime Marxist ecologists like John Bellamy Foster.[28] Moore sees in the contradictions of capitalism the possibility of an alternative political economy less ecologically brutal than capitalism and communism, and as a result, the rise of new social movements and even, to quote Moore, new ontologies for living after the era of "cheap nature."[29] Thinkers like Moore see the outline for the possibility of a more just future within the very terrestrial and ecological limits of the planet. Yet Moore, whose contribution I value because he sees the possibility of transformation not just in a radical political economy but also in ontology, cannot but help to fall back on the logic of "grave diggers." Even as the return to a historical conception of nature aspires to an ecological reading of history, the categories fall back into the same questions of productivity and profit: "The weird and dynamic process of putting nature to work on the cheap has been the basis for modernity's accomplishments—its hunger for, and it[s] capacity to extract the Four Cheaps: food, energy, raw materials, and human life. These capacities are now wearing thin. Industrial agricultural productivity has stalled since the mid-1980s. So has labor productivity in industry—since the 1970s."[30]

Thus, while Moore says he departs from the old logic of contradiction, his claim that "the breakdown of capitalism today is—and at the same time is not—the old story of crisis and the end of capitalism" is not a withdrawal from Marxist eschatology; it is an amplification.[31] The alteration is one in which ecological contradiction, that is, Moore's "cheap nature," is merely additional fuel for the dialectical machine of contradiction, revealing, and progress: a progress through negation. And so while Moore adds a number of forces beyond labor and profit to the frame of his historical analysis—something vital and important—in the end the "four cheaps" accomplish little more than "save" history as a Marxist analytic.

In a less dramatic way, Gopal Balakrishnan's rereading of contradiction follows a similar path.[32] Geopolitical conflict, civil war, and ecological crisis are folded into the crisis of capitalism as *accelerants* to the demise of the order, instead of as coconspirators in a larger order beyond the capitalist mode of production. Rather than distance themselves from the logic of contradiction, Balakrishnan and more subtly Moore have gone to great lengths to resuscitate this classical Marxist claim about the limits and transformation of capitalism via the drama of the ecological crisis. Like imperial and capitalist projects of world integration, Marxist drives for political unification by means of a polemicized historical necessity are still calls for homogenization. What I will try to respond to is the "upshot" of Balakrishnan's close and admittedly rich readings of the early and later Marx's clash with Georg W. F. Hegel. As I understand the payoff, Balakrishnan wants to argue that the later Marx finds himself again thinking about the ways in which a particular contradiction of capital, civil society, and state form "digs the grave" of capital—that the contradictions of capitalism, while not necessarily producing its transcendence, do create its demise. In so doing, argues Balakrishnan, Marx returns to an earlier debt to Thomas Malthus and the "dismal science" of demographic growth.

If we follow Balakrishnan to the end, we should infer the relevance of Marx on two critical points about our future. The first is that there will be a continuing decline in the rate of profit due to the limits of automation and efficiency improvements resulting in what Balakrishnan extrapolates as "the failure of new phases of accumulation to materialize."[33] And Balakrishnan's second extrapolation, that Marx's return to the problem of upheaval and civil war is inspired by the contradictions of capital, may be coming to maturity, historically speaking, in the form of an accelerating ecological crisis that reinvents Malthus's original limits of growth thesis. Balakrishnan argues that the ecological crisis, or what Marx at other times called the human "metabolic rift," was not fully transcended by Marx. It is back to accelerate, or at least intensify, the contradictions of capital accumulation and labor exploitation. In the end, Balakrishnan argues that capital will produce its own limit or crisis in the very failure to produce the economic and political form necessary to accommodate the scale of our coming crisis. In Balakrishnan's words: "However long its 'internal' structural contradictions take to play themselves out, no system so incapable of implementing the colossal scale of social planning that it would take to meet this impending challenge can be considered viable."[34] Therefore, according to Balakrishnan, capitalism should meet its end on the rocks of its lack of imagination, that is, the inability to find a new mode of accumulation, and drown in the incompetence of its political stagnation, or its inability to invent social planning sufficient for its contradictions.

I want to take up both premises as they make more explicit what is also central to Moore. First, I want to consider the question of capitalism as a failed social system planner, and whether the natural limits of the ecosystem represent an accelerant for capitalism's contradictions. The claim that capitalism is at odds with social planning is simply unsustainable. After all, geopolitics *is* a form of global social planning, and maybe the largest ever, as it is a system of planetary management. The problem is that we have a progressive functionalist bias in what constitutes successful social planning. The global networks of surveillance, multicountry military operations, mutual security alliances, and transnational border, seaport, and airport cooperation suggest that capitalism can innovate and implement global-scale social systems with high degrees of efficiency and effectiveness. What is left out by Balakrishnan is that one can have vast and unprecedented social and economic planning without benefiting the majority of the planet. Balakrishnan, to his benefit, makes note of the vast socialization of capital in the form of health care and education for the benefit of industry. However, more importantly, I would add to the cost of educating workers and maintaining the health of workers the huge price tag of the American-led geopolitical order. The cost of the wars in Iraq to maintain access to oil runs in the trillions of dollars alone. Yet even in regard to climate change, clean water, species loss, and migration crises, it is true that vast capitalist social planning is in the works: for instance, ecomodernist proposals for geoengineering that consolidate the hegemony of a few great powers. They do this via climate control and cooling northern countries while devastating subtropical and tropical countries by slowing the rain cycles necessary for agriculture; through capital-intensive desalinization, which ensures water security for those who can afford it; and through synthetic meat and food production, which represents real alternatives to land-based food production but extraordinary capital input. All these strategies for "adapting" to the climate and ecological catastrophe—while currently being pursued—are rejected by academics as seemingly unrealistic because they do not scale for 10 billion people. However, it should be considered that the visions of an ecomodern vanguard are less concerned with the speed and feasibility of scale as long as there is enough for them.

In fact, the most common critique by Marxists like Andreas Malm and neo-Keynesian ecologists like Naomi Klein is that such design schemes for cooler weather, clean water, and food security will leave potentially billions without the means to survive. From the ecomodernist perspective of elite blocks, that is, contemporary capitalism, so be it. Unlike Balakrishnan, Klein, and to a lesser degree Malm, the crisis of "failure" is just the next stage of what Klein calls elsewhere "disaster capitalism."[35] The fact that these solutions do not provide for everyone, or might create periods of tumultuous transition, are not critiqued within the standards of the narrow few pursuing them. Instead, the limited applications of industrial adaptation models are a virtuous form of managed scarcity and opportunity for the reconfiguration of governance. This transformation, while incipient, is for me a significant transformation between labor, contradiction, and political crisis. It envisions a world, and then engineers a world that can live, and even thrive, without the majority of people currently alive on the planet. The limited survival and making of a new fully manageable planet is precisely a "system-wide economic renewal."[36] The reconfiguration of the planet such that labor and resources play a significantly different role than previously experienced may not exactly be capitalism as we have understood it, but neither is it Balakrishnan's vision of a future created by Marx's understanding of contradictions.

Even the presumption of Foucault's schema of biopolitical capitalism is that somehow capital cannot live without labor, and therefore contradictions, that is, immiseration, will create turbulence for revolt, even if not revolutionary change.[37] In the biopolitical logic of labor, the threat of extermination is limited by the necessity of a population or mass to protect and to make live.[38] However, in the schema developing among the futurists of Silicon Valley, this presumption is simply false. Mass death at scales comparable and even exceeding the bubonic plague are no longer a problem for capital functioning. In fact, it is quite the opposite: losing one-half or more of the global population if sufficiently contained by drones, space weapons, and other long-range autonomous weapons systems provides a great benefit to a certain class of an increasingly cosmopolitan elite who look forward to automated forms of labor that make human exploitation instrumentally even if not morally obsolete. The limit of Balakrishnan's and for that matter Marx's vision is and was the presumption that there are limits to what humans can bear materially and morally. Like Moore, Balakrishnan is simultaneously too optimistic and too pessimistic. These thinkers are too pessimistic about the creativity of the Eurocene, which historically has departed from classical capitalist logic and structure in important ways, while maintaining the consistency necessary to continue. Furthermore, they are too optimistic about the ways that consciousness and political action change under increasingly sadistic material conditions.

A project of homogenization also hides behind the problematic logic of the contradiction of capital creating its alternative. There is something quite dangerous in how this logic interiorizes resistant ontologies or uprisings against capital as "new." Of the many examples given by Moore and hinted at more obliquely by Balakrishnan, the insurgent ontologies, whether rural or indigenous, are exceedingly old. I do not mean in the sense that indigenous peoples are outside of time but in the sense that Eduardo Viveiros de Castro has used when he talks about "extra-moderns,"[39] that is, forms of life that have creatively survived, so far, the onslaught of homogenization under the

Eurocene. To classify these extramodern forms of life "new" or contingent upon the "contradictions" of cheap nature is to resign their cause and constitution to the logic of capitalism itself. And as such, this logic drags along with it a project of homogenization as a way of thinking, whereby the resistance or even indifference to the Eurocene is created by the Eurocene. Not unlike Marx's insistence on proletarization as a necessary precondition of revolution, interiorizing struggles like "Idle No More" or the farmer revolts in India to the logic of contradiction is to rob them of what autonomy they have carved out against the dictates of development and modernization in both capitalist and Marxist iterations. These insurgent ontologies against extractivism are continuous across centuries of rebellion against the appropriation of land and resources. They are not a recent phenomenon created by the end of cheap nature. What is new, potentially, is that those of us consonant with a modern form of life have begun to care.

Modern concern for extramodern struggles may well be because of the glaring crisis caused by lives built on and from cheap nature. However, to conflate that concern with the cause of those who provoked our concern is politically and ethically dangerous. Like the logic of the Anthropocene whereby the power to break the world somehow suggests the power to engineer the world, the interiorization of indigenous and rural struggles against the Eurocene enables many of the same assumptions about the grounds and legitimacy of global-scale governance, even if such forms of governance would be more indebted to cosmopolitan solidarity rather than elite geopolitical control. In the former, the difference is more sentimental than material, as the cosmopolitan solidarity of a new, more just ecological order would itself also be a project of incorporation and homogenization, except this time done in the name of the marginalized irrespective of their input or shared governance. Thus, while it is all well and good to declare allyship for struggles like Idle No More against the destruction of the planet, we should be careful that the grounds for agreement are not a prelude to a new enclosure.

As an alternative, we could consider what Rosie Warren has described as Marxist pessimism.[40] This begins with the invitation to consider that the depths of human sadism and indifference are vast and even bottomless, a particularly important move as automation is on track to replace as much as 50 to 60 percent of all forms of labor in the next twenty-five years.[41] Furthermore, the automation and even autonomy of military labor, that is, modes of destruction rather than modes of production, open up a dark array of new economies where death, human labor, and capital accumulation take on a more horrifying plasticity.

From this pessimistic perspective that I am trying to develop here, the advent of Donald Trump is not the result of contradictions—that is, the dying gasp of capitalist class or a new populism—but the full cynicism of state, elite, and geopolitical control. The future is never fully determined by initial condition, but for me the idea of a movement of history, a limit imposed by contradiction, is at best an echo of naïvely hopeful confirmation bias. There is no sense of justice or development to which history arcs. There is no providence or historical force in favor of the better over the worse. However, that does not mean we cannot see the ways the present is developing and projecting itself into the future. If I were to make a prediction based on the emerging ecology of transnational capitalism, I would bet that Trump, rather than exacerbating the contradiction of the current order, will instead find new ways to weaponize it. Barack Obama's veneer of multicultural sensitivity will be replaced by an austere and sinister defense of brute force for whatever is needed. In this sense, I believe Trump will fulfill the truly postpolitical dream of neoliberalism, not in the pathetic governmentalities of the World Trade Organization (WTO) or Trans-Pacific Partnership but in what Saskia Sassen has called "the savage sorting of winners and losers," where large swaths of laborers and consumers can be ignored or liquidated, and where even once valued classes like the middle class can similarly become irrelevant.[42]

In this emerging epoch of capitalism, we will witness en masse and at the end of the barrel of a gun Glen Coulthard's reworking of "accumulation by dispossession" as rare earth minerals, dwindling petroleum supplies, and water all become significantly more important than human labor. Coulthard's point in *Red Skin, White Masks,* which we should take quite seriously, is that "primitive accumulation" was never primitive; it was ongoing particularly in settler societies, and it is now accelerating.[43] In this diagram of resource- rather than labor-intensive capitalism, we end up with Achille Mbembe's necropolitics, or the affirmative and productive industrialization of death and annihilation.[44]

Contra Balakrishnan's hope of falling profits, the liquidating rather than proletarizing of populations can still produce capital accumulation. This can be accomplished by selective displacement and murder such that new infrastructures for flows like oil pipelines or access to the rare earth minerals necessary for technological transformation become available to support cognitive economies less constrained by labor. What populations remain, driven mad by anxious consumption, are sufficient to maintain adequate consumer demand for increasing profits. In this diagram of capital accumulation, we have dispossession and platform transformations for new means of consumption, from the internet, to the internet of things, to the projected internet of spiritual machines—it stacks platform on top of platform. In what Benjamin Bratton has called the *black stack* or *stack geopolitics,* the successor of Donald Trump is the slicker and more sophisticated Elon Musk or Peter Thiel, for whom inventing gadgets and electric cars is already being projected out to interplanetary schemes for asteroid mining, Martian colonization, and a universal income guarantee for the few who will follow the intergalactic pathway of human development and commerce.

What we can already see in the excitement over Donald Trump by the alt-right wing of tech enthusiasts is precisely this ruthless disregard for human life in the name of getting things done. And this has been a long time coming. In 1989 Félix Guattari had this to say about Donald Trump: "Just as monstrous and mutant algae invade the lagoon of Venice, so our television screens are populated, saturated, by 'degenerate' images and statements. In the field of social ecology, men like Donald Trump are permitted to proliferate freely, like another species of algae, take over districts of New York and Atlantic City; he 'redeveloped' by raising rents, thereby driving out tens of thousands of poor families, most of whom are condemned to homelessness, becoming the equivalent of the dead fish of environmental ecology."[45] We should repeat Guattari's social ecological judgment of Trump for Musk, Bill Gates, and others whose toxic ecology is now pursuing an interplanetary scale of conquest.

Peter Sloterdijk, Vilem Flusser, and Lewis Mumford, in response to the cybernetic zeal for the future, refer to what they call posthistory. Each of these thinkers is attempting to understand a culture that is built around the idea that a particular race of humans, moderns, has escaped Marx's warning that "men make their own history, but they do not make it as they please."[46] Posthistorical humans believe that they do have the power to determine their own circumstances. Posthistory then is not meant in a Hegelian way. For each of these thinkers, the post marks an aspiration and state of exhaustion that ensues from the failure to make good on its promise. Euro-American world making—terraforming—has reached a limit with the seeming permanence of global interconnectivity and programming. As Flusser puts it, moderns now face the problem of programming where the capacity to program returns each of us to the question of who or what programs us. The aspiration of totality eats itself but continues anyway. Even the dark spots that periodically emerge in the world without exteriority are at best interruptions—wars, catastrophic accidents, acts of nature. Seceding, much less disappearing, is no longer possible as the globe is currently enacted and perceived. There is nowhere to hide. “History brings about the catastrophe of local ontologies.”[47] Mumford adds to this formulation the concern that the state of exhaustion and the presumption of a programmed order simultaneously inflates the hubris for global-scale management and creates a sense that there is nothing to be lost or gained as everything is transferable, malleable, useful or not. For Mumford, with the eclipse of animism and the sacred, we also lose the capacity to understand value beyond instrumentality. In this sense, we are done with history because what does happen is not historical, not an event; everything is modulation. Therefore, for all three thinkers it is possible that we are not a “we” in any meaningful moral sense but that we are nonetheless stuck: a global condition without a global people.

Those not completely alienated by the state of affairs swing to the other extreme, hell-bent on renewed expansion. Elon Musk’s desperation to take globalization on the road to Mars is the result of the same stuckness, but he rallies resources for a vicious exit strategy. Just as those ground up, lost at sea, or stolen for labor were, Mars is the horizon of possibility for another great age of exploration. Whether anyone makes it to Mars or if most of us are left behind is secondary to the redoubling of Euro-American terraforming.

Posthistory, stuckness, dreams of planetary and species transcendence—this is what I have in mind for this book’s subtitle, “Geopolitics at the End of the World.” Transcendence in this industrial and instrumental register seeks another savage ecology, a new planet to saturate, another surface to render spatial at the cost of regions and places of contour and difference. Whether life is discovered on Mars or not, the aspirations of colonization are dreams to once again transform “lifeworlds into locations.”[48] For Sloterdijk, the global approach transplanted from one planet to the next still captures the difference between the metaphysical age of antiquity and the modern age in the geometric difference between ascending and flying. Ascending was the imagination of escape velocity—to leave Earth and continue on and away. Flying requires mapping and following a surface, making a planet by flattening the planet epistemologically. Even those who wish to ascend to Mars actually want to fly, that is, resurface another sphere rather than cast off into the mysterious void of space. Mars is desired because it is useful; it is what is next. Mars is an effort to postpone the end rather than begin again. And so the pursuit of a savage ecology continues well beyond the contradictions of terrestrial capitalism.

If neither the planetary limits nor the limits of capital accumulation hedge against capitalism’s expansion, then we cannot take seriously Moore or Balakrishnan’s even half-hearted hope that contradiction will produce “grave diggers," or that civil wars may return to fracture capital. There will be grave diggers, but they will be automated by an algorithmic hunger for which there is no satiation. To put it another way, civil society, humans, and the political are—for a capitalist metabolism run on minerals and regulated by lethal automated force—luxuries, not necessities. The cozy relationship of Google and the state, as well as the vast network of joint ventures between defense departments and technology firms around the world, suggests that the state has new forms of innovation and control that do not require either Hegel's or Marx's visions of social order and social control.[49] The horror show of the next century, if not derailed, will be entrepreneurs and resource tyrants all the way down. In a world of necropolitical accumulation by dispossession, the reproduction of capitalist social relations may matter in the short run but not significantly in the longer term.

Labor automation in both economic and security sectors, vastly augmented by heuristic machine learning, can quite literally live off itself. This is assuming "the self" can continue to expand to asteroids and nearby planets. The limits and the catastrophe that we have been reduced to hoping for may be temporally and spatially out of reach. For those in what McKenzie Wark has called the vectoralist class, there is no catastrophe.[50] The ecological population growth apocalypse is an opportunity for an upgrade. The vectoralist class, or those for whom interest and benefit is not directly limited by the logic of capital, is even smaller than the dwindling size of the labor class. Further, unlike labor, they are better prepared for adaptation and reinvention than Marx would have suspected possible in the nineteenth century. As such, Peter Thiel and other paleo-accelerationists who funded and now celebrate the election of Donald Trump are coldly indifferent to the possibility of race wars, ecological collapses, and territorial displacement.

So rather than serve as obstacles, contradictions, for the vectoralist class, can be leveraged like any other hedging strategy pursued during the financialized epoch of capital. It is not a coincidence that Peter Thiel, with Elon Musk, is the founder of PayPal, the vast online system for deterritorializing and denationalizing capital flows, as well as Palantir, the data surveillance and logistic firm that does all spying for the National Security Agency (NSA). Like PayPal's ability to make money on the movement of money, Palantir makes billions of dollars from surveillance itself because the data can accrue interest and benefit in excess of its financialization.[51] Thiel, Musk, and their strongman Donald Trump are evidence of a new elite class native to a resource rather than labor-dependent logic of capital and equally invested in the dark possibility of a capitalism not constrained by its contradictions. In this future, we can now begin to see the horizon of how capitalism works precisely because it is broken. Thiel, Musk, Gates, and Trump are Joseph Schumpeter's creative destruction inside out. Accumulation by dispossession represents a logic of destructive creativity that requires little from social relations or the people who would populate them. Instead, the surveillance, displacement, and even massacre of those people is itself profitable.

If Marxism is meant to do *something* rather than merely satisfy our nostalgia for when intellectuals were truly dangerous, then the accelerating mutations of capital have to be considered, and the geopolitical capabilities of containing and even leveraging those contradictions must also be reflected upon more seriously. To do so means watching the horror unfold while also trying to understand its contemporary logic. The fact that this horror show seems bigger than our political imagination, or more than can be overcome by the force of *history*, is no reason to look away and hope the old tropes of contradiction and declining profit will hold.

Only Walmart Can Save Us Now

Other neo-Marxists dissatisfied with waiting for capitalism to undo itself have begun to develop strategies for intensifying the very means by which capitalism has succeeded in an effort to bring about a more just future. The logic captured best by Steven Shaviro is one in which "the only way out is through."[52] Unlike the waiting game of contradiction, the accelerationists hope to bootstrap historical agency—becoming historical subjects rather than subjects of history—through the creative and experimental technological transformation of supply chains and cybernetic governance platforms.

The left variant of this renewed futurism is much more interesting than that of the ecomodernists but also problematic in an age hell-bent on homogenization. Designated under the hashtag #ACCELERATE, thinkers such as Nick Srnicek, Alex Williams, and Benedict Singleton see in an accelerating rate of exponential technological innovation the possibility of a postscarcity social order. Admirably, the accelerationists demand a "recovery of lost possible futures, and indeed the recovery of the future as such"[53] against the resigned elitism of Thiel and others. Furthermore, the accelerationist future is enlivened by an ethos that is variably described as experimental, open to the outside, and creative rather than as a flat, all-digital singularity like the vision of Kurzweil.

Unfortunately, the ethos of experimentation is overshadowed by a commitment to "a modernity of abstraction" coordinated by a barren and homogenous "technology" and "globality." Contrary to the "outside" the manifesto[54] is dedicated to, all forms of life at odds with this abstract abstraction are characterized as "a folk politics of localism."[55] The arrogance of such a claim is particularly dangerous because the accelerationist accounts of technology are about as diverse as Monsanto corn. Like the aristocracy of the Habsburg empire, the insular and parochial trajectory of technological thought is bereft with recessive traits. While I share the view that the "left must become literate in . . . technical fields," I am less convinced that "big data" will be as important as Hawaiian sustainable fish farming or the capacity to proliferate wilder forms of life in the intensifying apocalypse of our time. At issue for me is a fundamental disagreement that "technology," as is synonymous with an accelerationist modernity, can, in their words, "win social conflicts."

While humans are often the source of technological mutations, whether or not the mutation "works" is rarely up to the designer. The ecology of each technic, whether the spear or digital consciousness, determines whether a particular arrangement of things is possible, and what effect or event it can enter into alliance with. For this reason, technology evolves, but so far its evolution has been tethered to the epigenetic structures of other animals and, primarily but not exclusively, humans. Therefore, it is necessary to think of each technical object as a kind of exogenous expression of an actual and virtual arrangement of things in a milieu in which humans experiment with and replicate but do not control or engineer. We are their DNA but little more. The expressive intensity of their existence could quickly leave us behind. Like the Tibetans who inherited the capacity for living in a thin atmosphere from Denisovans,[56] or the relationship and knowledge of reindeer herding that has enabled the Tofa to live through thousands of long Siberian winters, diversification and *not* singularity is the history of planetary creativity. The loss of planetary dominance in favor of another intelligence is possible. Irreversible catastrophic changes are certain, but extinction is unlikely. What *we* stand to lose as a species in this current apocalypse of homogenization is unimaginable, not because of the loss of life but because of the loss of difference. Who and what will be left on Earth to inspire and ally with us in our creative advance is what is uncertain. If the future is dominated by those who seek to establish the survival of the human species at all cost through technological mastery, then whatever human "we" manages to persist will likely live on or near a mean and lonely planet.

I believe it is a narrow view of industrial and mechanistic technology—from the perspective of many forms of life, whether Mohawks, phytoplankton, humpback whale, Tibetan, or Missourian—that is *the* social conflict. Abstraction then, for accelerationists, is stuck in the tight orbit of mechanistic conceptions of data and emancipation, whereby conflicts are reducible to narrow utilitarian and abstract conceptions of need such as hunger or other resource demands. That forms of life may themselves face a crisis of scarcity—social conflict—in their ability to flourish under the pressure of global supply chains and the extractive economies necessary for the infrastructures of materials science and big data is rendered secondary. In this sense, the sandbagging of the means by which one "wins" the narrowly defined social conflicts of acceleration aids and abets the crisis of homogenization.

I still hope that the dismissal of localities as "frozen" cultures in favor of accelerating a new global modernity is the result of limited worldly experience rather than theoretical necessity and that it may loosen up as accelerationism gains footing outside the narrow confines of Europe and the United States. However, at some level the tension will always remain between the converging operational transformation of the species envisioned by accelerationists and the possibility for many different forms of life. When faced with the contradiction, the question is this: Who will be empowered to choose one side of the contradiction or the other? For me, to write off the varied human animal and nonhuman animal forms of life throughout the world that run contrary to the late-modern technical epoch would be an unimaginable loss. This is a fate, as it were, worse than an unpredictable future for modern forms of life already pushed so far to the front of the line in the history of the Eurocene. The prospect of shiny and hopeful technologies that are being held out to break the deadlock of political thinking is all too intimate with the flattening of earthly life. I agree that there is no technology as such that can be distinguished from nature. I do not think we can "go back" to the Rhine or the Black Forest, but it is a categorical error to conflate the iPhone and global logistics infrastructure with all forms of liberating technology. We have no way to know if it will be Fredric Jameson's communist Walmart, or technologies of nomadic living practices, or urban farming, or some as-yet-unseen configuration of life—human or otherwise—that will offer creative and less cruel possibilities for another ecology.[57]

Therefore, we should be skeptical of a world "set free" by Western contemporary technology. In the contest between the accelerationist vision of experimentation and adventure, and the historically overdetermined tendency of these systems to necessitate vicious forms of management and control, on what basis do we expect a different outcome? While accelerationists certainly represent a somewhat different trajectory of thought to utopian dreams of consciousness without bodies and the neoliberal megacities of ecomodernism, the underlying understanding of social conflict as material scarcity makes me wonder if accelerationism could possibly end in any global order that was not coincidental with the singularity of ecomodernism. While I want to affirm the accelerationist commitment to experimentation, I want to push for a wilder and less managerial ethos of experimentation that is willing to risk stepping outside the boundaries of operational modernity, even if it risks the end of so-called human progress. The accelerationist wager that a renewed hyperfunctionalism can escape the deadlock of social antagonism falls all too easily back into a survivalism. Even if this survivalism is pursued in the key of egalitarianism, it carries the danger of a sovereign decision vastly overdetermined by the creation of the Eurocene. In the end, who will decide and for whom, and at what scale and character the technical transformations will take place, is anything but technical.

If accelerationism cannot in some sense make both a more modest and a riskier wager, then its experimental abstraction will join in the long legacy begun by Marx and carried on by state Marxists of cooperating with industrial state capitalism to deprive discontinuous forms of life, and in particular indigenous people, of full, meaningful lives. For Marxists, annihilating other forms of life is carried out so that nonindustrial populations can be liberated in the name of an abstract notion of equality that few, if any, people asked for in the first place. The poverty of Marxist thinking about other forms of life and value outside of species-being and labor is an ethical and practical hazard.

What if things are so bad that it is try or die? Isn't a Marxist egalitarian modernity preferable to a world ruled by Trumps and Thiels? I suppose I am still struck by the way we consistently are able to stage such dichotomous choices to reempower a geopolitical vanguard to speak on behalf of others, and in pursuit of a global future that may in fact do little if anything to disrupt the unequal distribution of pain and suffering native to the Anthropocene. It is hard to think some other platform, whether data-silicon, mega-urban, or luxury communist, would make much of a difference to those populations so easily left behind or outright exterminated by the last several iterations of modernity. I suppose my point is that there are worse things than death, there are worse things than apocalypses, and survival at such a sadistic cost is not something I want to try.

That there is currently no alternative is not a reason to invent another reality where nineteenth-century Marxism still works by technological miracle—Jameson's communist Walmart. Rather than castigate those who lack faith in a global transformative revolution to undo capitalism, we could take seriously that things might just be this bad. If this all sounds *too* pessimistic, I would counter that there is no normative guardrail to prevent reality from being *too* catastrophic. There is, of course, the danger of negativity overwhelming what political options still remain. However, the reverse danger is also possible—that the inability to break from consensus reality comes to support a naïve assessment of the world in place of evidence, and a naïve faith in history or technology. Against this logic of "it can't be true because it is too terrible to be true," we can join Marxist Rosie Warren in developing a realist pessimism that resists the precious nihilism of darkness, and the moralized insistence of optimism. Warren warns and insists:

> There is, of course, a danger of bending the stick, becoming rigidly pessimistic, fetishising pessimism tout court in just as evacuated a position as those so eager to accentuate the positive. Salvage is not interested in pessimism for pessimism's sake, in prolonging our pessimism any longer than is justified by our analysis, and aches for a time when pessimism is no longer necessary. . . . Pessimistic is just another word to describe those who fear we might be doomed but are fighting anyway, those who don't have a lot of hope but plenty of hate and heartache, plenty of yearning for something more, who have no certainty about the way forward except that it cannot be this.[58]

If we follow Warren, if we fight anyway, the fight should not come at the cost of other ways of life and other forms of life, much less other species of life. Otherwise we are left with yet another vision of transformation no less cruel, despite its Marxist commitment to egalitarianism, than the capitalist iterations of transformation that preceded it.

Vision III: Encountering the Third Offset, or a Vision of Geopolitical Transcendence

On November 13, 2016, I was driven to West Point Military Academy for an event on the fiftieth anniversary of the publication of Thomas Schelling's *Arms and Influence*. Along with me were a retired lieutenant general who was heading up the overhaul of the U.S. Army Cyber Command and an army intelligence officer recently stationed at the U.S. Pacific Command, only a few miles from where I live in Honolulu. As we wound up the Hudson River, the conversation was a bit stilted at first. The two career officers were polite but skeptically curious about what I was doing in the van. The driver, a West Point dropout and retired NYPD officer with the same crew cut he left the academy with, shifted into tour guide mode and began to tell us the facts and figures of the West Point property. As we approached the campus, he pointed to a wide bend in the river and told us that this was where General George Washington had planned his last stand against the English. A giant iron chain had been forged so long that it could stretch across the entire width of the Hudson. The great chain had in fact never been used, but if it had been deployed the English were to have met chaos and the full force of the river after becoming entangled in it, and that fantasy of victory is what is remembered. There was a palpable sense of awe in the van.

Despite only the hypothetical value of the technological innovation, over the weekend the chain returned time and time again to stand in for the ingenuity and capacity of the American military spirit. We heard very persuasive presentations by academics, seasoned veteran soldiers, and policy makers of the highest levels of military experience about the declining value of force, the nonfalsifiable history of the so-called success of nuclear deterrence, and the increasingly complex world of what now is called "multidomain warfare," that is, the next in a long line of euphemisms meant to contain the fog of war. Yet the offers of hope for U.S. hegemony were unflagging. The final report of the event returned to the faith in Washington's chain. According to the leading voices at West Point's Modern War Institute, the future of U.S. supremacy was to be found in the "Third Offset." The term *offset* was originally meant to capture the capability to maintain supremacy despite numerical inferiority in troop levels. More recently it is meant to capture the various ways the U.S. military plans to adapt to asymmetric warfare, and the unpredictable geopolitical future of climate change.[59]

During the 1950s, the cornerstone of the offset strategy was nuclear superiority in order to fill in the gap or offset the superior numbers of Soviet troops and tank divisions. Tactical nuclear weapons in particular were sent to NATO countries so that "usable" nuclear weapons could be quickly deployed to prevent a full-scale Soviet invasion of Western Europe.[60] Similar conversations are beginning again as Russian and U.S. competition in the Baltic countries escalates.[61] During and in the aftermath of the Vietnam War, there were heavy investments in offset strategies such as intelligence, surveillance, and reconnaissance capability (ISR), along with advances in precision weapons that now define contemporary combat, from the first Persian Gulf War to drone strikes in Yemen, Afghanistan, and throughout Africa. This is often referred to as the Rumsfeld Revolution in Military Affairs, or the turn to net-centric warfare.[62]

The renewed interest in offset strategies, commonly referred to as the Third Offset, is less defined and even more ambitious than earlier iterations. As it has been described by military planners and think tank researchers, the

Third Offset is based on an honest accounting of U.S. advantages over the vastly larger Chinese and Russian adversaries.[63] The argument, following in the spirit of Washington's chain, is that the U.S. possesses an inventiveness or creative capacity that, historically, geopolitical competitors are unable to match. So rather than see a particular weapon (nuclear weapons) or infrastructure (ISR and precision capability) as critical to military supremacy, the Third Offset attempts to leverage innovation itself as the strategic advantage of the U.S. military. What will offset the vast numerical disadvantage vis-à-vis either of the major U.S. competitors—much less the combination of the two, or the new nonstate threats, or the complex humanitarian-security crises of the climate refugees, or disruptions in the global food system—is the American "can do" spirit combined with the industrial capacity of the United States.[64]

Critiques of technological optimism, particularly in the U.S. military, are familiar and even hackneyed as the naïve exuberance for new toys among the Defense Advanced Research Projects Agency (DARPA) and arms manufacturers provides an easy target. However, the Third Offset comes with an unprecedented twist from earlier claims about technologically led revolutions in military affairs.[65] Because the cutting edge of supremacy is invention, or the speed of thought according to the Third Offset thesis, the U.S. needs a breakthrough in thinking rather than the individual outputs of thought. Therefore, the cornerstone of the Third Offset is now quite casually referred to as "general artificial intelligence."[66] If, according to this thesis, the U.S. possessed the capability for fighting cyberwar at the speed of computers rather than the speed of hackers (another critical area of "numerical insufficiency),"[67] or could compute scenarios, or deploy and coordinate thousands and thousands of micro drones, all faster than the most anomalous human brain or team of brains, then the U.S. will have broken the proverbial sound barrier and weaponized its greatest asset. The key move is to lock in the ability to think faster than humans, and to innovate and fight at a new velocity of thought.[68]

One cannot help but see the transformational optimism of Ray Kurzweil, Peter Thiel, and Elon Musk here, who all have greatly influenced a turn toward the fantastic in the strategic thinking of U.S. defense culture.[69] However, the plans for development and deployment are much more modest and much more immediate than the grand vision of spiritual machines. In the interim, while the DOD waits for its investments in quantum computing, neuroscience, and human–machine interfaces such as the Centaur program to pay out, existing machine learning and near-term advances in robotics and autonomous machines are meant to bridge the gap between what can be imagined and what can be deployed.[70] In the words of one of the members of the Army Research and Development Command, "We have all this technology, and we're trying to figure out how to integrate it."[71]

One of the more novel ways interim research attempts to cheat "the intelligence barrier" is by borrowing from nature. Like the Human Machine Interface (HMI) project, biomimetic and cybernetic research has produced everything from preimplanted and controllable butterflies to cyborg sharks.[72] Bootstrapping the operating systems of existing humans and nonhuman animals as a kind of hybrid intelligence platform is already being utilized in the development of new generations of drone technology. To quote a researcher from the Intelligence Advanced Research Projects Activity (IARPA) after completing a comprehensive review of military AI and drone research, "I realized Hollywood has it all wrong. The future of military robotics doesn't look like *The Terminator*. It looks like *Planet Earth II*."[73] The hybridization of nonhuman animals follows a longer history with the arming of dolphins and other highly intelligent animals for combat.[74] Related to and in many ways mirroring the autonomy of general AI, the modeling of insect and animal intelligence forms the foundation of control algorithms and expands the capability and collective action of fully synthetic autonomous machines.[75]

Like the bold vision of the army itself, theorizing about these trajectories means considering things only now possible or virtual rather than actual. The big breakthrough may never happen; however, if we take intelligence to be a fact of nature rather than a gift of some other realm, there is no reason not to assume a breakthrough will eventually take place. Whether or not it will resemble the desires of the researchers is what is up for speculation. Even if the outcome is not what is envisioned by the Department of Defense, the outcome will appear to us as unprecedented.

To consider the possibility of worlds not yet present requires developing critical faculties for thinking through the future that neither affirm the control fantasies of the Third Offset nor dismiss them as impossible. This is the task of what Rudy Rucker calls transrealism. Following a speculative trajectory, I would like to consider seriously that the military's neo-Hegelian joyride in search of artificial intelligence and networked everything might come to a grinding halt, reverse, or advance in unpredictable and unsettling ways. In considering these possibilities, I want to develop a mode of inquiry aversive to the net-evangelical gestures of Silicon Valley, as well as the military strategists inspired by them and who see technological convergence as preordained. And at the same time, I do not want to rule out the possibility for the new to emerge without warning or sufficient initial conditions, even if the new that enters the world is truly horrifying. To do so means that we cannot—as many neo-Kantians do—insist that artificial intelligence is unattainable because somehow consciousness cannot be possessed by machines and things. To reduce all thinking and critique to a single transcendental maneuver—the human as the only condition of possibility for thinking—is to doom us to remain in the confines of what is likely rather than explore what is possible. Furthermore, an overzealous faith in the uniqueness of the human will not prevent the vast resources of the world's militaries from proving Kant wrong. Military researchers may not discover what they set out to find, but research into automation and intelligence may nonetheless unleash something new in the world.[76] Instead of the false safety of Kantian prediction, we could instead follow Quentin Meillassoux's argument that we should ditch the correlationism of presuming that there is a kind of dependence of the cosmos on the human. Contra Kant, Meillassoux insists that rational critical thought must be open to the possibility of a cosmos that is indifferent and exceeds humans: "To identify rationalism with the eternity of natural, deterministic, or frequential laws is to render thought powerless before originary phenomena, and ultimately to resign oneself to acknowledging a transcendent foundation. Reason teaches the exact contrary: laws have no reason to be constant, and nothing entails that they will not contain new constants in the future."[77] Instead, rigorous investigation can be speculative in an orientation to the very real possibility of novelty, or what Meillassoux calls advent: "The advent of life is not the necessary effect of a material configuration (such claims have never made sense). Instead, it is the contingent and conjoint creation of a Universe of qualities and material configurations that were both inexistent until then."[78]

The advent is an eruptive difference that, like quantum probability, exceeds initial condition, and therefore in a Kantian sense has no condition of possibility at all. This is the character of the wild thought I would like to add back into critical security thinking, to describe conditions of possibility whose conditions are not yet probable (not the same as impossible) and only virtual. Given the dark audacity of military research, our critical thinking must keep pace with the undetermined character of the world that inspires technical research that seeks to engineer "new constants."

And yet at the same time, I would like to dump a bucket of cold water on the naïve optimism of Silicon Valley and its obscene correlationism, which sees not just Earth but the whole cosmos as a providential home for transhumans, and therefore "believes" the singularity is a necessary conclusion to history for the benefit of transhumans. To disrupt the growing consensus on a transhuman future, we can dramatize the discordant chunks (like inhuman intelligence emerging) or remainders that do not so easily reincorporate into the vast cosmic journey from cave paintings to intelligent silicon-based life (like the vast ecologies of intelligent waste emerging from endless battle zones). Jussi Parikka calls these chunks of deep time *future fossils*.[79] Similar to Timothy Morton's *hyperobjects*, these future fossils themselves demonstrate a temporality that is antagonistically noncorrelationist.[80] Future fossils and hyperobjects like nuclear waste, atmospheric carbon, plastic, and, as I argue, nonhuman emergent intelligence, inhabit a present that reduces empires, civilizations, and even the human itself to fleeting moments rather than long arcs of progress.

Rather than a dialectic or resolution to the seeming antimony between these advental objects and the doldrums of a future already fossilized, they can persist in torsion with one another as a mode of inquiry. In so doing, speculative analysis in a transrealist genre seeks out possible futures rather than probable futures that dramatize the multiple layers of time and change often exceeding human time frames and certainly exceeding human control.

Pursuing transrealism in the bizarre world of the U.S. military's transformational vision of the Third Offset would be, as Meillassioux says, hyperrational, but not in the confines of a human rationality. Rather it would consider the indifference of change, advent, and the virtual as contrary to a providential human destiny. Thinking about the problem of emergent autonomous weapons might give us a sense of how the homogenizing and cruel refrain of the Eurocene could continue beyond the humans that made it, and vastly exceed the strategic intentions of those invested in maintaining the existing geopolitical order.

The Human Abstract or What Might Be Next

"I was just figuring," said Montag, "what does the Hound think about down there nights? Is it coming alive on us, really? It makes me cold."
"It doesn't think anything we don't want it to think."
"That's sad," said Montag, quietly, "because all we put into it is hunting and finding and killing. What a shame if that's all it can ever know."—RAY BRADBURY, *Fahrenheit 451*

Either war is obsolete, or men are.—R. BUCKMINSTER FULLER, *Utopia or Oblivion*

Every day Fuller sounds more and more prescient, but the increasingly likely outcome of his prediction is that men are obsolete rather than war. In our moment in the Eurocene, war appears more durable than the human. So to understand the truly inhuman possibility of the Third Offset as a transformation of the global order, we need an analytic paradigm shift. If life is as simple as it seems—appetite, satisfaction, and replication—what new trajectory of creative evolution are we becoming? It is time to stop talking about general artificial intelligence as a kind of mirror of what humans think about their own intelligence and start talking about the possibility of a more general and nonhuman artificial life. There are a few reasons for this, but first there is the question of probability; the basic characteristics of life are already attainable and are vastly more dangerous than the narcissistic human obsession with general artificial intelligence or superintelligence.

Second, the possibility of a *breakthrough* in machine awareness may not be planned. Instead, the emergence of something truly novel—a new kind of intelligence—seems more likely to come from the mountains of war's rubble and remnants than from a laboratory. The world is already drowning in e-waste and surplus military technics.[81] What if those flows of waste become purposeful or aware? For waste to become brilliant would require little more than the intelligence of *E. coli* with its limited sight and strategic capability. Imagine something as simple as cockroach intelligence or ant intelligence creeping out of the ecology of broken lethal things filling the landscapes of Afghanistan, Iraq, or Yemen. Brilliant waste is what happens when weapons and other lethal forms of life persist purposefully. Land mines and IEDs are a primordial version of this form of waste as lethal things, with limited but definitive awareness of their environment.

In such a machinic ecology, humans would return to a state of predation variably similar to Paleolithic and early Neolithic eras but significantly distinct as well. Megafauna certainly posed a threat to humans during these eras but few, if any, animals targeted humans specifically. However, the ecological lineage of brilliant waste suggests the possibility of a particular appetite emerging for *Homo sapiens*. Aware-weapons-cum-brilliant-waste would be human-specific in their primordial design, and as likely to remain path-dependent along this trajectory of development, and potentially in the worst-case scenario, homo-mimetic. Consider already the weight limits of smart mines, the radio wave triggers of IEDs, and the pattern-of-life targeting system of drones. All these evolutionary structures, ranging from the simple to the complex, represent an appetite for humans.

The competition environment of increasingly long wars, now stretching decades rather than months or years, is ripe for the emergence of such forms of life not unlike the already too-present emergence of child soldiers or battlefield-created antibiotic-resistant microbes.[82] Conflict zones now last for generations. Think of how few denizens in Iraq can remember a time before a U.S. invasion. Beyond the ongoing combat with ISIS, current generations of Iraqis now suffer from war-induced air pollution and genetics-damaging heavy metals in the water table. War has, on a human timescale, permanently changed their zone of the atmosphere and terrestrial environment.[83] The long war is, for many, an endless war as they were born in war and will likely die before a conclusion is found. The stories of J. G. Ballard and Joe Haldeman are no longer fictional for many in the Middle East, as well as parts of Africa and Central and South Asia.[84] As William Gibson said in an interview on NPR, the future is already here, just unevenly distributed. This is particularly true of the future of war.[85] Thus contemporary warfare waged by U.S. and NATO forces is a form of sustainable warfare: slower casualty rates; geographically dispersed targets; automated and subcontracted violence; and zones of indistinction between war, crime, development, and humanitarian intervention. All render the otherwise effective means of war termination ineffective.

Sustainable warfare means a near-permanent state of war to incubate new forms of life: organic, mechanical, and digital as well as every possible hybrid permutation. What comes out of such an encounter between war, life, and creativity is something like the weaponization of life itself. Sloterdijk has considered at length the weaponization of the lifeworld in the history of gas weapons and incendiary bombing.[86] However, I have something a little different in mind. Rather than the weaponization of the *umwelt* of life, I want to think through the weaponization of the *process* of life during the five-hundred-year ecological experiment of the Eurocene. What artificial life confronts us with is the possibility of weaponizing evolution, or the conditions that reproduce and create new forms of life. I do not mean the "weaponization" of life in the banal metaphoric sense but the actual harnessing of creative evolution in the making of weapons. This is a selective breeding and husbandry of war made possible by the collision between the faltering and expanding strategic aims of Eurocene geopolitics, and autonomous or creative forms of emergent artificial life.

The coming-to-life of lethal technics with an appetite for "us" and our "others" is not as fantastic as the history of robot-paranoia-inspired movies like the *Terminator* series would suggest. After all, humans and other nonhuman animals such as dogs are already evidence of this phenomenon. Humans and human cognition coevolved with and through war and technologies of war.[87] According to Sloterdijk, with the emergence of "long-distance weapons and tools, the hominids managed to escape the prison of bodily adaptation."[88] Therefore the novelty of the approaching future is a kind of second-order weaponization of life or, maybe more appropriately, the vitalization of weapons. What would constitute such a "vitalization"? And what would the consequences of this vitalization be? In a philosophical sense, it would be a world of military surplus that insists on its own superlative existence.[89] Are weapons seeking their intended use as an end in itself? Objects for themselves are creeping out of the bifurcating ecologies of brilliant waste; the descendants of land mines and IEDs and cluster bombs are thriving in the data-rich nutrient of the always already militarized internet of things. Images of these vital weapons, of mechanical-machinic life, should also challenge our concepts of artificial intelligence and how AI might succeed at modeling human consciousness. That is, human consciousness is itself already a weaponization of a brain, appendages, thumbs, eyes, and coordination, and therefore a martial consciousness by design.[90] To successfully model human consciousness would be to create machines capable of sadism, torture, and murder rather than instrumental killing.

And what of the automation of us?[91] The long-term corrosion of human life resulting from the global automation of war, peacekeeping, and policing is impossible to predict in its particularity but less difficult to predict in its philosophical outcome given current trajectories of research and deployment. The ethical catastrophe of making war and surveillance easier and cheaper, while at the same time automating ascending layers of decision-making, transforms the thin amalgam of cosmopolitanism and global rights claimed by contemporary interventions into a perpetual motion machine. Although algorithms currently automate technical procedures such as take-off and landing, object recognition software makes possible the automated acquisition of targets, the killing of those targets, as well as the risk and value of collateral damage. The real revolution in automation is happening as object recognition software and its capacity to recognize objects such as tanks or AK-47s gives way to more sophisticated (or pseudo-sophisticated) capacities of gait and behavior recognition.[92]

Following Banu Bargu's work on necroresistance, this form of targeting is necropower par excellence as it returns the entire field of battle to the physiognomy of the body itself. According to Bargu, "The insurgent's body becomes the concrete battleground of domination and resistance, subjugation and subversion, sovereignty and sacrifice."[93] In 2016 Turkey's armed forces foundation company ASELSAN began construction of fully autonomous machine-gun turrets called "smart towers" to use at the Syrian border with the capability to target and kill without human oversight.[94] Politically, the decisionistic character of sovereignty to pursue war becomes a mathematical fact rather than a human judgment of enmity automating politics, if not fully automating any one machine or human soldier. Furthermore, the developing drone capabilities referred to as signature strikes that choose targets based on behavior rather than identity move beyond questions of citizenship or even identity common to war and violence. Instead they rely on metrics for identifying and judging behaviors as worthy of eliminating.[95] From the ecology of autonomous and increasingly aware weapons, a new ontology of the enemy emerges as a technical procedure well beyond the political landscape of enmity.

Rather than simply automating the "hunt" for enemies chosen by political processes, so-called signature strikes signal a shift to the automation of the political decision of who is and is not an enemy in the first place.[96] The identity that corresponds to the enmity of combat is replaced by an algorithmic definition or function of dangerousness. This shift from what Human Rights Watch has termed "human on the loop" practices to "human out of the loop" practices pushes the posthuman character of war further into the Cavellian nightmare zone in which everything is an object to be targeted but never to be encountered or recognized.[97] An object either is or is not dangerous as its temporally specific and targetable function. The cause or reason behind those attributes is no longer relevant. These changes will not be political events. The switches will be flipped by military planners or software developers without accountability.[98] If we continue on this trajectory, practicality will replace both strategic and moral thinking. Further, that practicality will be habituated by bodies that also cease to think about the quaint human algorithms of morality and duty. The fora in which such decisions will be made if at all are likely to constrict, as secrecy predominates in an environment charged by a dangerous mix of paranoia and real danger. And then it will all be modeled, explicated, and encoded.

In such a world, the event of machine awareness will parallel the loss of forms of human awareness. To take a more speculative tone, what will a close encounter with nonhuman intelligence do to force a "persisting us" to rethink the use to which we have put machines in the pursuit of what we ourselves have been unwilling to do? The answer is hard to consider because the "we" that remains at such a point may appear to us now as alien or inhuman as the machines I am speculating about.[99] We should take seriously the full spectrum of possibilities of what artificial life will look like if it emerges from the current world of surveillance and war that drives its evolution. To put it simply, what happens when we make artificial life in our own image or our martial image refracted through the physiognomy of beetles or sharks? What emerges will be a form of life native to the Eurocene.

Conclusion: An Elegy for Human Transformation

Rather than treat the current drive for homogenization as inevitable, or throw our weight behind one of the visions of transformation presented here, we could instead seek wilder visions of the future that start with a healthy sense of skepticism about the promise of technologically and algorithmically enhanced forms of human-centric control. Images of the future that take the Eurocene as their starting point are not the only possible futures. Making his case for minor languages, K. David Harrison questions the virtue and probability of a techno-singularity future.

> Perhaps we will grow plants in greenhouses and breed animals in laboratories and feed ourselves via genetic engineering. Perhaps there are no new medicines to be found in the rainforests. All such arguments appeal to ignorance: we do not know what we stand to lose as languages and technologies vanish because much or even most of it remains undocumented. So, it is a gamble to think that we will never avail ourselves of it in the future. Do we really want to place so much faith in future science and pay so little heed to our inherited science?[100]

Rather than see reindeer-herding practices of Siberians as "localities" or "folk" knowledge that must be cleared away for some as-yet-unknown transformation to come, why not see the very complex practices of the Tofa people as a sophisticated science for living in an extreme environment? There is a danger here too, though. The Tofa, as mediated through the work of Harrison, could become yet another noble savage, a mimetic stand-in for what "we" moderns do not like about ourselves. Viveiros de Castro's advice on the subject is quite useful. To take up the other forms of life as provocations for thought means "refraining from actualizing the possible expressions of alien thought and deciding to sustain them as possibilities—neither relinquishing them as the fantasies of the other, nor fantasizing about them as leading to the true reality."[101] There is no guarantee, but we also should not rule out that it may be the Tofa or the Tibetans who will provide the techniques to live on a cold moon with a thin atmosphere. A longer historical view would keep in perspective that fossil fuel–based technics are fizzling out after a measly two centuries, whereas the technologies of the Tofa have been sustainable and innovating for thousands of years. If the goal is a "jailbreak" from this planet, such an escape will certainly require some combination of techniques that can persist for scales of time that far exceed, by several orders of magnitude, the five-year lifespan of a laptop computer or cell phone.[102] Try making a wafer-thin CPU on an ice moon.

I share Harrison's skepticism. The technics of our technological rut are fragile unto death. The third and fourth industrial revolutions depend on sterile labs and rare earth minerals, which when assembled for computation are fatally allergic to heat and water, and entirely dependent on luxurious amounts of electricity. In a world that is getting hotter and wetter, and where energy is scarce, one would hope that other technologies as well as other forms of life are possible. Instead, disposability hovers over all modern technologies. That each object "innovated" reaches obsolescence before the close of any given financial quarter and is replaced by a nearly identical but slightly improved object follows developmental thinking as much as the consumer products revolutions of the mid-twentieth century.[103] The prospect of a new hypermodernity, then, is homogenization.[104]

The danger of homogenization is not new to grand modernization projects. However, the prospects for a human future are so fraught that the resources and constituency necessary for another transformational vanguard do seem possible as fewer and fewer people are necessary to have a global impact. Yet any such global unification of humanity is now so materially unsustainable that its inevitable failure appears equally likely. As with earlier projects for global unifications, the question will be what violence is unleashed when those leading the charge refuse to accept failure. It is unfortunate that advocates of ecomodernism, the singularity, and accelerationism so thoughtlessly join in the long legacy of depriving discontinuous forms of life the ability to determine their own fates, much less other, often mutually exclusive visions of full, meaningful lives.[105] The race for the singularity, even a failed race, will be extremely resource intensive. In some projections, the singularity is proposed to survive by consuming the entire sun in a Dyson sphere, a hypothetical device meant to completely surround a star so that all the solar energy can be captured.[106] Even then there is unlikely to be "storage" space for everyone. The ecomodernist variants of modernization are no less catastrophic for obstinate forms of life that are put in competition to fight to the death over what little resources remain for those not entrepreneurial enough to dominate other populations for their enrichment. These visions of the future do not present an alternative or way out but instead offer a completion of the geopolitics of homogenization.

Could it be otherwise? Of course. Futures are always unknowable. Furthermore, humanity is not at the helm of history. The many other possibilities we debate in laboratories and parliaments may reflect little more than the provincial plans of a small fraction of just one planetary species. However, what is certain is that "unleashing" the potential of technics, as promised by the singularity and other visions of human transformation, is a dead end. If we are to truly accelerate human adaptation, it must include the minor and incipient techniques of a world of new and old alliances across cosmologies, creature affinities, and even organic and inorganic forms of life. Any sufficiently significant change in the nature of life will be a multispecies and multicultural endeavor. And still most of us will not make it.

To be clear, I am not suggesting that the 'way out' is to throw our weight behind Idle No More rather than Elon Musk. I also do not want to suggest what is suggested too often, which is that the resurgent practices of indigenous people are our saviors because 'they' (timeless natives) can guide 'us' (moderns) back to nature. Indigenous peoples struggling against extractivism are not outside of history. They are lines of flight, minor traditions, who are inventing new futures, not relics hiding in the past. There is something perverse in the way these indigenous futures are being raided by scholars and policy makers. The success of indigenous struggles against the current catastrophic outlook of global politics is undone when wayward adherents to the Eurocene try to jump onboard. It seems improbable to me that eighty percent of the earth's biodiversity survives in indigenous territory because of secret heirloom squash, or mindfulness practices that could be adapted for a new iPhone app. Indigenous sovereignty and governance is the secret. However, I find it doubtful that those interested in indigenous agricultural products or subsistence water management techniques are ready to give up settler sovereignty in exchange. As a result, the selective incorporation of techniques of indigenous resurgence will be little more than another round of homogenization until the excitement over poaching indigenous knowledge and practices is eclipsed by a corresponding zeal for indigenous sovereignty.

Beyond the disgust I feel watching the beneficiaries of the Eurocene going back to the all-you-can eat genocide buffet in search of a sustainable way of life, I also find this logic dubious. There is an assumption made by bioprospectors and scientists seeking indigenous historical climate knowledge that indigenous peoples retain some natural nobility or authentic connection to the earth. It seems more likely that the indigenous, black, and brown people of Earth are the cutting edge of political struggle and livable futures because they have had front-row seats to the making of this crisis. There is expertise that comes from surviving an apocalypse. But that wisdom is soaked in blood, and those that wish to 'scale-up' indigenous solutions for modern living owe an unpayable debt and ought not be allowed to pillage the survivors of European expansion under the guise of allyship, or worse yet, the venal claim that *we* are all in this together.

Certainly, the Eurocene is not synonymous with all people of European heritage. Many people found in Europe or its colonies continue to live under the harsh violence of the vanguard that leads the worldwide project of Europeanization. There are many minor traditions within the Eurocene that resonate with the need to oppose rather than expand the current geopolitical order. These minor lines of invention immanent to the Eurocene, if truly resistant to the homogenizing tendencies of Europeanization, do not need to parasitically feast on indigenous forms of life. These other becomings of Europe, as well as those who find themselves scattered between different communities and forms of life, could instead find comfort and possibly inspiration in knowing they are not alone rather than instrumentalizing others in the desperate pursuit of survival. How will we know who to trust, who to oppose, who to have sympathy for, who to turn our back on? There is no safe, sovereign position from which any one of us can make a determination of who is in and who is out. The categories are vague and too frequently exploited for cynical and sinister ends to build a new program just yet, or maybe ever. We can neither let go of difference nor give into the desire for identification. Yet the categorical murkiness between the Eurocene and particular Europeans, as well as other forms of life, only intensifies the need to rethink the responses to the eco-geopolitical crisis as a politics that demands a new inequality. Anything like a way through will require the ability to creatively constrain those who continue to viciously pursue incorporation and homogenization at the cost of everything and everyone else.

PART III. MUST WE PERSIST TO CONTINUE?

8. APOCALYPSE AS A THEORY OF CHANGE

We should calmly ask ourselves, however, if the world we have conceived in accordance with reason is itself a viable and complete world.
—GEORGES BATAILLE, *The Accursed Share*

But what if you discover that the price of purpose is to render invisible so many other things? —JEFF VANDERMEER, *Acceptance*

Everybody wants to own the end of the world.
—DON DELILLO, *Zero K*

As I find little inspiration in the technological optimism of the singularity or a cosmopolitan future and no hope in the inevitability of power politics, I want to try to make more visible where I see the possibilities of making a life amid a dying or worse yet expanding Eurocene civilization. What I offer here is not an alternative world order or a new categorical imperative. Instead, I want to sketch out what possibilities I think might exist in the terrain of apocalypse and war for those of us moderns no longer interested in being along for the ride.[1] The possibility of catastrophe, while always present, is more or less open to creative intervention. Even if there is no way out, so to speak, sadism is not the only condition for persisting. So rather than start from the position of how to end war or transcend the Eurocene, I would rather think about what other becomings are in the neighborhood of the Eurocene's martial order but are in flight away from this epoch's cruel trajectories. This puts me in stark contrast to many thinkers in international relations (IR) who most often seek peace.

This is because rather than attend to these subtle and deeply ecological practices of war and homogenization, liberal international relations theory, whether it be the democratic peace theorist or the providential tone of cosmopolitans, tries, like Kant, to expel war from the world while maintaining a modern order entirely indebted to it. Despite Kant's predictions for an end of war as part of "nature's secret plan," or as the cosmopolitan "desire" of man, or Hegel's completion of the liberal democratic state, global war and apocalypse are not things that can be outlawed, regulated, and governed out of existence as they are too intimate with the very order these visions of the world hope to elevate over war and competition.

To understand global war and apocalypse as a becoming is not only to lay bare the facile and destitute liberal understanding of peace and the future but to open up explorations of a becoming otherwise than the Eurocene. Such a becoming is likely to be illegible to the current indexes of progress and global order. The normative markers of peace—the absence of conflict—need not define the limit of possible becomings other than war. Becoming agonistic, becoming active, becoming rage, becoming justice, becoming quiet, becoming still, becoming disobedient, becoming graceful, becoming kind, becoming indifferent, becoming defiant, becoming gentle, becoming sacrifice, becoming fire (as many monks in Vietnam did and at least three individuals in the United States have in the face of the Iraq War), becoming generous, becoming courageous, becoming feral . . . The restoration of belief in the world requires an affirmation of being in excess of a regulative or repressive model of peace and progress.

War and the drive to homogenization endemic to the Eurocene cannot be disowned or expelled. They must be diverted by other incipient becomings. Other forks must be taken. This does not require that the world slow down. It might require that we unblock certain flows corralled by the arborescent strategies of fortress state craft. Redirecting the affective economies of war toward other attachments—arguments, justice, compassion, forgiveness, politics, resistance, grief, art, beauty, the world—cannot be accomplished by repression or separation; that is a recipe for ressentiment.

To understand the processes of becoming that enliven and rigidify the Eurocene is to understand the possibility of becoming something else. If we externalize or banish the Eurocene to the place of evil or outside ourselves in the name of some new alliance, we fail to understand just how indebted the modern form of life is to the Eurocene. In this moment—returned to us by a kind of attunement to depth of this catastrophe—we might find other practices, bodily dispositions, emotions: grief rather than rage, compassion rather than revenge, determination rather than resignation. For some the otherwise will only give contrast to the power of hate or rage to overcome other impulses, but in others it may spawn other directions; new questions, alternatives to the dissatisfaction, or burnout from rage, hate, and revenge. I am a pessimist but I am not a nihilist.

In fact, desperation may not always lead to the same result if returned to a fork in the stream of becoming rather than the inevitable requirements of the stultified responses of bombing, killing, starving, incarcerating, deterring, sanctioning, hating. New machines can be released into an assemblage, new cutting edges, new transfigurations and modifications—metamorphoses. Each of the current geopolitical options is a commitment to the human as being rather than becoming. Each fails to see possibilities contained in a body that evolves and carries potential to continue evolving.

This view of evolution is not determinism. It is the condition and insistence of modification and change, each modification confronting the possibility of multiple directions, trajectories, lines of flight, new practices, and experiments. Like all experiments, from winged reptiles to speech, some fall flat on their face and others produce sonnets. But at each moment of modification, time forks, slows to a near halt, like a drop of water just before it separates from its source. In complexity theory, such a moment is called a bifurcation. Manuel DeLanda explains:

> Furthermore, even if we are destined to follow the attractors guiding our dynamical behavior, there are also bifurcations, critical points at which we may be able to change our destiny (that is, modify our long-term tendencies). And because minuscule fluctuations in the environment in which bifurcations occur may decide the exact nature of the resulting attractors, one can hardly conclude that all actions we undertake—as individuals or collectively—are irrelevant in the face of these deterministic forces. Bifurcations may not be a "guarantee of freedom," but they certainly do provide a means of experimenting with—and perhaps even modifying—our destinies.[2]

For those attuned to such possibilities—the succession of moments passing from one alteration to the next—the inevitability of the next moment cascades into a set of possibilities: the Israeli soldiers who suddenly will not pull the trigger; the flinch of a silo captain when confronted by an incoming nuclear missile; saving the world from a nuclear war almost triggered by an unusually rapid weather balloon rocket launched in Finland; food sovereignty movements; the inexplicable generosity of an Algerian Jew who returned the hatred of anti-Semitism with the impossible generosity and affirmation of deconstruction rather than the self-destructive drive of Zionism; love among state enemies; the impossible gesture of the African National Congress refusing to expel Afrikaners who once tortured and murdered them; career military officer William "Fox" Fallon, who sacrificed his prestigious position as head of Central Command because he would not go along with the plan to attack Iran; the cascading events of the Arab Spring.[3] The miracle need not be transcendent—coming from outside the world, from a god—but the incipient chaos of possibilities contained in every moment of becoming expresses my belief in immanent miracles or unpredictable moments of bifurcation.

Such moments of possibility are obscured by the towering authority of normative theories of order and progress. Experiments, practices, new media, drugs, social arrangements, habits, irrationalities, bizarre affinities, creativities that attend to these fleeting punctuations in historical movement sometimes allow us to become otherwise than what is expected, planned, prudent, pragmatic, realistic, or ordained.

Global war is the condition of the Eurocene and does not appear to be getting any better, but the martiality of the Eurocene is neither in our nature nor contrary to our nature. The current epoch is simply the "so far" of a particular human evolution setting the condition of possibility for the next move—coextensive with other minor becomings not yet fully emerged, still emerging, or incipient. That the list of surprises we can enumerate seems paltry in comparison to the list of horrors need not be entirely discouraging. For in each case, the aleatory surprise of becoming otherwise than anticipated was seemingly undeterred by the quantity of data to the contrary.

To this end, ethics in this next section is defined as the means to intervene in the vitality of becoming, not to steer its course as captains of our destiny but as attempts to drag our feet in the water in hopes of going productively off course. It requires only a little drag, a slight dynamic difference for an object in motion to change its course. One discovers through aerodynamics and friction that as the speed of an object increases, the effects of slighter and slighter variations in drag are magnified. The slight movement of a rudder or flap on a plane can cause it to loop or spin out of control given the right speed. Slight changes in shape can slow down or speed up a vehicle without having to alter the mass of the vehicle. The drag of affirmation in the face of "bodily" inevitability or providence is the kind of drag I have in mind.

I think this view of ethics makes contact with Deleuze's insistence that the task of the contemporary condition is to "restore belief in this world."[4] The "moral of the story" is not a simple *assertion* of belief. This dimension of belief slides much more deeply into bodies, tendencies, and dispositions. One cannot either deny or accept the current tendencies of the Eurocene. The apocalypse is part of the world, and thus something that cannot be resented, and is yet so hard to affirm. To this end there must be a style of affirmation, or an ethics of affirmation. As Deleuze says of moralities of the ass and the ox, "they have a terrifying taste of responsibility, as though one could affirm only by expiating, as though it were necessary to pass through the misfortune of rift and division in order to be able to say yes."[5] It is surely not the case—given the complexity and interpenetrating nature of the Eurocene—that the practice of affirmation I have in mind could be called autonomous or sufficient to the issue. How we prepare ourselves for moments of bifurcation matter. Attunement or care for the world can alter the affective dispositions or primed response toward less hateful or resentful responses to dynamism and unexpected change.

This is how experimentation can proceed,[6] with a sense of texture and malleability that says to go slowly, generously, but still experimentally, with care and attention, pursuing an attunement for what passes. The addition of care cannot but conflict at some point with many readings of becoming, but it should not be read as reticence or as opposed to becoming. The development of an ethos of affirmation is neither a call to "slow down" nor an insistence on revolution. Instead, understanding the Eurocene as emergent or as a field of immanent relation requires experiments that provoke people's bodies to betray them. This should be the goal of all new political strategies! Such experiments are vital to the question of becoming something else as we increasingly find ourselves resonating with different phyla of human species.

For Georges Bataille, something like posthumanism already came and went. It was a brief moment for upper Paleolithic man in which our equality among animals was attenuated by works of art. From that moment of consciousness, things went another way. The species enslaved, declared war, and left the adventure of consciousness behind for the pursuit of a narrow instrumental reason. I am not sure I can quite go all the way with Bataille. There is a little too much nostalgia in his recovery of artistic man. But I am intrigued by other becomings of possible human trajectories that do not, as Bataille says, culminate in the atom bomb. What of true holidays? Not holidays for wars, nations, or order but "half-divine, half-demonic" events of celebration.[7] Can the inhabitants of the Eurocene still go feral, not as a return but as a form of speciation, a breaking point in which some of us diverge from this particular dead end? If so, it is worth considering what such a pursuit would look like if only to glimpse what might have been and what still could be. How such possibilities will come about, when they will take place, will be neither predictable nor rational. Change of this magnitude will exceed *us* and may even end *us* but it is unlikely that it will end everything. Change at the scale of epochs, or evolution, is always more and less than an extinction; it is an apocalypse.

Life, therefore, has been often disturbed on this earth by terrible events. . . . Numberless living beings have been the victims of these catastrophes. . . . Their races even have become extinct, and have left no memorial of them except some small fragment which the naturalist can scarcely recognize.—GEORGES CUVIER, qtd. in Stephen Jay Gould, *The Structure of Evolutionary Theory*

It turns out I suddenly find myself needing to know the plural of apocalypse.
—RILEY, character in *Buffy the Vampire Slayer*

Hold tight, we're in for nasty weather.—TALKING HEADS, "Burning Down the House"

In *The Structure of Evolutionary Theory*, Stephen Jay Gould puts to rest the idea that life slowly and continuously developed from the careful selection of fittest life forms. Despite the near consensus of neo-catastrophists like Gould, Michael Benton, and many others, the gradualism of Charles Darwin continues to infect our political and philosophical thinking. As the debates about the emergence and diversification of life focus more and more on the symbiotic and cooperative elements of behavior, environment, creativity, and non-genetic memory (epigenetics), we find that the image of a teleological and continuous climb toward human sentience is less and less credible.[8] In fact, major turning points in the development of life may have been punctuated and provoked by events wholly exterior to the inter- and intraspecies competition thought to drive Darwin's theory of evolution.[9] Save for an asteroid or unusually explosive supervolcanoes, dominant and maybe even sentient life could have been represented by reptiles or cephalopods.

Political theory and history follow similar stories of development and continuous upward mobility. Claims to the ubiquity of Western political forms such as the state, universal theories of rights and norms premised on provincial Kantian and Hegelian traditions, chauvinistic species claims to natural resources in Locke and Marx—all these traditions find refuge in presumptions of crypto-providence, that is, the idea that success of a particular way of being is selected by nature for its superior functionality or character. In "Idea for a Universal History from a Cosmopolitan Point of View," Kant calls it "nature's secret plan."[10] Furthermore, the idea that life, ideas, and ways of life improve as time moves forward, and that outmoded forms of life are culled or left behind, has as its tableau an image of nature that is vicious but consistent, such that selection can drive a grand dialectic forward toward improvement. If accidents, exogenous interventions, and unforeseen and meaningless alliances of organisms and environments determine the fate of living things, then what security can be found in rectitude and superiority?

None. According to Gould, "we grant too much power to the calm of daily life because we live within its immediate surrounding pervasiveness. We therefore fail to realize that rare and unusual events set the basic patterns of history."[11] This does not mean giving up on order altogether. Gould, following Georges Cuvier, argues that the fossil record is catastrophic but that "this sequence of catastrophes imparted a directional history to earth and life. . . . Life's vector of progress records an increasing adaptation to harsher climates of a cooling earth."[12] So life is more complex and this striving for complexity can be seen intensifying over time, what Gould calls a "vector of progress." Yet the history of mutation and adaptation cannot be mechanistically reduced to a process of natural selection. Instead, there is a dynamic, natural history of creativity, selection, catastrophe, alliance, convergence, divergence, and real chance at work in the emergence of the human estate. Following Gould, I want to consider how we might theorize differently if our attempts at making sense of the world accorded with the actual world we have inherited. What onto-ethical adventures might come out of the tumultuous and catastrophic history of our planet? And why, despite our claims to intellectual progress, do we humans remain so indebted to a gradualist image of thought that refused, until the middle of the nineteenth century, to even accept that there had been a single extinction of an organism?

Unlike many of his peers, political theorist William Connolly has become interested in the turn away from gradualism to catastrophism in contemporary geology and evolutionary theory. Although geology seems a strange touchstone for a political theorist, the history of the world—deep geological time—confounds much of our inherited wisdom about the relationship between humanity and the planet that creates the condition of possibility for humanity's existence.[13] In a blog post at *The Contemporary Condition*, Connolly and I note how even secular stalwarts of political theory such as John Rawls follow a very literalist Christian view of the planet.[14] The formative geological events responsible for the creation of Earth from the big bang to the cooling of the planet's surface took place in the "beginning" and have since been replaced by the imperceptible and thus politically insignificant cyclical behavior of a "mature" planet. There was creation and now there is the age of man.

The geological history of our planet and even the more recent history of life on our planet tells a very different story. Geologists and paleontologist describe at least five great extinctions generally defined as the loss of more than 70 percent of the species on Earth. In almost all cases, exogenous events such as asteroids, or in one case the emergence of a new mountain range, set into motion a series of amplifying feedbacks that accelerated too rapidly for the majority of the planet's life forms to adapt.

In this light, Alfred North Whitehead's characterization of life as a war against entropy takes on a more startling and dramatic character.[15] Not only is the world not promised to humanity; it is not even guaranteed to be hospitable to organic life. There are no promises. For Connolly and Whitehead, this is the opening for the possibility of freedom and ethics. If life, human flourishing, and planetary systems were in some sense irrefutable, it would also mean we lived in a mechanical, law-driven world devoid of the possibility of novelty, which is a precondition for something like freedom.[16] Contrary to a world of tight and perpetual equilibria, life is, in the final instance, novel and not reducible to initial conditions. It is here that we can also see Connolly's attraction to the "teleodynamism" and "teleosearching" of biologist Terrance Deacon as it resonates too with Whitehead's concept of aim, or creative struggle, in the evolutionary process.[17] Reality is not path dependent. Precisely what makes catastrophe possible is also what makes creative evolution possible, or the capacity to effect change that is unprecedented, novel, and therefore unpredictable. This is as true for the innovation of the eye as it is for the rise of U.S. hegemony or industrialized animal slaughter. Therefore, Connolly and Whitehead's interpenetrating open systems from microbe to cosmos have to be capable of catastrophe—that is, not self-correcting—if something like *real creativity* is to exist. This is the speculative wager Connolly makes in *The Fragility of Things*, which I hope to push further into the thinking about the many apocalypses of humanity and earth.

An Apocalyptic Tone

Apocalypse is a touchy subject even for those of us in critical traditions prone to question developmentalist and teleological theories. We often respond to the possibility of catastrophe with skepticism. The practiced intervention is to criticize those proposing the possibility of apocalypse with critiques of eschatological thinking or to argue that representations of the end-time stem from cultural malaise or a reactionary romanticism for simpler times. Cultural theory has long since been enamored with the underlying psychic and discursive explanations of the fear of human extinction. Much of this work located the advent of nuclear weapons as the zero point for a renewed sense of apocalypse. In general, this work—such as Martin Heidegger's dread in later works like *The Question Concerning Technology*, Jacques Derrida's "Missives and Missiles," or the vast troves of literary theory on science fiction—

focuses on what apocalypse represents in the sense that it does not actually represent the possibility of apocalypse. It must be something else that we are obsessing about.

There are good reasons to be skeptical of apocalyptic thinking, particularly because it has become an entertainment industry in its own right. After all, much of the genre of American apocalypse horror, from the disaster movies of the 1990s such as *Deep Impact* and *Armageddon* to the popular TV series *The Walking Dead,* depends on the narrative adventures of mostly white privileged people having to live like most of the rest of the world does on a daily basis: no food security, the risk of being forced from one's home, unpredictable access to basic things like medicine and emergency care, terrifying people or zombies or robots coming to get you in the dead of night. There is something undeniably precious about this vision of apocalypse where people with perfect teeth pretend to be terrified at the possibility of killing and preparing their own food.

But the genre is also dangerous because images that depict the loss of a manageable world do not remain in the world of fiction. The fear that is amplified and given form by these immersive experiences of doom find their way into the major budgetary and strategic decisions at the U.S. Department of Defense as well as many other military agencies across the planet.[18] Since the second term of the George W. Bush administration, the DOD has been the most outspoken division of the U.S. government on the dangers of climate change as a driver for apocalyptic upheaval.[19] The fear of security analysts is that U.S. citizens will require military repression to maintain order during the inevitable tumult of sea-level rise and agricultural disruption resulting from erratic weather and seasons.[20]

In the case of many environmental advocacy groups, it is the apocalypse that is coming for the poor and the marginal that will be most impacted by the storm surges and food shortages. Given the high density of low-income populations near costal zones, this will likely be true. However, there are as many reasons to believe that what is really animating the intensity of both military and environmentalist fears is that climate change will bring a particular way of life to an end. It is a way of life that is as threatened by peak oil or any of the other shortages of minerals or capital that are necessary for the predictable routines that many Americans and Europeans have grown accustomed to, undoubtedly at the expense of the rest of the planet's population of human and nonhuman Earthlings.[21]

So why study apocalypses? In part because we can learn a lot about the Earthling condition from how that condition has and will be punctuated by events far beyond our control.[22] We live in a world sensitive to perturbation, prone to turbulences of various kinds, and it is out of that noise that creativity can be cultivated even if only by alliance rather than willed individualism. So to come to grips with apocalypses means also to think about the scales of action and efficacy with which we can participate while also cultivating attentiveness to what kinds of living things we want to intervene with and on behalf of. This is the mess that we find ourselves in. Transformation is possible but its possibility may be indifferent, or at least inured, to our existence.

This puts the emphasis on how to live and how to die rather than whether we live or die. This is, I think, also present in the cacophony of apocalypticisms. There are minor strains of what, much more than Kant's sense of enlightenment, we ought to call maturity. This time it is not the knowledge of our unique capacity for reason that should be championed and cultivated but the limited hold we have on this world and just how vulnerable we are to forces beyond our control. Maturity as humility and tragedy bares the marks of what many have called the Anthropocene much more than the particular consequences of sea-level rise in the course of any one human life. Whether our current trajectory toward climate turbulence succeeds in mass extinction cannot exclusively cause or prevent the apocalypse before us. The confrontation with the Earth system, its fragility, and its capricious grip on life will irreversibly change what it is to be human. So there must be both concern and sanguinity in preparing ourselves for what is already happening. We need to find an immortality, what Whitehead called perishing, worthy of the event of humanity. What this means is that we should not be fighting so hard to avoid perishing as a species, if that even means something, but rather we should be trying to perish better.

This is a dangerous endeavor. For all the reasons Connolly's work has fought so hard against negative critique and the debilitating stupor of Theodor Adorno, Giorgio Agamben, and other followers of the dark arts that see this life as damaged or in need of redemption, we have to find a place to take the catastrophism of the universe seriously while also following Deleuze's invitation to intensify belief in *this* world.[23] This might ask too much of words and ideas. Connolly warns that catastrophes "shatter the bond of trust in the world that had tacitly bound you to humanity and the world."[24] Or even worse yet, apocalypses may, as Bataille writes, "conceal a possibility of enticement."[25] I hope that we, particularly in the extravagant and luxurious countries of the world, are reaching a point of saturation in which apocalypse is becoming so obvious as to no longer paralyze or entice but to finally provoke. I suppose we will see.

So how do we begin? In a time of big data, predictive modeling, and renewed positivist hubris, Connolly invites us to be seers. What Connolly calls an "endemic" complexity requires of us, he says, an attention, at multiple levels of perception and experience, to those "protean moments of incipience."[26] The seer for Connolly is caught up in practices of speculation informed by and in conversation with science but not curtailed by current findings. Instead, with each new set of experiments and scientific insights, new speculative opportunities become possible. We can see in Connolly's work, certainly since *Neuropolitics*, an adventurer's fascination with the opening up of scientific research to bolder forms of cosmological and even metaphysical speculation. Complexity theorists such as Ilya Prigogine and Stuart Kauffman, neuroscientist Giacomo Rizzolatti, and more recently biological anthropologist Terrence Deacon have all made significant contributions to the way William Connolly reads Friedrich Nietzsche, Henri Bergson, William James, Gilles Deleuze, and maybe, I think, most importantly Alfred North Whitehead. It is hard to say what first attracted Connolly to Whitehead. Undoubtedly it was not the writing style of *Process and Reality* or Whitehead's partnership with Bertrand Russell. In many ways, Whitehead is aesthetically out of synch with the thinkers of becoming who drive Connolly's contemporary thinking.

I believe, in part, that it was Whitehead's conversion from a follower of a rationalist mathematical universe that he had described with his student Bertrand Russell, to a creative and transexperiential universe of process that caught Connolly's attention. Whitehead's leap into speculative philosophy was notable and resonant with Connolly's own thinking because it took inspiration from the crises in knowledge taking place in early twentieth-century physics and biology. Whitehead's attentiveness and then risky pursuit well beyond his expertise is kindred with what, for me, is Connolly's most exciting intellectual phase.

However, the affinity between Connolly and Whitehead is, at first cut, tricky. Connolly is foremost known as an affirmative thinker. His engagements with the tragic character of the world have always been followed by an expanding universe of fecundity and plenitude. Even the "turbulent" and "emergent" character of the world described in his final chapter in *A World of Becoming* pairs "joys" with "risks" and "jumps and bumps" with "real creativity." *The Fragility of Things* leaves none of this joy behind, but the *real* of creativity comes at a cosmological price that is closer to Whitehead's speculative thinking. Novelty and creativity exist because of—not in spite of—the fragility of existence. In *The Fragility of Things*, Connolly's nonprovidential view of complexity and creativity, like Whitehead's epochal cosmology, lets in a darkness I do not think is fully present in his earlier work.[27] The Lisbon earthquake, cataclysmic climate change, confrontations with the melancholia of planetary extinction, and the risky character of global politics find a place in a world that is creative but makes no guarantee that its creativity is for us. At a similar pitch of thinking, Whitehead described his own work as follows: "Almost all of *Process and Reality* can be read as an attempt to analyze perishing. . . . We can see the universe passing on to a triviality. All the effects to be derived from our existing type of order are passing away into trivialities."[28] Similarly, *The Fragility of Things* marks the making explicit of an ethical and ontological landscape we negotiate in a universe passing into triviality. Creativity and triviality—this is the hard problem Connolly is confronted with and that he must push through in order to imbue the world with a sense of possibility rather than futility. After all, for Connolly and Whitehead this is not an exclusively human world. Martin Heidegger's world of the call of Being in which humans are unique is replaced by an age, the Anthropocene, that is merely distinctive. We live in a moment where it is the whole Earth system—its climate patterns, ocean currents, and hydrolytic cycles—that is teaching us just how not alone we are in this cosmic adventure of real creativity. At the right conjunction of events, everything and anything from cyanobacteria to asteroid collisions can be world forming.

In an attempt to give a little depth to my pairing of creativity and fragility with apocalypse, I will illustrate a few of the many apocalypses that our cosmos and planet still bear the marks of. I then bring into sharper relief the incipient apocalypses that while still virtual pose an end to what we know as the modern human, *Homo sapiens*. I hope this investigation might wake people to what Tristan Garcia has called the epidemic of things.[29] I want to amplify and dramatize that the deluge of waste, dying animals, eroding shores, flooding streets, dwindling diversity, expanding wastelands, and the geopolitical arrangements that enforce the unequal distribution of destruction and precarity ought to leave us rattled but too easily becomes normal. A certain apocalyptic tone is necessary if we, as humans, quite drawn to oblique and selective appreciations of the world, are going to wake up from our pathetic slumber.

"Time falling away. That's what I feel here," he said. "Time becoming slowly older. Enormously old. Not day by day. This is deep time, epochal time. Our lives receding into the long past. That's what's out there. The Pleistocene desert, the rule of extinction."
—DON DELILLO, *Point Omega*

A catastrophe must have terminated one world and initiated another.
—STEPHEN JAY GOULD, *The Structure of Evolutionary Theory*

An exhaustive list of apocalypses is not possible; however, such a list would certainly include a number of geological and human events. There is no way to know where the beginning would be. However, there is in the middle of things a definitive turning point in our planet without which *our* existence is inconceivable. It is speculated by scientists that two billion years ago, the earth was a warm cozy sea of methane and rust. One can only imagine what such a planet looked like. It certainly was not the pale blue dot for which I now have such an affinity. The atmosphere of this early world not yet Earth was almost entirely free of oxygen gas (O_2), also called dioxygen.[30] Dioxygen when released was quickly captured by minerals such as iron and deposited as rust and other oxygen-mineral compounds. The magnificent banded iron formations running through the sedimentary rock of Minnesota and Western Australia are the remnants of the "great rusting" of this planetary epoch. The irony of our current view of the world in which CO_2 is the enemy and O_2 is the savior is that all the experiments trying to replicate the chemical evolution from which the simplest forms of life emerged have failed if oxygen is present.[31] There is strong consensus that had there been free dioxygen in the atmosphere, early planet life would have never emerged.[32] Organic monomers or building blocks of early cellular life would have been oxidized or destroyed by free oxygen. So the first inklings of life on our planet were "obligate" anaerobic, meaning they could not live in the presence of oxygen.

A creative prokaryotic (lacking a defined nucleus) cellular organism changed all that. An early form of cyanobacteria began to capture sunlight to convert CO_2 and water into energy. The exhaust of these newcomers was O_2. The amount of O_2 released exceeded the mineral capture cycle that had kept the anaerobic world thriving for half a billion years. The oxygen-free ecosystem of the early planet thrived for ten times as long as primates have been on the planet. This was not a fleeting stage of planetary existence. What followed was the greatest climate catastrophe in the history of the planet. Obligate anaerobic life was torn asunder, dismantled by the rapid oxidation of the planet. A very few of these early forms of life survive deep in the ocean near volcanic vents or in other oxygen-free pockets around the planet. That was the extinction. Ninety-nine percent of all life on the planet died. But there was more; this was an apocalypse. In this new era, the photosynthetic life we now call plants began to thrive.[33] The ocean belched oxygen into the atmosphere in such concentrations that it threw the climate system into chaos. Free oxygen broke up the high concentrations of atmospheric methane. As those who follow climate change debates in the contemporary era will know, methane is a rapid warming agent. The loss of methane rapidly cooled the planet and resulted in the longest "snowball Earth."[34] Despite this 400-million-year glacial period, the variety of life increased and a planet more closely resembling Earth began to emerge.[35] Geologists now credit plants with the invention of mud, a vital component to emergence of a more complex and varied ecosystem. Plants also created rivers, which had not previously existed. In fact, much of the topology of the Earth is contingent entirely on plant life.[36]

The next billion years witnessed an explosion of life and the blooming of the blue-green planet we call home. The world would never be the same. Earth scientist Martin Gibling says that the oxygen catastrophe should call to our attention that "plants are not passive passengers on the planet's surface system. They create the surface system. Organisms tool the environment: the atmosphere, the landscapes, the oceans all develop incredible complexity once plant life grows."[37] Plants are earth forming, and for the pursuit of life that is more important than world forming.

The oxygen catastrophe was not the last of the great extinctions. In fact, most accounts of the five great extinctions do not even include the oxygen catastrophe on their lists. The fecundity of plants overshadows their violent disruption of planetary life. While there is an understandable bias for sentient oxygen-breathing organisms, we cannot underestimate just how much the seemingly inert vegetal form of life was once the great destroyer.

In the intervening 3 billion years between the emergence of plants and our entry onto the global scene, cells would gain a nucleus; life would differentiate fronts and backs (bilaterians); fish, amphibians, and plants would creep onto the land to finish the great project of terraforming; insects, reptiles, and then, in the geological equivalent of a nanosecond, mammals entered the scene roughly 200 million years ago.[38] From there it would only be another 188 million years before creatures we might recognize as other humans took shape.

It is easy to recount this timeline and see, as did Darwin, the smooth curve of evolution upward toward complexity, sentience, and then intelligence. But this was not so. That smooth curve was shattered and set off into radically divergent tributaries of biological development by at least five more events that have qualified as extinctions. The definition used to distinguish between extinction and speciation, or species loss, is that extinction is characterized by the loss of more than 70 percent of all living species on Earth. However, in reality this is much more than 70 percent of all living things, as it means entire forms of life disappear forever, which changes the capacity for what can return and flourish in the aftermath of a given crisis. The dinosaurs are a good example: an instance when a dominant and highly differentiated form of life containing many species disappeared from the planet almost entirely.

In the case of the so-called Great Dying at the end of the Permian era, 96 percent of all ocean life and 70 percent of all terrestrial life became extinct.[39] In my Cold War childhood, it was common knowledge that the only thing that would survive the gigadeath of all-out nuclear war was the cockroaches. The Great Dying is the only extinction in which there was massive insect extinction. Eighty-three percent of all insects came to be extinct. The Great Dying is instructive because it demonstrates just how capricious our planet can be. The initial chaos was caused by what is suspected to be an unusual spike in the activity of supervolcanoes. The sulfur and ash created a primordial nuclear winter. Sunlight was reflected by atmospheric SO_2 and the planet froze.

As destructive as this rapid-onset ice age was for life on the planet, it was likely the one-two punch that made the event so catastrophic. As the lava spread out over the planet, frozen methane was released and combined with a spike in CO_2 to rapidly warm the planet. Like our contemporary impending climate turbulence, there is not a linear outcome set by initial conditions. Climate turbulence created by warming can flip as a result of unforeseen feedbacks and send the planet spiraling in the other direction in terms of temperature. What the Great Dying teaches us is that the diversity of life is less dependent on the temperature of Earth than on the speed with which that temperature changes. The changes in climatic extremes happened too quickly for the overwhelming majority of life forms to adapt.

When time contracts and grows more familiar to our historical imagination, apocalypses scale at different degrees of extension and intensity. The human world as a subset of Earth—two intercalated open systems—has its share of apocalypses too. Some of these apocalypses have been influenced by the tumultuous character of Earth, but they were also driven by a number of other forces more microcosmic than supervolcanoes or planetary terraforming. Some thirty thousand years ago, human residents of Europe and much of Asia lost their last peers in the hominid world. Neanderthals and Denisovans disappeared from the world, and their existence can only be reconstructed through careful excavation and speculation.[40] The accepted wisdom typified by Jared Diamond's *Guns, Germs, and Steel* was that *Homo sapiens* had been unwilling to tolerate the presence of other upright and intelligent hominids and wiped them out with superior lethal technics.[41] There may have been such conflicts. However, that was not all.[42] The narrative of *Homo sapiens* leaving Africa and laying waste to those hominids in this way is belied by more and more genetic and archaeological evidence to the contrary.[43]

Most humans with predominant ancestral roots in Europe have at least 3 percent of their DNA from Neanderthals.[44] Similar percentages exist for humans from East Asia in terms of Denisovan DNA. What else was gained from these peer hominids is more speculative although not without evidence. Neanderthals were also artists and may have imparted other skills for surviving in the new colder climate of Europe to their new guests from Africa. There is compelling evidence that it was intermarrying with Denisovans that made it possible for Tibetans to survive at higher and higher altitudes.[45] What else Tibetans gained from Denisovans to survive so close to the edge of our oxygen-rich atmosphere is not known. In all these cases, it is not really accurate to say what "we" gained from "them." The first generations of humans out of Africa could make such claims, but those of us today are "them" as much as we are the children of the humans from whom we gained our genetic material, as in each case the first generation of Neanderthal-human children were 50 percent of each form of life. They were, genetically speaking, something new—for which most of the planet now is. Those who remained in Africa are the only truly "pure" *Homo sapiens*. It is hard not to think about this evidence in light of the violent history of racial thinking. Further, the hominid apocalypse calls into question just how plastic the human species is and how porous the various species categories are even beyond the so-called human race. That there was planetary "miscegenation" among hominids, not to mention now-humans from every corner of the planet, shows just how tragically stupid racial pseudosciences were and continue to be.

We will probably never know how tragic the hominid apocalypse was. There may have been intense violence or no more than is experienced among humans today. What we do know is that the global spread of humans, the first great wave of globalization, was enabled by joining with other species. The hominid apocalypse also demonstrates how insufficient extinction is as a definition of apocalypse. The loss of the other three hominid species represents a fraction of a fraction of a percent of the total number of species on the planet thirty thousand years ago. However, the spread of human life to every corner of the world and the acceleration of that global circulation would come to be of planetary importance. There would be no Anthropocene if the adventure out of Africa had not succeeded. It seems like an overstatement to think that 3 percent of human DNA could matter that much or that interbreeding could make that much of a difference in the success of the human race. However, during this period of human history thirty thousand years ago, there were as few as one thousand humans in Europe and Asia.[46] I have more Facebook friends than there were total humans on the planet. Every birth mattered. If even only one hundred of those one thousand lived because of fertile Neanderthal or Denosovian parents, that could have been the difference between flourishing and extinction. The irreversible convergence of hominid life reorganized the entire ecology of the planet, and even the atmosphere, in just under thirty thousand years. The results are mixed to say the least but it is not possible to say that they were inconsequential simply because we did not lose more than 70 percent of all species. The importance of any given apocalypse is only to be found in the postapocalypse.

Subsequent waves of human globalization have been punctuated by apocalypses more clearly tragic. At the end of the fifteenth century, the offspring of European Neanderthals and African *Homo sapiens* set off in search of trade routes and landed in the Americas. Unfortunately the European humans did not come alone. With the waves of conquistadors came the microbial world of Europe, which was transplanted to the Americas. The results were catastrophic. Some estimates put death rates at 90 percent for the North and South American continents. Historian David Stannard puts the death toll at approximately 100 million.[47] Whole civilizations disappeared in a matter of a few generations. For good reason, the American apocalypse is remembered most for the cruelty unleashed by the volitional actions of Europeans who set out to conquer these newly "discovered" lands. However, this version of events is insufficient. While we should never diminish the wanton violence committed by the European invaders, the vast majority of deaths in the Americas would have happened even if the conquistadors had been hospitable visitors. This should, I think, give us pause for two reasons.

First, the complexity and fragility of life is not solely under the dominion of human decisions and intentions. We are all part of vast assemblages, and the costs of change and adventure can be unforgivably high. This should be remembered when considering the sixth great extinction underway. Even often progressive visions of humanity as a global species that is universal in its character and for whom Earth belongs can risk unexpected dangers that, when actualized, are violently dismissive of human and nonhuman life alike. The impending amphibian extinction is likely the effect of humans casually tracking fungus all over the planet. The bubonic plague that devastated Europe and India has a similar accidental history.

Second, the self-congratulatory myths of European intellectual and technological superiority are verified in many ways by narratives of American conquest that focus exclusively on the human violence committed against indigenous populations. Europeans were not superior in number or technique. Europeans were contagious. Given how much our current era still suffers under the arrogance of developmentalist logic, it is worth foregrounding that it was not the ascending, smooth geometric curve of European technological advances that resulted in the industrialization and rapid economic growth that underwrites the contemporary Euro-American global order. The "great acceleration" blamed for the current climate crisis required primitive accumulation on a global scale, and then the subsequent transplanting of that industrial form of life to every corner of the planet. It is impossible to imagine the small bands of European invaders succeeding if it had not been for the "Great Dying" of the Amerindians.

Similarly in Africa, the Maafa beginning in the sixteenth century and continuing into the nineteenth century killed nearly 4 percent of the entire human population on the continent, collapsed ten empires, and permanently altered the economic sustainability of the continent. This devastation emerged from an assemblage of the demand for "free labor", in the mostly emptied lands of the Americas; the economic intensification that emerged in the triangle between Europe, Africa, and the Americas; and the brute cruelty of racial chauvinism. A short five hundred years after these settler colonial and slave relations were canalized by mercantilism and then capitalism, we are discovering that human apocalypses and geologic apocalypses are differences of neither degree nor kind. Each is implicated in the other. There is cruelty in the concentration of atmospheric CO_2, and that concentration of CO_2 will intensify new cruelties that traverse the boundaries of species. Similarly, the conceptual and temporal boundaries of apocalypses are frustratingly diffuse. Are the Americas and Africa part of a larger European apocalypse? Is industrialization an apocalypse of mechanization disfiguring the holism of animism as Lewis Mumford and others have suggested?

These divisions cannot be made in the final instance. Instead, I want to think about apocalypse as a concept in the Deleuzian sense. The function of a concept is not one of demarcation or clarity; rather, it is for building bridges that complicate and drive thinking further into the complexity of things. What I am trying to capture with the idea of apocalypse is that fecundity and destruction cohabit. History is neither a series of creative destructions nor a series of destructive creativities, as if either creation or destruction could be given primacy as the driving tendency of existence. Our cosmos is fragile, and fragility is the condition of possibility for either or both novelty and destruction. Thus the relationship of novelty and destruction is neither oppositional nor dialectic. Furthermore, the history of our planet and our form of life as contemporary humans belies any view of life or our world as necessarily ascending from simplicity to complexity, or from chaos to order—much less from savage and unjust to civilized and pacific. Nothing is reversible and nothing is permanent.

In the present age of technological wonder and extreme gadgetry, we are often under the misapprehension that humans have exited history, or at the very least nature. Even climate change is often discussed as an engineering problem that can be solved with human management. We should, however, pause and put into perspective just how far our technological advance has not come. Consider the suspected causes of each of the five great extinctions as potential threats today. Supervolcanoes, asteroids, unpredictable and emergent forms of life, the loss or addition of novel atmospheric gases, cosmic rays, rapid warming or cooling—none of our revolutionary technology could contain or combat any of these formidable cosmic or planetary-scale opponents. In fact, even the human apocalypses of the Americas and Africa still seem daunting.[48]

Antibiotic-resistant bacteria are flourishing in the most sophisticated hospitals in the world, and so it is certainly conceivable that we may confront a pathogen that we cannot stop from wiping most of us out. In the case of Africa, all our collective political development has been incapable of preventing, much less reversing, the horrors of continued colonial exploitation. We have not innovated our way out of the genocidal practices of warfare and territorial domination. Since the first Gulf War, the U.S. has managed to kill or deprive of life more than 10 percent of the Iraqi population. African Americans legally freed of slavery some 150 years ago still find 50 percent of their male population living in a state-run cage during some period of their life. Europe has a similarly unimpressive track record. So the hubris and ignorance that presumes humans are the captains of their destiny ought to attend more carefully to our planetary heritage before declaring "The Age of Man." Even the critical application of Anthropocene as a geological category may risk presuming too much. In an essay on Kafka's *Ape Odradek,* Jane Bennett prods us to consider the limits of the Anthropocene. According to Bennett, Earth is marked by humans, "but not in the sense that earth has become thoroughly humanized. There are multiple creatures, shapes, misfits, simulacra, doodles, and vapors afoot, colliding, entangling, co-operating, competing, lurking and crashing."[49]

9. FREAKS, OR THE INCIPIENCE OF OTHER FORMS OF LIFE

The human figure, to be grasped, must be *read*. To know another mind is to interpret a physiognomy, and the message of this region of the *Investigations* is that this is not a matter of "mere knowing." . . . The human body is the best picture of the human soul—not, I feel like adding, primarily because it represents the soul but because it expresses it. The body is the field of expression of the soul.
—STANLEY CAVELL, *The Claim of Reason*

Among all of the things that can be contemplated under the concavity of the heavens, nothing is seen that arouses the human spirit more, that ravishes the sense more, that horrifies more, that provokes more terror or admiration to a greater extent among creatures than the monsters, prodigies, and abominations through which we the works of nature inverted, mutilated, and truncated.
—PIERRE BOAISTUAU, *Visions of Excess*

He judges not as the judge judges but as the sun falling round a helpless thing. . . .
His thoughts are the hymns of the praise of things.
—WALT WHITMAN, *Leaves of Grass*

The danger in building such an apocalyptic tone as this is that it could embolden reactionary forces and desires. My concerns regarding the end of the human, the rise of autonomous machines, and the bifurcations and speciations of humanity are not meant to foreshadow a restoration of humanism. These changes are irreversible. What comes next is an effort to build capacities and intensities attuned to novelty that can remain political while also finding satisfaction and even inspiration in whatever comes next. I want to find a way through the apocalypse before us that is generous and affiliative rather than cruel and isolationist. This is easier to say than do, particularly as we come in contact with new or radically different forms of life than our own. The track record of *Homo sapiens* is mixed, at best. However, our Neanderthal DNA suggests the possibility that not all encounters with radical difference ended in annihilation. Speculatively one can imagine that the difference was one of what Stanley Cavell in this chapter's first epigraph calls *reading.* Cavell sees problems like enslavement and other (in)human horrors as a kind of aspect blindness such that a "darkness" can be projected onto the other.[1] The other becomes or is illegible; the other is a physiognomy that exceeds our imagination.[2] As we dilate what we mean by the species or even move beyond any recognizable face of a human other, we are confronted by a kind of expressive illiteracy. There is expression—what Cavell calls a soul—but there is no capacity for us to experience it. However, I do not believe this is an either/or proposition. The careless victim we cannot encounter, the enemy that cannot be more than our relationship of enmity, the incipient intelligence not yet capable of hailing us, the form of life too incongruous with our own, the useless forms of life already discarded—each of these categories is learned and habituated rather than given to us by the world as such. Therefore we can, I think, cultivate techniques of sensitivity that change our affectability and our sympathy such that a novel encounter may produce something other than disgust or, worse yet, the nonencounter that lacks any structure of care and produces nothing at all.

Freaks live in this interregnum between disgust and indifference. They are forms of life that bear some connection with the others capable of being encountered but simultaneously possess an incongruity that offends our physiognomical literacy. Freaks live among us as horizons or closures to a world of possibility between the normative somatic human and the monster. In Bataille's short study of the freak show, he says: "A 'freak' in any given fair provokes a positive impression of aggressive incongruity. . . . And, if one can speak of a *dialectic of forms*, it is evident that it is essential to take into account deviations for which nature—even if they are most often determined to be against nature—is incontestably responsible."[3]

The tension between the ontological real character of everything "under the concavity of the heavens" and the illegibility or outright inversion of normative nature forms the focus of this chapter. Emerging and existing freaks invite us into the becoming of the world and demonstrate just how little regard history has for our platonic fictions of form and eternity. Like many of the figures in this book, I want to avoid romanticizing the freak as an authentic or noble form of life. Instead I want to do the labor to be capable of a broader sense of wonder that enables encounters well beyond the mimetic boundaries of experience.

The agent provocateur of this line of thinking for me was a cinematic encounter with Tod Browning's 1932 film *Freaks.* Browning's film is often remembered as one of the first horror films, but I think that is only true if we take a very expansive understanding of horror in the transrealist sense

Rudy Rucker suggests, as noted in the introduction to this book. The horror of *Freaks* is not to be found in the introduction of supernatural ghouls or sadistic violence but in the experience of our own limitations of experience. In fact, what inspired the censorship of the film was not gore or violence but the subversion of normality. Browning's narrative is one of a community of freaks held together by deep bonds of trust and kindness. The "horror" that was found to be objectionable was that at the end of the film the antagonist—a blond normal—who betrays the community in an attempt to swindle one of the freaks out of their family inheritance is physically transformed into "one of them." What was inconceivable to the community standards applied by the censorship board was the idea that someone normal would be made freakish. It is hard not to also suspect that the offensiveness of the film's content was amplified by a general desire to hide the actors of the film, themselves also freaks. Evidence to this point can be found in the visceral response of F. Scott Fitzgerald, who for a time shared a commissary with the cast of the film. Two sisters with an anomaly called microcephaly sat down at the table to eat with Fitzgerald. Fitzgerald reports that their mere presence caused him to vomit and retreat from the lunchroom.[4] It is telling that Fitzgerald, one of the greatest storytellers ever, a talent requiring a great deal of attentiveness for the world, lacked the sympathy to see the sisters as colleagues or even fellow humans. Browning, relatively minor by comparison to Fitzgerald, possessed something more profound as an observer and storyteller.

Unlike many exploitation films and horror films, which rely on the conflations of monstrosity and evil, Browning's film stages the possibility of a sympathetic encounter with freaks. The narrative is already quite sympathetic. In the story, a generous community of outsiders readily accepts a "normal" into their family, culminating in a famous scene of the members of the freak show chanting "one of us" as the normal is inducted into their family with little or no skepticism. That the normal then trades on this generosity to steal from them leaves the audience with little remorse as she is rewarded with her false request, that is, to become one of them. For me, the artistic genius of the film and true ethos of curiosity and generosity is in the naturalistic breaks from the narrative. In one such vignette, Prince Randian—also known as the living torso—gracefully uses his mouth to pick up a cigarette, pick up a matchstick, strike the match, light the cigarette, spit out the matchstick, and take a deeply satisfying drag of the cigarette. Each time I see the scene, I am left with a sense of amazement. Unlike Bataille's freak, it was not the incongruity that struck me but a congruity and continuity that was not my own. I was in some sense both incapable of reading Prince Randian's limbless body and drawn into his form of life at the same time. I suppose at some level I am also a gawker at the freak show, but the difference between wonder and disgust is, I think, a resource in developing the capacities of sympathy capable of new literacies beyond the human. These vignettes of life throughout the film are in sharp contradiction to the "warning" issued at the beginning of later editions of the film that speaks of the tragedy or shame of the abnormal. Browning's original film gives us moments of a life suffuse with sympathy and entirely devoid of a sense of the pathetic or shameful. In an older vocabulary, I would say there is a profound humanity in the characters of Browning's film but that language will not be sufficient to the futures before us. Of course Prince Randian is still a face, a remnant of my normative somatic order, and in this chapter I want to see how far we can go beyond that order to resist the reactionary humanism so common in left and right critiques of human evolution and change.

The appended warning in later editions of Browning's film also misunderstands our freaky future. The disclaimer states quite matter-of-factly, "Never again will such a story be filmed, as modern science and teratology is rapidly eliminating such blunders of nature from the world."[5] For better or worse the teratology now known as genetics that the film speaks of is as likely to create freaks as prevent them. The algorithmic teratology of neural networks, and machine learning too, will produce more freaks. Tay (bot), Microsoft's artificial intelligence "chatbot," has already come and gone because of her fascist and racist rants. Tay (bot), like many freaks, was intolerable because she too closely reflected an inhumanity we could not stand in ourselves but lacked the standing or capability to hold her ground, so she was *destroyed* like Old Yeller for being rabid.[6]

Freaks as an Empirical Fact

Genetics, computer science, neuroscience, and nanotechnology are converging. Increasingly what can be fabricated is limited more by the imagination than technical capability. Many already speak of a singularity—a convergence of organic and inorganic life and physical and biological sciences—in which the capability to intervene in the course of human history will experience a sudden and dramatic scale shift. Whereas scientific discovery has previously accelerated the pace of politics, economies, warfare, or medicine, synthetic biology means the ability to intervene in the very conditions of existence, even radically to alter the trajectory of human evolution or to create new sentient beings—the dream of an Anthropocene by design rather than folly. In response to this opening, I am going to try my hand at what William Connolly calls the political theorist as seer. The task: to read the entrails and portends of scientific, technological—both organic and inorganic—evolution and to look for the "pluripotentiality inhabiting . . . such temporal tiers."[7] To what end? The concept of critical responsiveness as developed throughout Connolly's work has been received in strictly human terms, not always either man or woman but always human.[8] As such, a certain intelligibility or logic of recognition underwrites the application of Connolly's deep pluralism and its struggle to acknowledge previously unintelligible parts of newly emergent identities. The presumptive generosity that is extended has at least an inkling of where to look or listen for the incipient, but not yet fully public or political, demands of those whose minor tradition is not yet audible. Thus, Connolly's concept of generosity and critical responsiveness has been circumscribed by a certain humanism when placed in the context of other concepts such as the public sphere and democracy. Nevertheless, the potential of these concepts has not been exhausted by the humanist frame, and these concepts must be pushed beyond the accepted limits of the species community.

The purpose of this chapter is not to prove the existence (or inevitability) of artificial intelligence or of newly emergent posthuman forms of life, any more than to declare the end of *man*. My hope is to consider the possibilities and *limits* of a moral order grounded in what we now call the human species. In the first section I lay out some actual and possible trajectories of social beings that have not been welcomed into the species family with open arms. The second section considers the concept of the species in the moral theory of Jürgen Habermas. The final section considers critical responsiveness or what Jane Bennett calls sympathy as a strategy for pursuing generosity without the presumptive boundary of a common humanity. This is ethics without a net; there is nothing to reassure ourselves that our duty has been done or that our generosity is sufficient. The moral calculation of where our commonality begins and ends is, from this prospective, an alibi for indifference and even cruelty. Fortunately, Connolly is not a theorist of moral actions or duties but of ways of life, ethos, reflective vigilance, and care. I argue that these are the resources we need in a world of material and political becoming. Tumult need not be cause for panic and resentment.

Hostility toward Variation and the Emergent

In the early part of the twentieth century—beginning as an offshoot of mind-body debates—some materialists began to describe the human qua human as a machine. A debate between machinists and antimachinists thus ensued.[9] After twenty years of academic speculation, the introduction of the first thinking machine, Electronic Numerical Integrator and Computer (ENIAC), provoked a slight shift in this discourse. Rather than simply debating whether "Man" was a machine, the question was inverted: could machines become human? One particular antimachinist, Paul Ziff, denied the possibility that machines could ever do more than process data. In particular, Ziff asserted the inability of machines to acquire feelings and thus (according to Ziff's logic) consciousness.

In response to Ziff's 1959 essay "The Feelings of Robots," Hilary Putnam questioned the "inhumanity" of inorganic life. However, he did not then posit the "humanity" of robots. Putnam instead concluded that "there is no correct answer to the question: Is the robot conscious?"[10] The result of this unambiguously uncertain conclusion set off a torrent of articles asserting the exceptional character of human consciousness and claiming that no "artificial" machine could ever do better than mimic that consciousness.[11] Putnam's answer was taken as an attack on "our" place in the chain of being. While these arguments were not explicitly religious in content, they were uncharacteristically—for the analytic tradition from which they emerged—religious in tone. They were strikingly reverent. Their intensity is curious, given that the modern microchip had not yet been invented at the time and that computers still filled rooms and barely had the computing power of today's cell phones. It seems that years of sci-fi films and comic books filled the imaginations of an otherwise *sober* lot of academics. Of the eight articles that responded to Putnam, seven staunchly disagreed. The only person who agreed, Dennis Thompson, did so because it was not "such a radical claim."[12] Thompson did not really see the point; we *were already* machines in his estimation.

For my purposes, Thompson misses the point entirely. Putnam's claim is not interesting because he sided with the "machine theorists" but because he concluded that there may be no way to ever resolve the question. As Putnam says, "the question calls for a decision not a discovery."[13] The instance of a decision confronts the otherwise purely rational enterprise with an ethical choice. Where his contemporaries buried the unknown possibilities of this new technology under centuries of tired arguments regarding mind–body dualism and humanist claims that "we" are the sole possessors of consciousness and perception, Putnam *decided* the following: "If we are to make a decision, it seems preferable to me to extend our concept so that robots *are* conscious—for 'discrimination' based on the 'softness' or 'hardness' of the body parts of a synthetic 'organism' seems as silly as discriminatory treatment of humans on the basis of skin colour."[14]

Putnam's decision represents an atypical response in his community of philosophers—perhaps even an *inhuman* response. As is shown by the work of Masahiro Mori and others, many people instinctively fear that robots will in some way challenge the human race. Putnam's response is different: in the flurry of attacks on even entertaining the possibility of artificial life, Putnam could not help but see a connection to the racial injustice that was present at the time of publication in 1964 and thus refuses to repeat the error.

Foregrounding the fear of robots is a long history of human mutation, having been the focus of the various iterations of the eugenics movement. Before that, it faced the wrath of superstition. As such, questions of political rights have never been divorced from biological or more specifically species considerations; there exists, rather, a biopolitics of citizenship and in regard to not just the polis but also the species.[15] Whereas classical politics could speak of the organization and governance of subjects, the advent of a theory of biological evolution that included humans introduced the possibility of governing the production of subjects, not just in a legal or discursive sense—citizenship, caste, class—but in the "fitness" or biological character of its subjects. Race as a biological concept extends politics from the demographic questions of reproduction and health—as has been reconstructed in Michel Foucault's *Security, Territory, Population*—to the intrinsic character of the babies born. The notion of the survival of the fittest lent the epistemic supremacy of science or objectivity to earlier moralized discussions of worker productivity or even the spiritual origins of industriousness and laziness. Nascent theories of racial superiority and inferiority came under the purview of governmentality in the form of Malthusian public health initiatives: birth control for the poor; sterilization of and experimentation on "incompetents" and racial minorities, in particular Native Americans—practices that did not end until the 1970s.[16]

A brief explanation of recent legal and political responses is worth noting. In 1927 the U.S. Supreme Court decided in *Buck v. Bell* that "for the protection and health of the state," forced sterilization of imbeciles and other infirmed or abnormal people was not a violation of fundamental constitutional rights. The official position of the courts has not changed. In 1981 the courts decided in *Poe v. Lynchburg Training School and Hospital* that the eight thousand women forcibly sterilized in the state of Virginia had not had their constitutional rights violated.

In the last year, an email alert was sent to attorneys who serve *ad litem* for foster children to publicize that the Environmental Protection Agency (EPA) had adopted guidelines for testing known environmental toxins and carcinogens on unwanted or unknowing children. According to 70 FR 53857, "the EPA proposes an extraordinary procedure applicable if scientifically sound but ethically deficient human research is found to be crucial to EPA's fulfilling its mission to protect public health. This procedure would also apply if a scientifically sound study covered by proposed § 26.221 or § 26.421—i.e., an intentional dosing study involving pregnant women or children as subjects—were found to be crucial to the protection of public health."[17] The explanation and scope of this decision was focused on children who "cannot be reasonably consulted," such as those who are mentally handicapped or orphaned newborns: these groups may be tested on without informed consent. It also stated that parental consent forms were not necessary for testing on children who have been neglected or abused. As was the case in the original case of *Buck v. Bell*, being unwanted or otherwise downtrodden was made synonymous with being genetically deficient. To be included in the political community of constitutional rights, one has to be—by this logic—capable of demonstrating an "understanding" of those rights and one must be wanted by that community.

As the EPA's proposed guidelines demonstrate, the century-long effort to eradicate human variation has not in any way eliminated unwanted or "subhuman" individuals. Now birth "defects" tend to mark class and national origin. Despite the best efforts of modern science—or because of the best efforts of modern science—public fear of mutation and the specter of genetically engineered beings and artificial intelligence has joined the ranks of the abject and unwanted. There is a recurrent hostility toward forms of life that do not narrowly fit the definition of humanity—a kind of somatic fundamentalism that insists that the genetics, phenotype, and manner of expression all conform to a norm of what it is to be human.[18] As Georges Canguilhem argues, norms require a certain abnormality or pathology in order to take on meaning.[19] But the abnormal is not merely an index for the normal; the abnormal becomes a moralizing category for measuring and finally determining what *is* human.

In the case of mutation or variation—of either natural or artificial origins—deviating from the human image is what inspires revulsion. Robots and androids create the same feeling but through an inverse movement: they intrude upon the species by transgressing into the proprietary capabilities of consciousness, language, and other monopolies claimed by the human species. Japanese roboticist Masahiro Mori identified the phenomenon and proposed a hypothesis called the "Uncanny Valley."[20] The theory is that humans are fascinated by, even attracted to, robots as they gain human qualities—eyes, ears, and an identifiable face. Then, once robots become visibly or unmistakably humanlike, the fascination and attraction turns to disgust. The human participant in the experiment becomes agitated and uncomfortable. In recent experiments, the human respondents refuse to allow the robot to stand or move behind them.[21] "Movement is a sign of life" and as such seems "wrong" for a machine.[22]

Although history is rife with the exploitation of other races (colonialism), nonhuman animal species (mechanization of animal husbandry and slaughter and animal experimentation), or subhumans (eugenic policies toward abnormal human development, including the poor and the sick), artificial life is a newly emerging horizon. Even if not "conscious," the existence and increasing importance of "intelligent" machines confront us with the horror of the automaton while the genetically modified human presents us with something also "not quite right." The possibility of artificial life in all its forms seems to provoke a response somewhere between atypical human bodies and inanimate objects. The possibility of artificial life treads in both the forbidden zone of challenging human superiority (in this case because it may exceed it, whereas mutations malign it) and the more traditional uncanny provocation of living objects. Therefore, life that does not resemble the norm of human life thus far, whether artificially created or naturally variant, will be met with the same violence and ignorance that those differently abled have faced from eugenicists unless the narrow definitions of life and the fear and ressentiment that inspire those definitions can be altered.

This, I argue, requires what Connolly names cultivation. Connolly sees the "visceral attachment to life" as a resource for deep pluralism, one that hopes to transform the fear and loathing of variation into the "preliminary soil from which commitment to more generous identifications, responsibilities, and connections might be cultivated."[23] But I will add to this point that the "Uncanny Valley" not only exists for all those beings that stray from the normative boundaries of the species—whether biologically or synthetically divergent—but also represents a formidable obstacle to the cultivation of connections and generous identifications. This logic holds, in particular, if one's visceral attachment to life is an attachment to a "human" life. Connolly's immanent naturalism actively resists the temptation to circumscribe generosity to human subjects. Throughout *Neuropolitics*, Connolly insists that gratitude and generosity find their inspiration in an "attachment to the earth and care for a protean diversity of being."[24] To read this statement alongside early works such as *Identity/Difference* or *The Ethos of Pluralization*, one may wrongly assume that the "protean diversity of being" refers to a human being. However, what is clear in Connolly's contestation of the nature/culture opposition is that an "attachment to the earth" complicates what Connolly means by *being*, as examples of communicative bacteria and participatory chimpanzees and crocodiles demonstrate.[25] From this standpoint, the danger of falling into an anthropological limit is apparent. Reading Connolly's deep pluralism only in the human terrain of traditional democratic theory obscures many of the sources of the gratitude and "earthiness" that inspire the necessary ethos for a deep pluralism.

David Howarth's thorough and generous piece "Ethos, Agonism, and Populism" is emblematic. While affirming Connolly's concept of ethos of agonistic pluralism, Howarth argues that agonistic pluralism presupposes "a common symbolic order" and a "democratically organized public space" so that "those who are 'othered' [can] be cultivated, respected and brought into the public sphere."[26] As a result, Howarth's inclusion of an ethos of agonistic pluralism in his rendering of democratic politics is actually an exclusion of the grounds for Connolly's ethos. What Howarth sees as a "populist politics" in Connolly's theory circumscribes the depth of pluralism on the basis of those who can be brought into the public sphere. Nonlinguistic forms of life or forms of life that simply cannot be "cultivated" sufficiently to be recognizable in the public sphere represent an irreversible limit as long as the population of a populist politics is underwritten by the image and norm of "Man" assumed by Howarth's reading. The "limits" of deep pluralism, the assumption that deep pluralism's agonism takes place in a human and narrowly linguistically driven public sphere, results from ignoring the expanding jurisdiction of Connolly's notion of life as becoming as it develops in *Neuropolitics* and later works that contest the hard distinction between nature and culture. Connolly insists in *Neuropolitics* that human culture is made up of "essentially *embodied* beings" and that once theorists understand "the corporeal layering of language, perception, and thinking in human life," the discrimination against nonhumans or subhumans that currently underwrites the borders of the public sphere begin to break down.[27]

The subsequent grounds of culture and politics can better be described as an assemblage of nonhuman, living, nonliving, and human agents alike rather than in terms of "individuals" or "human rights" as Howarth does.[28] Thus the agonism that Howarth describes as requiring the cultivation of others such that they can enter the public sphere takes place in a much different terrain in my reading of Connolly, one that resembles neither the strictly human public sphere presumed by most democratic theorists nor a theory of cultivation that is exclusively human or agent driven. The complexity of human and nonhuman assemblages alters the expected provocateurs as well as tactics of cultivation necessary for participation. For Connolly, relying on an "accordion theory of language" that constantly redefines communication and agency to suit the limits of anthropocentrism denigrates the "nontheistic reverence for an abundance of being" that is necessary to inspire affirmation rather than cynicism and ressentiment.[29] For Connolly, agonistic respect and critical responsiveness require the nutrition of such an ethos of abundance, which is much more than a traditional democratic ethos that would be defined by Howarth as "a respect for the common rules of the game" and the requirement of "a common symbolic order."[30] This understanding of ethos is certainly necessary but not sufficient. But what counts as "the game" as well as "playing" must be inflected with a gratitude and openness to other forms of life and participation that are not quite so dependent on the commonality of communication and public space.

Reading Connolly this way suggests that while Howarth is right that cultivation and respect will be needed, by Connolly's account, neither ought to be limited to or require a "common symbolic order." Therefore, evaluations of Connolly's concepts of critical responsiveness and agonistic respect will necessitate an account of publics, agency, and language that does not take for granted the often assumed anthropocentrism of the democratic theoretical landscape. Otherwise efforts to increase "inclusion" and "respect" in the "public sphere" and the "symbolic order" will fail to attend to the inhuman, the subhuman, or the insistence of *things* that exceed their status as objects because those efforts will focus, as Howarth has, on subjects that can negotiate or be represented in a public sphere. My fear is that Howarth's commitment to "foster and encourage" "the emergence of new identities" will be confounded by norms of "negotiation" and "representation," which are in some sense off the table for contestation when the "plurality and heterogeneity" of the public sphere is defined by a "common symbolic order."[31] The source of the ethos that Howarth and Connolly agree must animate politics will not be found in Howarth's account. Democratic theory must go deeper beyond the multilayered experience of "humans" to the multilayered experience of life more broadly as an "attachment to earth and the protean diversity of being," what will be discussed in the final section of this chapter as Deleuze's notion of *a life* or the creative machine of abundance that far exceeds the provincialism of humankind.

Jürgen Habermas is also concerned with the eugenic impulse and with robots, but in *The Future of Human Nature* he focuses on the distinction between humans who are "naturally" born and humans who are the product of scientific intervention. The distinction between artificial and natural is the basis for Habermas's defense of the human as a species. The goal of insisting on this difference is to guard against the invasion of science by declaring artificially modified humans not human at all and thus moralizing the results of any scientific intervention into life. For him, the concept of the human being as God-given and unalterable—something that he transmutes into a biological fact rather than a religious one—is being disgraced by the next phase of eugenics, genetic intervention, and by research aimed at producing artificial intelligence. Habermas argues that posthumanism and "self-styled Nietzscheanism" threaten to turn humans into objects, such that we no longer have bodies and instead are bodies. Echoing Adorno, Habermas warns against the instrumentalization of human beings and a permanently damaged life.

Habermas rejects the religious image of humans as sacred, and seeks instead a postmetaphysical means to ground his challenge to the objectification of human life. He proposes a species ethic: the hope is to understand humans as a species constrained by particular guidelines to produce morally appropriate laws regarding interventions into life. For Habermas, beings can only be human if they enter life: "as members of a species, as specimens of a community of procreation," and only if they participate in "the public sphere of a linguistic community." If these two requirements are met then and only then is it possible to "develop into both an individual and a person endowed with reason."[32] Moreover, and inversely, both requirements are important to his argument: species membership is a prerequisite to participation in the linguistic community. This comes as little surprise given the presumption of many that humans are the sole possessors of language. This, however, is not the basis for Habermas's position. What is the distinction between those *Homo sapiens* of natural birth (who "owe" no one for their traits) and beings who result from human intervention (whose "abilities" are not their own but caused by a scientist)? For Habermas, this difference alters the basis of human responsibility. Genetically altered humans cannot answer for their actions or capabilities because they are "determined" from the outset.[33] What Habermas fears is that genetic engineering will become a means for instituting determinism writ large. The irony of this position is that it misunderstands the determining power of genetics and, in addition, it clings to enlightenment concepts of freedom and autonomy that Habermas's fears seem to invalidate. If it is possible to determine human behavior and freedom via genetics, is it not the case that we were always already determined? In this regard, Habermas's terror regarding the loss of autonomy reflects his own lack of faith in its existence, while Habermas then transmutes that very insecurity into the instrumentalizing intentions of genetic scientists. The apocalyptic tone reaches an apex when he concludes that genetic intervention would result in a new species of life that existed in "a moral void, a life not worth living."[34]

Habermas's other concern is more paternalistic. While he believes the new beings would not be properly human, he also fears for the treatment of the damned, the mutated subhumans. Whereas Mori sees in the Uncanny Valley a social phenomenon that can be overcome, Habermas tends to render the hatred of difference natural and inevitable. He even points to this (naturalized) reaction as proof of the "immoral" existence of altered beings: "Symptomatically, it is the revulsion we feel when confronted with the chimera that bear witness to a violation of the species boundaries that we had naively assumed to be unalterable. This 'ethical virgin soil,' rightly termed such by Otfried Hoffe, consists of the very uncertainty of the species."[35]

Contrary to Habermas's goal to prevent a new era of eugenics, this is the very logic that animates both the antagonism toward robots and the eugenic response toward human variation. Is it not the fact that they fall outside, and even offend the boundaries of the species, that first defines their subhuman status?

Despite the hope to provide a rational ground to protect the human, the affective charge of sacredness betrays Habermas's "postmetaphysical disposition." In several places he refers to artificial insemination as "perverse"; he describes those who entertain the possibility that machines could possess anything approximating humanity as "engineers intoxicated by science fiction," as agents of "adolescent speculation," or as "self-styled Nietzscheans."[36] Habermas is attached to a concept of the human outside of evolutionary time. He can imagine historical change but not biological or, in the case of cybernetics, nonbiological becoming as defined by alteration. For Habermas, the human is transcendent and timeless even if not religiously sacred. However, there is no reason to believe that human "nature" is not just as contingent upon genetic variation and selection as, say, the behavior of an antelope. Thus, Habermas's "humanity" is not *categorical* or *intrinsic*, for which he can make universal determinations of value or equality. Humans continuously change over time.[37]

The problem is that this requires a static and valorized concept of the human that is consonant with the very animus toward difference that motivates eugenics. How could strengthening and clarifying our definition of the species *not* exclude those at the margins of biological and social intelligibility? This effect is further compounded by the purely linguistic approach Habermas takes to participation in the species. What redoubles the logic of *Buck v. Bell* are the countless lives that can participate in neither the linguistic construction nor the procreative construction of the species. Many people who are categorized as autistic would not be part of the species by the linguistic definition, and those who are not heteronormative or fertile could not participate procreatively. This is not to say they cannot reproduce. After all, a new method of extracting stem cells from bone marrow and inserting them into artificial sperm seems to enable lesbian couples to have children without male participants.[38] Other obstacles to reproduction such as infertility or genetic variation like hermaphroditism can be overcome using other technological methods. Or, more radically, one could consider the way artists and thinkers procreate; it is hard, when reading Beat poets like Allen Ginsberg and Jack Kerouac, not to think of them as Walt Whitman's progeny. But these methods are by Habermas's definition perverse or at least insufficient and thus not constitutive of the human species: the offspring would not be equal in birth to the rest of the human race and would lack the foundation for moral freedom and autonomy.

It is thus not surprising that Habermas would favor one exception to his opposition to human modification: "therapeutic" gene elimination, as in the case of monogenetic conditions such as Down syndrome. The ability to distinguish between artificial modification and therapeutic genetic intervention is only possible because the latter restores a child to a normative image of what it is to be human. This is of course the very core of the eugenic spirit: not improvement or evolution of the human race, as he accuses posthumanists and Nietzscheans, but purity and maintenance of an already superior strain of humanity.

It would be easy to follow this critique with the accusation that Habermas's use of species carries a racist tone. One could note that species membership was the backbone of European colonialism and race science up to and including the Nazis. But this critique, which celebrates multiculturalism, merely shifts the line between what is and is not a moral being deserving of the full rights and duties of a political subject. The more interesting ethical question comes after one partially grants the premise of Habermas's argument. What if a radical difference does exist? What if the entity that confronts the human species, however defined, exceeds a certain kind of moral or mirrored intelligibility? What is just on the other side of the mimetic divide of species membership? This is the question that drives the problem of becoming in the age of the Eurocene. The first act of drawing the boundary of the human raises the second question of *why* "we" treat those who fall outside that boundary so badly. In this respect, the question is not why do we treat previously effaced subjects like objects, but why do we treat objects or quasi-subjects so terribly?

What Else Could a Species Be? The Human Refrain and the Politics of Becoming

> I believe robots have the Buddha-nature within them—that is, the potential for attaining Buddhahood.—MASAHIRO MORI, *The Buddha in the Robot*

> The point is to discover and restore belief in the world, before or beyond words. What is certain is that believing is no longer believing in another world, or in a transformed world. It is only, it is simply believing in the body.—GILLES DELEUZE, *Cinema II*

The rejection of Habermas's regulative ideal of the human species should not be interpreted as a wholesale endorsement of genetic engineering or attempts to create artificial intelligence. Instead I will use Connolly's concept of critical responsiveness, and more generally the ethos of immanent naturalism that attends to life as becoming, as a means for thinking through the motivation for eugenics and the paranoia of *natural* and engineered freaks. Connolly speaks of a *politics of becoming* as "the paradoxical politics by which new and unforeseen things surge into being" and *critical responsiveness* as "the form of careful listening and presumptive generosity to constituencies struggling to move from an obscure or degraded subsistence below the field of recognition, justice, obligation, rights, or legitimacy to a place on one or more of those registers."[39] These, he insists, require the cultivation of creativity and the infusion of generosity such that "principles are not doled out in a stingy or punitive way."[40]

Unlike Habermas, who believes that the current trajectory of scientific development can be arrested, or even regulated, to the point of being abolished, Connolly sees in this desire to slow down a kind of ressentiment—one not that different from the hatred of the world that prompts Habermas's reaction in the first place. In his criticism of Sheldon Wolin, Connolly argues that it may in fact be "a quick tempo of life, to put it bluntly, that sets a crucial condition of possibility for the vibrant practice of democratic pluralism. [Connolly's] wager is that it is more possible to negotiate a democratic ethos congruent with the accelerated tempo of modern life than it is either to slow the world down or to insulate the majority of people from the effects of speed."[41]

The warning about the alternative orientation toward change is that the fight to slow the world down will come at a cost. The nostalgia for a simpler or slower life often inspires a rogue's gallery of enemies and scapegoats to blame for the failure of restorationist movements.[42] In part this is because the pace of life is not solely under the dominion of human control; life has a life of its own. Failure to constrain life's unpredictability and acceleration does not lead those who demanded slowness to concede: instead, their energies of ressentiment are redirected toward something they can accomplish—the vilification of those identified with the acceleration of life. Across the political landscape, one can observe the right blaming queer lives for the breakdown of "stable" families and from the left the scapegoating of technophiles and scientists for destroying nature and human authenticity.

This seems true of Habermas, who, in his chosen perspective of the "future present," can consider evolution and change only in negative and provincial apocalyptic terms. The fear of change and of the unpredictable expresses a kind of revulsion toward life. And life is nothing if not mutable and inclusive of the aleatory. Connolly's political theorist as seer also attempts to peer into the future, but the seer looks for incipient *possibilities*, not catastrophic certainties. Instead the political theorist as seer "reviews forking moments, not apparent to most participants when things are still open."[43]

This distinction between Habermas and Connolly as fortune-tellers is not as simple as optimism versus pessimism. Connolly is certainly skeptical of the acceleration of life and even experimentation with life, but he is also open to the possibilities of new conditions for action created by the alteration of and by the world even when humans are not the sole purveyors of political action. He holds no nostalgia for a static human species that never really existed in the first place. Connolly is concerned with *what* "holds things together" but not in *holding* things together. Unlike Habermas, his nonprovidential, immanent naturalism has faith in the world but does not require that that faith be in what Habermas limits to the human world. It can sustain its belief in the world, its attachment to principles, without the species concept that Habermas is terrified of losing.

What is required to develop a sense of gratitude for the abundance of life? It certainly means a gratitude for the unhuman or what resembles life but is not quite human. As Deleuze argued in a discussion with Antonio Negri, "becoming isn't part of history; history amounts only to the set of preconditions, however recent, that one leaves behind in order to 'become,' that is, to create something new."[44] I contend—without a sense of drama—that the image of man held so tightly by Habermas may be that bit of history left behind in order to become. And if gratitude requires belief in this world, not "another world, or in a transformed world," as Deleuze says, then it is necessary to search beyond the current confines of species community. After all, a belief in elsewhere would pit us against the world that we have. It is hard not to see ressentiment or hatred as what animates the affective charge in Habermas's outright dismissal of new forms of life whether they are conscious life, nonhuman animals, or even man-made humans, as in the case of assisted reproduction. Habermas cannot help but use terms like *perverse* and *narcissistic* to describe these interventions because they disrupt the image of "man" on which all his values rest. Connolly's sense of gratitude need not require the meager subsistence of a species in order to find fulfilment or satisfaction in the vital becomings that precede and exceed the parochial limits of man in favor of what Donna Haraway has recently called oddkin, those queer becomings that traverse species, temporal, and blood boundaries in favor of something messier and more generative of affirmative heterogeneities.[45] Or, to put the project somewhat differently, I am referring to a world where kinship is made rather than born.

In this way, Connolly's invocation of abundance and gratitude may illuminate the possibilities of Deleuze and Guattari's concept of the refrain. The refrain is already latent in Connolly's situating of abundance and gratitude within an understanding of a nonprovidential chaos, such as those theorized by Ilya Prigogine in the sciences and Nietzsche in philosophy, but a return to the refrain further illuminates what is at stake in amplifying the attention to the assemblages or interfaces with other species such that the coherence of the species is only loosely present and can give way or itself participate in relays with other forms of life.[46]

Deleuze and Guattari "call a refrain any aggregate of matters of expression that draws a territory and develops into territorial motifs and landscapes."[47] They identify the means by which musical birds can mark territory with song. The mobile assemblage of bird songs requires not just one singer but corresponding rhythms of multiple birds' songs. The territory takes on a sonorous shape: a shape in sound. A territory defines the species and it is also a sonorous species that holds together the territory. But this is not reducible to a species ethic or a fixed identity. In Deleuze and Guattari's view, "Territorialization is precisely such a factor that lodges on the margins of the code of a single species and gives the separate representatives of that species the possibility of differentiating. It is because there is a disjunction between the territory and the code that the territory can indirectly induce new species."[48] This is because the differentiating possibilities promote variation. Deleuze and Guattari write, "One launches forth, hazards an improvisation. But to improvise is to join with the World."[49] And so a territory can be held together with different refrains—aggressive, violent—refrains of security, order, purity; or with rhythms of grace, generosity, and gratitude. We are not wholly in charge of our rhythm much less the refrain, but there is room for amplification and addition or subtraction of sounds. This is the cultivation undertaken by an immanent naturalist. According to Deleuze and Guattari, territorialization is both creative and destructive, and "the rhythm itself is now the character in its entirety; as such, it may remain constant, or it may be augmented or diminished by the addition or subtraction of sounds or always increasing or decreasing durations, and by an amplification or elimination bringing death or resuscitation, appearance or disappearance."[50]

Connolly's amplified rhythms are gratitude and experimentation, and the style in which he reads these terms emphasizes the necessity to proceed, as he says in *Neuropolitics*, "thoughtfully, modestly, experimentally."[51] Connolly is not modest in the sense of being timid or cautious or apprehensive about the world. Instead he is careful in the sense of *caritas*; he applies care to his investigations to look for those as-yet-unheard or unrecognized voices. This is where he parts ways with Deleuze and Guattari's ambivalence or near indifference to the cutting edges of change that can be violent and dismissive of the suffering of others, and yet he affirms the becoming that punctuates life chaotically.[52] We have a paradox and a danger—neither of which it seems Connolly would want to avoid.

Through Connolly's attention to the unthought—in our experience with time, politics, and the suffering of becoming—we are reminded that, like the birds Deleuze and Guattari speak of, an ethos or theoretical disposition can be either musical or nonmusical. Connolly's disposition is musical; Habermas's is not. Or rather, Habermas marches to the meter—consistent staccato—of a Kantian march. According to Deleuze and Guattari, "Meter is dogmatic, but rhythm is critical."[53] Evolutionary biologist Brian Goodwin identifies the critical edge of rhythm with a new form of biological science: "Relationships are primary in understanding the type of order that can emerge, whether . . . cascades of symmetry-breaking processes that give rise to biological form in developing organisms, rhythmic activity . . . that both engender and depend on the creative activities of persons. . . . A science of qualities is a science of holistic emergent order that in no sense ignores quantities, but sees them as conditioning rather than as determining aspects of emergent process."[54] According to Goodwin, it is this concept of assemblage and shifting connections that best defends against the atomistic view of species and thus the "biology of parts" becoming a "medicine of spare parts, and organisms becom[ing] aggregates of genetic and molecular bits with which we can tinker as we please."[55]

Brought together, a science of qualities animated by Connolly's spirit of generosity and gratitude is necessary if we wish to listen to the emergent life before us. *It* will not march well enough to stay in line with Habermas's species meter; in fact, in some cases *they* (maybe a community of *its*) will lack the legs to do so. And those of us already included will grasp for pronouns to describe them. But critical responsiveness and the ethos of an immanent naturalist can listen for the not yet audible, or see the legible emergence of the songs of the legless, soulless, even those who have not yet learned to even hum but merely whir as they plod through the jobs designed for them by humans. The forms of life that confront us on the fringe of our species will need rhythm, not meter, lest they face the violence inflicted by those who especially hate objects. Robots, cyborgs, hermaphrodites, mutants—all categories otherwise than normatively human—will require the work of musicians, not marching band leaders, who judge those not yet issued the uniforms of the human species as examples of a "life not worth living."[56]

If there is something deserving of reverence about the anthropological moment in the vastness of *universal history*, it is not to be found in our rules or morals. It is even less likely to be found as some sacred or permanent configuration of our bodies as if they were already baked loaves of bread. If there is something to affirm and extend, it must be wrested from the newly discovered ethos from which we define our relationship to the world and ourselves—something that (while impermanent) is nonetheless persistent in the continuous alterations that will define, reinvent, and at times disprove the grounds from which we proclaim the rights of man. This is the insistence of becoming: becoming not as a deterministic pessimism but becoming that affirms that compassion, generosity, and care are not under the exclusive dominion of *homo ratio*. Such virtues can exceed the interior relations of the *human subject* and the human species. The crisis is not the future of humanity; it is the necessity—which has always existed—to engage in profound acts of courage that defy the crass politics of survival (species or otherwise) and affirm instead the dissonant harmonies and plural agonisms of life. Whether we feel the warmth of care in what many have called *dark or damaged times* is not dependent on the ability to distinguish or define species-being but to cultivate new ears to listen for the insistent moments of resonance across the lines of kingdom, phylum, genus, species, culture. Objects and patterns of life, geological shifts, architecture—they all affect the evolution of organisms; they are like an exogenetic helix, externally and collectively ontogenetic rather than the individuality of one's own DNA strand. Thus how we pattern the world will become its heritage even if it is not directly, genetically inherited from us.

Habermas finds this entire line of analysis absurd: "In everyday living, we don't think twice before distinguishing between inorganic and organic nature, plants and animals, and, again animal nature and the reasoning and social nature of man."[57] And in part he is right, as those ordinary category distinctions mark differences. But upon closer inspection at the boundary of any of these categories, the choice between them appears arbitrary or, at best, a compromise of pragmatic necessity. Contrary to Habermas's common sense, we can observe in contemporary debates over the initial transition from geological formation to evolutionary biology the breakdown of the distinction between organic and inorganic. Attempts to theorize the emergence of the first living cell (chemical evolution) and resolving the leap from structure to content ("phenotype" to "genotype") have foundered on this sharp categorical difference between cause and effect and elided the degree to which each theory has attempted to explain the transition or event of life as internal to a single organism—the individual—ignoring the inorganic milieu from which life emerged.

A. G. Cairns-Smith eschewed the focus on the production of a particular gene sequence in hopes of discerning a more complex relationship between structure and its generative cause, DNA. In political theory terms, he thumbed his nose at identity, breaking out of what Brian Goodwin calls the "genocentric" biological model.[58] Cairns-Smith sought to identify the interface rather than an ontological divide between organic and inorganic existence. The divide—the proverbial and primordial chicken/egg problematic—presupposes that for something to develop in evolutionary terms, it must have a means of passing on the information (DNA) of its more competitive or innovative structures. However, to develop such structures, it must have had some means for recording them. Despite evidence of their mutual interdependence, the prevailing assumption was that DNA must have preceded the structure so that the structure would have a record on which to base its developments and a mechanism for recording subsequent changes. The problem with this model is that it had no way of explaining the cause of the DNA itself, which also would have needed some prior recording mechanism. Life needed to precede itself.[59] Cairns-Smith took a different line of thought: what if the content was the structure? After all, the distinction between content and structure elided the materiality of the process being described. Genetic information is not like spirit; it is a molecular structure. Thus what he calls "naked genes" (genetic information without a wrapper or organism) may have had particular structures that allowed them to survive and replicate simply because of their shape and organization.[60]

We have a deconstruction of Derridean proportions—there is no longer a need to distinguish sharply between phenotype (structure) and genotype (content)—but this does not obviate the need for a cause of the initial naked gene. For this, Cairns-Smith has to look outside the jurisdiction of biology and organic chemistry. He found an explanation in an encounter between organic and inorganic material. The initial organization (the phenotype) of the genetic information—the assembly of basic molecules into more complex structures like peptide bonds and nucleic acids (which together form RNA or a single strand of DNA)—came together as a result of chemistry enabled by an otherwise inert or unreactive substance, clay: "Often clay minerals that are produced from weathering solutions seem to organize themselves fortuitously, in a rough and ready way, into the kinds of things that might be needed for primitive organisms."[61] The crystalline structure of clay was the catalyst, a pattern for which the otherwise simple components could assemble into something more complex. The engine that drove development against the grain of entropy was not vital in the sense of active or dynamic. It was crystalline, a pattern for life.

The search for the fundamental component of life (water, carbon, etc.) was misguided. No one component, or even combination of components derived from breaking apart and analyzing the current composition of the human, was capable of explaining the transition from nonlife to life. It required an event, an encounter, an interface between organic (carbon based) and the definitively inorganic (the silica crystals of clay) for life to emerge. The point of each of these digressions into the zones of indistinguishability is not to dismiss the categories of human, conscious, living, or organic; it is to loosen "our" grip, to disrupt the certainty that dismisses the emergent or the as-yet-unclassified identities, entities, and other new patterns—between life/nonlife, sensory/inert, conscious/unconscious, linguistics/autistic—as insignificant because inhuman.

As an alternative to the panic represented by Habermas and other somatic fundamentalists, the politics of becoming suggests an enhanced attentiveness to materiality and the chaos of becoming. Connolly and Bennett suggests the need to experiment with experience in ways that draw attention to the world as it is rather than the world as we want it to be.[62] We do not have enough experience with being-uncertain or its more radical possibility being-thing—at least not in ways that are not negative or violent. *Thingness* need not be characterized by stasis or the inanimate.[63] Instead we can acknowledge moments that continue of their own accord, irreducible to a subject-centered consciousness. Try, for instance, giving up and allowing the cross-current of the ocean to drag you down shore, pay attention to the moment just before you fall asleep and the moment you wake up when your body is too heavy to move, enjoy the thrill of falling when you cease to be afraid of hitting the water, and allow yourself to be touched rather than always touching. Ruminate on those moments when muscle memory takes over and you cannot miss a jump shot or fail to hit the right note.

These all seem essential experiences—whether it is an actual bodily experience or just an encounter with a scientific debate that disrupts basic "common sense"—toward becoming material and thus learning to listen to the unexpected forms of life that continue to emerge. The terror of becoming-thing or being not *all* human cannot help but contribute to the animus felt toward objects or emergent forms of new subjectivity. Each of these entities questions our monopoly over the experience of being an active and free agent. Thus the moral or good life—in Habermas's case the species ethics—has wrought as much fear, resentment, and retribution as positive grounds for justice. A species ethic provides little sustenance to a life in flux, in the face of eroding boundaries in which what we value most about the human moment in time seems to be giving way to something else. Restoring belief in the world necessitates a certain attunement toward mutation and the possibilities of other forms of life, and so generosity and faith need not end with the particular arrangement of patterns and structures currently called the human. The politics of becoming can instead be animated by the ways such a refrain can continue, or hold new patterns together. The ethical space of becoming may consist in acts of generosity and belief that animate other becomings and forms of life. To give up on strict or tightly defined nature/culture and inheritance/heritage binaries can help us learn how to pass on certain refrains without the supposed prerequisites of human nature or human genomics.[64]

It is possible to pause and listen to the various relays with the world, and to practice what Connolly calls a "double entry orientation to interpretation, oscillating as a matter of principle between critiques of consolidated interpretations and the production of positive accounts that connect cultural life robustly to the domains of biology, neuroscience, climatology and evolution."[65] How to affirm the more "volatile image of being" is, according to Deleuze and Guattari, the vital task: "It is no longer a question of imposing a form upon a matter but of elaborating an increasingly rich and consistent material, the better to tap increasingly intense forces. What makes a material increasingly rich is the same as what holds heterogeneities together without their ceasing to be heterogeneous."[66] After all, the lessons of evolution and becoming are that we may not make it as we are. The human qua human may face literally what Foucault may have written figuratively: "The wager that man would be erased, like a face drawn in sand at the edge of the sea."[67]

Toward this end, Putnam's conclusion about robots is helpful: science can challenge our ways of thinking, but in the realm of ethics we are confronted by decisions, not discoveries. I contend that Putnam's generosity required a particular intimacy with uncertainty such that he could welcome the uncanny rather than suppress it. Likewise, those of us looking to practice critical responsiveness need an uncertain and curious orientation to seek out the locations to listen to most closely. As Connolly says of Deleuze, the generosity of an immanent naturalist requires a "fugitive disposition on the visceral register susceptible to further cultivation."[68] Organic life was given shape and existence lapping over and over on beaches of inorganic clay. The pattern or refrain of inorganic material, the crystalline structure of clay, in turn gave form and organization to organic life. We still bear that pattern even though we contain no actual clay in our content. The same could become true for the human refrain.

One can only hope that the human face drawn in the sand irreversibly alters the pattern on the beach; what is unknown is which refrain, which catalyst, we will leave behind. Heritage need not be instrumentalized by the somatic fundamentalists; it need not be inheritance in the genetic sense. Values such as courage, generosity, belief, and gratitude for the abundance of life—even if not wholly human—can be continued even if *we* do not persist. Put another way, why settle for a species ethic when a particular human refrain can return with a rhythm that gives new life to the characteristics we now recognize as worth saving? I believe this is what Connolly means when he says, "immanent naturalists pursue an orientation to ethics that resists entangling it from the outset in simplification and cruelty."[69]

To return again to Gibson, the future is already here, it is just unevenly distributed.[70] In March 2016, in Seoul, South Korea, a signal from the future was received. Move 37, game 2, Google DeepMind's Alpha Go artificial intelligence platform surprised world champion Go player Lee Sedol with a move he had never seen before. Alpha Go went on to defeat Sedol. After the match Sedol said to *Wired Magazine* reporter Cade Metz. "It's not a human move. I've never seen a human play this move."[71] Some observers said they were filled with sadness for humanity. They thought they were witnessing a coming obsolescence of their species or the possibility of a confrontation with a potentially hostile intelligence. Other players were angered or embarrassed by the rise of the superior gaming machine. Sedol just kept repeating "so beautiful." Sedol witnessed something new enter the world and he was in awe.

CONCLUSION. RATIO FERITAS FROM CRITICAL RESPONSIVENESS TO MAKING NEW FORMS OF LIFE

We are not unique, we are merely distinctive.
—WILLIAM CONNOLLY, *The Fragility of Things*

Is this new civilization being replaced by another? . . .
What has a beginning can have an end.
—STANLEY CAVELL, *Claim of Reason*

The biggest problem we face is a philosophical one: understanding that this civilization is already dead. The sooner we confront this problem, and the sooner we realize there's nothing we can do to save ourselves, the sooner we can get down to the hard work of adapting, with mortal humility, to our new reality.
—ROY SCRANTON, *Learning to Die in the Anthropocene*

The end of the world as we know it is not the end of the world full stop.
—DARK MOUNTAIN PROJECT, "Uncivilisation"

In a short book titled *The Function of Reason,* Alfred North Whitehead sets out to describe an urge or force that he believes distinguishes creative living things from other organizations of matter. Whitehead is unsatisfied with the functionalist Darwinian account of life in which accident merely selects organisms as fittest to survive. Whitehead sees neither fitness nor utility in life but creativity in the face of shocking fragility. According to Whitehead, from the perspective of deep time, life represents not the fittest of forms but the most unlikely. As he sees it, if the cosmos was determined by the ability to endure the ravages of time, then it was rocks, not organisms, that were the obvious winners. Furthermore, even within the kingdoms of plants and animals, surely complexity bears little survival benefit. In fact, unlike some bacteria, viruses and fungi that can live indefinitely as complex organisms are much more vulnerable and persist in comparatively smaller populations. So rather than seeing an ascending line of organisms growing more complex to outcompete simpler adversaries, Whitehead sees complexity as an outgrowth of a rare aim toward novelty. He names "reason" as this aim or struggle to break out of equilibrium and fight against the current of entropy. While reason is not possessed by all things, the capacity is highly diffused. More than a conatus to merely persist, reason is the "counter-agency" against the universal tendency of decay. According to Whitehead, "In the animal body there is, as we have already seen, clear evidence of activities directed by purpose. It is therefore natural to reverse the analogy, and to argue that some lowly, diffused form of the operations of Reason constitutes the vast diffused counter-agency by which the material cosmos comes into being."[1] For Whitehead, reason is what accounts for the existence of complexity.

What Whitehead is trying to describe is a weak but determinative force at work in those arrangements of things that strive toward greater degrees of complexity. Reason is not a necessary force; it is only a possible force. If it were a necessary condition of life, then we would be back in the realm of mechanism or physical law. For Whitehead, mechanism is a dead end.[2] A kind of "life principle" or pan-vitalism in all things does not make sense either. Whitehead sees the upward struggle of complexity and novelty as rare and precarious. Unlike Darwin, Whitehead believes that any species once struggling toward complexity can stall or even reverse. Even highly complex species like humans are capable of sliding back into a kind of brute repetition that Whitehead calls *fatigue.* For Whitehead, "fatigue means the operation of excluding the impulse towards novelty."[3] Therefore, novelty is dependent on an overcoming of fatigue. Connolly develops the concept of freedom along similar lines but adds specifications that, when placed in conversation with Whitehead, militates against the risks of a vulgar Lamarckianism. According to Connolly, freedom, or what Whitehead calls *aim*, cannot be willed directly. Whitehead leaves this point ambiguous, as his characterizations of *aim* at times suggest a self-conscious will even if not a human will. In order to get out of the humanist trap of negative versus positive freedom, Connolly argues that "creativity is a process in which we participate in uncanny ways rather than one over which we preside. It is therefore a process that upends the images of desire, will, agency, and intentionality often installed in negative and positive traditions of freedom."[4]

Therefore, we cannot simply *choose* to be creative. As Connolly explains it, "An agent, individual or collective can help to open the portals of creativity but it cannot will that which is creative to come into being by intending the result before it arrives. . . . The creative element is located somewhere between active and passive agency."[5] Interchangeably, Whitehead calls this thing that is "between active and passive" an *urge* or a *tendency.* Resonating with Connolly, Whitehead writes, "In the animal body there is, as we have already seen, clear evidence of activities directed by purpose. It is therefore natural to reverse the analogy, and to argue that some lowly, diffused form of the operations of Reason constitute the vast diffused counter-agency by which the material cosmos comes into being."[6] Whitehead and Connolly want to resist the positivist demand for a law so that we do not fall again into the trap of either mechanism or simple finalism. Instead Whitehead and Connolly are describing a minor tributary in the organization of matter that resonates and amplifies the virtual character of novelty such that change can erupt in the world as something new and not merely be the rearrangement of what already is under the blind determinism of regularity.

However, both thinkers also see in that precarious possibility for creativity its opposite. If there is no law of creativity to guaranteed novelty, then it is not ordained. For Connolly and Whitehead, life is rare, fragile, and unnecessary.[7] The lively surface of Earth is alone in its solar system. What exists beyond our galaxy is difficult to say, but even investigating how precarious and contradictory the emergence of life was on our planet suggests that the living are not a necessary outcome of matter. Connolly refers to this as the tragic possibility of the universe. For Connolly, "the experience of abundance . . . is marked by fragility and vulnerability."[8]

Whitehead sees in this tragic possibility different scales and moments of collapse and catastrophe. Novelty does not disappear from the cosmos but a particular form of life that pursued novelty can lose its "reason." According to Whitehead, "The urge of Reason, clogged with such inertia, is fatigue. When the baffled urge has finally vanished, life preserves its stage so far as concerns its formal operations. But it has lost the impulse by which the stage was reached, an impulse that constituted an original element in the stage itself. There has been a relapse into mere repetitive life concerned with mere living and divested of any factor involving effort towards living well, and still less of any effort towards living better. This stage of static life never truly attains stability."[9] As Connolly puts it, "The creative element of freedom is episodic rather than constant, and it is tinged with mystery."[10] Decline can gain an irresistible momentum whereby creativity disappears. In such cases, Whitehead argues,

> When any methodology of life has exhausted the novelties within its scope and played upon them up to the incoming of fatigue, one final decision determines the fate of a species. It can stabilize itself, and relapse so as to live; or it can shake itself free, and enter upon the adventure of living better. In the latter event, the species seizes upon one of the nascent methodologies concealed in the welter of miscellaneous

experience beyond the scope of the old dominant way. If the choice be happy, evolution has taken an upward trend: if unhappy, the oblivion of time covers the vestiges of a vanished race.[11]

Under the influence of Whitehead and Connolly, we see that the sixth extinction is not merely the loss of life. In our age of the Eurocene, the growing wasteland accompanies, but does not drive, the crisis.

In another of the many manifestos cropping up in the tumult of our time, a group of former environmentalists who refuse to continue fighting their governments and corporations have penned what they call the Dark Mountain manifesto. In it they propose an "uncivilising" of thought and art as an alternative to the anxiety-inducing obligation to save the planet.[12] They do this in part, they say, because "the self-absorbed and self-congratulatory metropolitan centres of civilisation" have wrought massive human animal and nonhuman animal suffering, and in so doing accomplished very little. The Dark Mountain Project suggests that the extremes of manicured suburban life and the meager existence of those living on trash heaps are not the only options. Instead, they say, there may still be time and possibility for something more interesting and less cruel on this planet, "somewhere on its wilder fringes." In a *New York Times* article profiling Paul Kingsnorth, founder of the Dark Mountain Project, the author, Daniel Smith, focuses on the collective's followers as former environmental activists who have lost their "faith."[13] Smith is wrong on this score, but it is important that he characterized the Dark Mountain Project's festivals and creative output as resignation. The article shows just how much a certain utilitarian rationality comes to dominate the valuation of forms of life. To underline the defeat of Kingsnorth and his friends, high-profile activists such as Naomi Klein are quoted in Smith's article as saying that Kingsnorth has "given up." Joining the chorus of blame, environmentalist George Monbiot calls the movement a "near criminal disavowal of one's moral duty" on the basis that the Dark Mountain Project gives up on traditional political action.[14] One wonders how Klein and Monbiot can continue to repeat the same exhausting gestures without achieving a different result. Do they really believe the power of ideological critique is at some tipping point of finally making a difference? Klein and Monbiot seem much too smart to be that naïve. So, Monbiot and Klein's dissatisfaction with the Dark Mountain Project is perplexing to me. If in Monbiot and Klein's assessment the lives of billions of people are at stake because of the failure of the current political order to even begin taking the current collapse seriously, then why keep demanding of that political order that it live up to something it seems utterly indifferent to?[15] In this sense, I wonder what counts in the current apocalypse as "doing something" and which habituated and empty demands for political action amount to doing nothing.[16] In particular, I wonder when Klein, whom I find erudite and compelling, will be overwhelmed by the fact that being so right makes so little difference.

Contrary to the morose profile in the *New York Times*, Kingsnorth and other Dark Mountain Project participants have created a series of festivals of mourning and celebration of those species and ecosystems lost to the great homogenization. Furthermore, Kingsnorth and his friends have committed themselves to learning to farm and feed themselves alongside adventurous and creative experimentation with artistic practices, particularly new forms of writing, that take all the species of the world seriously as inspiration and audience. In the words of Kingsnorth's "Uncivilisation," "Uncivilised writing offers not a non-human perspective—we remain human and, even now, are not quite ashamed—but a perspective which sees us as one strand of a web rather than as the first palanquin in a glorious procession. It offers an unblinking look at the forces among which we find ourselves. It sets out to paint a picture of homo sapiens which a being from another world or, better, a being from our own—a blue whale, an albatross, a mountain hare—might recognise as something approaching a truth." That such a bold and audacious experiment would be characterized as giving up says something about the current coordinates of ethical and political thinking. I think instead what the Dark Mountain Project represents is an evacuation from a set of practices, organizations, and alliances that have utterly failed almost all of us. Dark Mountain Project has set itself the task of learning to live and die well in this world, regardless of how this world turns out. The daring of Dark Mountain Project and others that commit themselves to this world but also to a form of life beyond the limits of what is currently seen as acceptably modern and maybe even human is that they take the fragility of the world as a provocation for something more interesting. Unlike those modernist projects that seek their fortune off the rock or beyond the confines of the human "meat suit," Dark Mountain Project digs deeper into the dark but unpredictable trajectory of planetary change rather than trying to escape it. I am not arguing that all is lost—although it may be. However, I am curious why inventing new forms of life that might live through the current apocalypse with what Kingsnorth calls dignity—a dignity recognizable beyond our limited Western humanist circle—is tantamount to giving up. Instead, I would like to consider what techniques for living creatively and with a greater sense of wonder for the diversity of life we might find in those "wilder fringes" Kingsnorth invokes.

The apocalypse before us is one of a great homogenization. It is the result not of floods, asteroids, belching mountains, and tectonic collisions but of sadism and fatigue. We are living in the shadow of an annihilating repetition that would, if successful, finish the process of operationalizing the planet in the image of the Eurocene. The question is whether the repetition of oil drilling, consumerism, primitive accumulation in the cruel territories of the postcolony, strip mining everywhere, and racial profiling at local and global scales has reached terminal velocity or if there are still nascent possibilities for new, wilder aims. In light of the heaps of burning cell phones and discarded computers, a common and dogmatic methodology of life is showing itself as exhausted. Contemporary warfare and ecological exploitation are first and foremost cruel but the cruelty is becoming tedious. In this state of exhaustion, the accelerationists' demand to restore the future, despite my deep reservations about the ethos of their future, is vital. And the Dark Mountain Project's endeavor to invent a wilder humanity is exciting and equally necessary. History has not come to an end, but much of humanity has stalled in vicious consumption of everything. The self-declared civilizational winners have neither a future nor a wild spirit. And we will need both if there is a point to persisting at all. And yet even Kingsnorth's thinking is sliding toward a flat-footed "green nationalism" that demonstrates just how fragile and uncertain these experiments are.[17] No one trajectory will provide the answer, much less a guarantee.

Venerated thinkers from Jesus to John Rawls have done little to prepare us for this creaturely life. To this end, we need a new social science, an uncivilized social science committed to a feral reason that is endemic to this world rather than the cold consciousness of a supposedly independent human mind or exclusively human social sphere.[18] It is time to think like the Earthlings we are.[19] Something is already beginning to take shape that is less enamored with its own humanity, something that cultivates a critical attunement to creaturely life. There is: in the work of Jane Bennett, Dipesh Chakrabarty, Donna Haraway, Brian Massumi, Steven Shaviro, Claire Colebrooke, McKenzie Wark, William Connolly, Catherine Malabou, Tristan Garcia, Eduardo Viveiros de Castro, and many more already committed to a social sciences for Earthlings. And many others have also begun to think in terms of an Earthling social science in the age of apocalypse. Claire Colebrooke's recent two-volume book on extinction sketches what a social science worthy of our apocalyptic times might require: "Perhaps something other than a discursive politics among communicating individuals needs to open up to forces that are not our own, to consider the elemental and inhuman, so that it might be possible to think what life may be worthy of living on. Such an approach would require a thought of the cosmos—of life and its durations—that would be destructive of the polity, that would not return all elements and forces into what they mean for 'us.'"[20]

I think Dark Mountain Project, thinkers like Connolly, and Whitehead, among others, are similarly oriented toward a cosmic life worth living. This would mean accepting "the world for what it is and to make our home here, rather than dreaming of relocating to the stars, or existing in a Man-forged bubble and pretending to ourselves that there is nothing outside it to which we have any connection at all."[21] Contra the dream of becoming data, or some other silicon life form, the problem is not the technological limitations of space exploration, geoengineering, or even digital existence. It is the belief that one of these options can escape this world. Such a desire for escape is at some level a hatred for this world. These various strategies of transcendence *will* extinction as their success. However, even the dream of digital or "spiritual machines" must cope with mortality.[22] The recent discovery of electron-eating bacteria is just one more reminder that there is no "jailbreak" from this mortal coil; it is decay all the way down.[23]

So rather than wish for the end as transcendent images of the future do, the wilder fringes should be in search of minor traditions, incipient practices, novel senses of belonging, and anachronistic forms of life, both futural and deeply old.[24] My senses are repulsed by the consolidation and homogenization of humanity and against the cherry picking of what forms of nonhuman animal life are useful. The task at hand is not aided by acceleration or transcendence but by differentiation. Those who see an eternal future in technological dominance or a digital life without death are like Friedrich Nietzsche's fools "who equate a philosophy of immanence and abundance with a mood of 'optimism.'"[25] Instead we must find our meaning in rougher waters. According to Viveiros de Castro, "to lead a good life (vivir bien) as it is said that Indians like to say—it is first necessary to enjoy living on the edge."[26]

If Earth's calamitous and creative history teaches us anything, it is that those who survive and thrive are not the fittest or even the survivalists. They are those creative forms of life that intensify their existence even if that intensity is only fleeting. After all, fitness is about fit, and fit changes without warning. In a creative cosmos, we must speciate often and wildly lest we find ourselves without reason to live, much less the ability to continue. To put it another way, we should fear fatigue, not oblivion.

To what end then? And how do we mobilize a wild creativity with the intensity of just how fragile we are? How could thinking take seriously the crisis of our contemporary condition without adopting the eschatological tone of a

Christian apocalypse? How do we go wild without the cruelty of indifference? That is what I am trying to begin: A search for a sober apocalypse, a slow apocalypse. A confrontation with perishing, finitude, and fragility but one that fills us with at least as much wonder as dread, more political energy than resignation, and takes seriously that apocalypses are not ends but irreversible transitions. These events punctuate our cosmic epoch. As events they are sometimes catastrophic, sometimes tragic and cruel, and sometimes generative. However, they are always more and less than an extinction. Apocalypses bear an ambivalent relationship to finality. It is the end of something but never *the* end.

So serious investigations of apocalypses have to get over the fascination with the idea of apocalypses. Apocalypses are not simply the climaxes of eschatology, even though eschatologies are inspired by and likely inspire the deep punctuations of real crisis.[27] Apocalypses are real in the sense that they have taken place before us, with us, and will continue to occur after us. We are involved in apocalypses but they are not for us.[28]

I do not want the attention of care for our apocalypse to be a scare tactic or even necessarily an exhortation to action but rather a way to bring into focus just how intimate a creative universe must be with fragility. I am quite fond of aspects of our species but I also see its limits and dangers to creativities outside our narrow trajectory of life. What I hope to do is push further out from the Eurocene and even the human estate in hopes that the trajectories of our becomings be more than simply components of the emerging apocalyptic transitions. Instead we need to propel forward those characteristics, those forms of life, those freaks that fill us with reverence and wonder. If every apocalypse is more and less than an extinction, then what will our heritage be? What trace can we leave on the future? What interventions can be made in the swirling incipiencies of our apocalypses that are gaining momentum? Apocalypses are certain and all things perish, but maybe the inflections of each passing and the conditions of each new beginning are mutually unsettled, underdetermined, and waiting for a creative, wilder nudge. This is my speculative wager. I am experimenting with the role of the seer in order to push further into the metaphysical fallout of cosmic fragility.

> *Truth as Circe.*—Error has transformed animals into men; is truth perhaps capable of changing man back into an animal?—FRIEDRICH NIETZSCHE, *Human, All Too Human*

> Tomorrow morning, he decided, I'll begin clearing away the sand of fifty thousand centuries for my first vegetable garden. That's the initial step.
> —PHILIP K. DICK, *The Three Stigmata of Palmer Eldritch*

#DIFFERENTIATE #SPECIATE

THE END VISIONS OF LOS ANGELES, CALIFORNIA, 2061

The camera pans down from the sky in a wide landscape shot; a plain, beige, flat-topped warehouse begins to fill the frame. As the camera pulls back, the prefab aluminum siding and corrugated roofing come into focus. Lacking any and all adornment, the clean, angular lines seem out of place in the arid desert.[1] The surrounding dust-scape is absent of trees or other distinguishing features. The land is parched and cracked.[2]

A simple razor-wire fence surrounds the building and a road winds to the entrance. Along the road are low piles of rubble, some that still resemble the simple dwelling of an unknown era.

Zooming in on the granular remnants of cinder blocks, the flashing red eye of a serpent drone winds up through the pocks and crevices of the once-forgotten homes still searching for signs of life.[3] Nothing has lived here for a long time; there is no water, and not even signs of water until you reach the salt-crusted beaches a few miles away.

The camera slowly pans again, a tracking shot, as the world turns around the perimeter of the facility. We can see around the corner of the building, and at the back of the building movement is visible.

A horde of upright corpses pushes against the chain-link fence of the facility's back gates. Automated surveillance dirigibles hover over the throngs of trudging bodies, some now only walking in place as space is quickly running out.[4] The blimps appear to be counting or scanning the horde. The preliminary numbers exceed 1.5 million but even the precision cameras of the drones have to estimate as the bodies will not stay still and the density of the crowd is intensifying, obscuring the possibility of an accurate count.[5] And still, more are coming.

Some are only pieces of humanity. Likely the targets of the first Gulf War's aerial campaign of smart bombs, others appear flattened or crushed, probably from the decade of so-called concrete bombs, steel-reinforced pylons meant for roads and development projects never built, dropped from thirty thousand feet to level whole neighborhoods.[6]

The rags still clinging to one corpse display a patch discernible as the Iraqi national flag. So many of the bodies have the winnowed look of lethal hunger. At least 500,000 of those are children starved by the deprivation of sanctions and the infrastructure targeting of the Clinton administration.[7] Other corpses are blackened likely from the incendiary weapons and cluster bombs that had flooded their homes.[8] Some corpses have bullet holes through the backs of their heads, having been killed execution style; many others had been maimed and murdered in unnamable ways by the venal creativity of war.[9] Still more corpses have distinctly American dog tags; their last injuries were distinctly self-inflicted.[10] The tattered remnants of mothers still clutching the rotting corpses of their children lurch slowly toward the distant warehouse. Tight-ribbed dogs, some wearing faded collars, tear at the ankles of those on the edges of the memorial procession.[11] Mechano-crabs scurry in and out of the open cavities of the lumbering dead. Switching to infrared inputs, phosphorescing smart lice can be seen pulsing and bleeding from every cadaver, almost appearing to give life to the moving history of injuries scaling the beach.

The camera pulls further back, going ever higher to get a sense of scale, but before the boundary of the horde is visible, the singularity of each body is lost in a sea of browns, grays, and bleached bone. Finally reaching sufficient altitude to capture the event, the horde becomes almost invisible, another feature of the land's tortured topography.

Even packed, shoulder to shoulder, the corpses file along 1,515 square miles of the coast, and there are still more dragging their feet across the sand, each emerging from the placid surf of a glassy ocean.

The video feed cuts again; this time it is an angular shot from the corner of a room, showing a rotund man in uniform, his black boots resting on the edge of a desk and his blank stare directed toward a monitor showing the gathering crowd outside the back fence of the facility. Behind him an entire wall filled with screens flickers in the eerie green of night vision infrared. On each screen is a tight shot of a body attached to a wall, in total darkness, head bagged, stripped naked, and twitching with the myoclonic jerk of profound sleep deprivation.[12] No other movement is visible or maybe even possible. Many of the bodies jerk together in time as if synched by some larger rhythm among them, a kind of perverse dance step to a song that was not.

One screen zooms in on the face of a body that still fights its fate, a new arrival. The bag has been removed and in the green darkness of the cell, you can see the vessels in the body's eyes begin to hemorrhage and spread from the intensity of its screams. The noise-dampening equipment and specially designed fiberglass tiles counteract any chance of sound being heard, even by the body screaming. The arms and legs buckle under the soft, padded restraints, and still no sound can be felt or heard. The lip movement recognition software records a phrase that escapes from the body: "we are not who we are."

The angle changes again, the view looking down on the uniformed man as he watches television feeds and eats a sandwich. Now visible is a badge just below the epaulettes of his gray, military-style shirt. The badge reads, "Securitas Corporation. Integrity first."

Back from commercial, the screen is filled with images celebrating the life of the now-deceased Rudolph Giuliani—pictures of smiles and handshakes with dusty New York firemen, graphs of plummeting murder rates, and then a clip of Giuliani at a podium, this time grave and serious: "The lesson of 9/11 is that America is truly exceptional. We withstood the worst attack of our history, intended by our enemies to destroy us. Instead, it drew us closer and made us more united. Our love for freedom and one another has given us a strength that surprised even ourselves."

He pauses, seeming to hold back tears. "For the victims and their families, every day is 9/11," he says. "Never forget."

The crowd erupts in cheers and applause.

From the line of reporters in the front row there is an inaudible question. Giuliani frowns, takes a beat, and looks directly in the camera as he says, "Well, revenge is not a noble sentiment, but it is a human one."

The shriek of falling tungsten rods breaking the sound barrier several times over can be heard even inside the facility, followed immediately by a muffled boom and the rattle of the prefab warehouse walls as if Earth itself were shaking from within. The blimps must have finished their count because kinetic kill vehicles once designed to take out nuclear bunkers and high-value military facilities have been targeted and dropped from outer space.[13] The Air Force's "rods from god" now rain down on the miles and miles of walking corpses.

The camera cuts back to the aerial view over the facility, slowly moving out over the horde flattened by the kinetic kill devices. But the irascible corpses are already getting to their feet before the dust has settled.

The camera follows the scrambling crowd toward the water, where the memorial procession begins and then disappears again where the water is too deep to see the bodies.

Over the ocean, the camera catches sight of itself in the water's reflection. Its smooth, long titanium wings outstretched from a narrow fuselage and bulbous nose cone are almost beautiful, elegant. The wings dip twice in a half roll as if the drone somehow had recognized itself.[14]

At the bottom of the screen, a gray bar slowly expands until the screen reads, "File uploaded, Los Angeles, California, September 11, 2061."

NOTES

INTRODUCTION

1 Union of Concerned Scientists, "World Scientists' Warning to Humanity."
2 Ripple et al., "World Scientists' Warning to Humanity."
3 Wittgenstein, *Philosophical Investigations*, 19, 23.
4 R. Manning, *Rewilding the West*, 3–10.
5 Agamben, *The Use of Bodies*, 208.
6 Deleuze and Guattari, *A Thousand Plateaus*, 358.
7 Levinas, *Totality and Infinity*, 21.
8 Whitehead, *Science and the Modern World*, 207.
9 Elsewhere I have described this cosmology as a martial cosmopolitanism. See Grove, *Target Practice*.
10 Guattari, *Schizoanalytic Cartographies*, 11.
11 Malabou, C., *What Should We Do with Our Brain?*, 2–3.
12 McKittrick, *Sylvia Winter*, 18.
13 McKittrick, *Sylvia Winter*, 31.
14 Connolly, *The Fragility of Things*, 49.
15 Shaviro, *The Universe of Things*, 55.
16 According to Latour, "the error is not that we trust Double Click—it's our whole life—but that we slip unwittingly from omission to forgetting." See Latour, *An Inquiry into Modes of Existence*, 275.
17 Schlosser, *Command and Control*, 245–47.
18 Kuletz, *The Tainted Desert*.
19 Law, *After Method*, 2–3.
20 Viveiros de Castro, *The Relative Native*, 51.
21 Muecke, "Wolfe Creek Meteorite Crater—Indigenous Science Queers Western Science."
22 Bataille, *The Unfinished System of Nonknowledge*, 115.
23 Whitehead, *Process and Reality*, 6.
24 Coates, *Between the World and Me*, 71.
25 Shapiro, *The Time of the City*, 21–23.
26 Rucker, "A Transrealist Manifesto," 2.
27 Valencia, *Capitalismo gore*, 30.
28 Valencia, *Capitalismo gore*, 33.
29 Valencia, *Capitalismo gore*, 31.

30 Rucker, "A Transrealist Manifesto," 1.
31 Rucker, "A Transrealist Manifesto," 1.
32 Guattari, *The Machinic Unconscious*, 36.
33 Valencia, *Capitalismo gore*, 67.
34 Rucker, "A Transrealist Manifesto," 3.
35 Rucker, "A Transrealist Manifesto," 3.
36 Cavell, *The Claim of Reason*, 354.
37 Cavell, *The Claim of Reason*, 354.
38 Cavell, *In Quest of the Ordinary*, 184.
39 Weizman, "Lethal Theory."
40 Huntington, Clash of Civilizations, 2011.
41 Network of Concerned Anthropologists, *The Counter-counterinsurgency Manual*; Chow, *The Age of the World Target*.
42 Hoffmann, "An American Social Science."
43 Mearsheimer, "Benign Hegemony."
44 International relations scholars do love nuclear weapons, but they spend most of their time studying the deterrence strategies and game theory of "great powers" rather than thinking about the long-term consequences of tens of thousands of nuclear weapons decaying around the planet.
45 I recently discovered a corresponding term to describe the kind of violence I am thinking through in the work of Sayak Valencia. In *Capitalismo gore*, Valencia describes these creative violence actors as "especialistas de la violencia," or violence specialists. Among those specialists for whom violence is the primary skill and technique of action, there are those she refers to as "emprededor/as," or entrepreneurs of violence. Like Valencia, I am interested in how these violence specialists are characterized by a "creative orientation" to "innovation, flexibility, and dynamism" (55–57). For Valencia, the drug cartel Los Zetas is emblematic of what she sees as the doubling of entrepreneurship from economics and politics into crime and spectacular violence. If there is any difference in our use of the term at all, it may be Valencia's emphasis on capitalism in the formulation of violence entrepreneurs. I see a much longer historical reach for the concept and describe at length how violence entrepreneurs played a key role in precapitalist settler colonialism. In a provisional kind of way, I would add that violence entrepreneurs play a significant role in the transitions between every "episteme of violence" rather than being idiosyncratic to the period of late globalization Valencia engages. Furthermore, although the term *episteme* is adapted from Foucault, I think Valencia would agree that epistemes of violence do not graph neatly onto the periods identified by Foucault, particularly as we dilate the geographic scope of the genealogy of violence beyond the provincialism of Europe. This history is detailed in chapter 3.
46 Anabel Hernández and her coauthors provide juicy details about how easily elites move between licit and illicit economies in the twenty-first century. See Hernández, Bruce, and Fox, *Narcoland*.
47 This idea of the "night sight" of thinking is borrowed from Eugene Thacker's work on pessimism. See Thacker, *Cosmic Pessimism*.

48 Physicians for Social Responsibility has made a compelling case that the U.S. global war on terrorism has killed some 1 million people. I say "for no reason" because the vast numbers of these deaths were the result of a fictional premise. There were no weapons of mass destruction. It is instructive of our moment in history that one has to make a "case" for murder and not the normal sort where the guilt or innocence of the perpetrator is being contested. Instead the scale of murder is such that groups have to make a "case" that murder has taken place. The numbers are too large to verify even when dealing with something as determinative as dead or alive, with estimates of dead equaling as many as forty or fifty thousand, a margin of error larger than many cities and far in excess of the total number of every person you have ever known. See Physicians for Social Responsibility, *Body Count: Casualty Figures after 10 Years of the "War on Terror" Iraq, Afghanistan, Pakistan*, 2015.

49 See Scahill, *Dirty Wars*, and Greenwald, *No Place to Hide.*

50 I am quite taken with Povinelli's characterization of the violence of late modernity as cruddy or eventless. See Povinelli, *Economies of Abandonment.*

51 Lauren Berlant writes about the left's obsession with the "cool facts" of suffering and how they become the "hot weapons" in the debates about agency that so often derail political change. See Berlant, *Cruel Optimism*, 101–2.

52 Coates, *Between the World and Me*, 116. Coates's lie is that of slavery and democracy but for me this is just the rotten core of the empire project of U.S. history. W. E. B. Du Bois referred to the vision of the U.S.-dominated UN Security Council as the "global color line." For Du Bois, U.S. foreign policy and empire were the Middle Passage inside out. See Du Bois, *Color and Democracy.*

53 Harney and Moten, *The Undercommons.*

54 Bataille, *The Cradle of Humanity.*

55 Halberstam, *The Queer Art of Failure*, 183–84.

1. THE ANTHROPOCENE AS A GEOPOLITICAL FACT

1 Sloterdijk, *Spheres*, 3:85.

2 Lyotard, *The Postmodern Condition*, 37–41.

3 Sloterdijk, *Spheres*, 3:25. See also Pickering, "Cyborg History and the World War II Regime."

4 Sloterdijk, *Spheres*, 3:65–72.

5 Sloterdijk, *Spheres*, 3:99.

6 Sloterdijk, *Spheres*, 3:101.

7 Sloterdijk, *Spheres*, 3:99.

8 Crutzen, "Estimates of Possible Variations."

9 Shapiro, *Violent Cartographies*, 14–16.

10 Hamilton, *Earthmasters*, 182.

11 The Annales school is the most developed of the attempts to understand the deeply ecological character of history. Fernand Braudel in particular added the distinctively geographic and climatic character of the rise of European hegemony and the system of capitalism that developed with it. Unlike other Marxist historians, for Braudel,

capitalism like European hegemony is contingent on a number of nonhuman factors and distributional effects of location. The result is a much broader understanding of materialism in the making of history beyond the political economic focus of traditional Marxist approaches. Braudel's claim that history requires all the social sciences and even the natural sciences is inspiration for this book. See Braudel, *A History of Civilizations*.

12 Crutzen and Birks, "The Atmosphere after a Nuclear War," 124.

13 Sloterdijk, *Not Saved*, 309.

14 Sloterdijk, *Not Saved*, 239.

15 Gabrys, *Program Earth*, 6–9; Litfin, *Ozone Discourses*, 178.

16 On different approaches to the dating of the Anthropocene, see Zalasiewicz, Waters, Williams, et al. "The Anthropocene Biosphere."

17 The American apocalypse is described in more detail in part III.

18 Alfred Crosby makes the point that motive is less salient than the assemblage of technology, plants, and animals that made the terraforming of "neo-Europes" possible. Crosby, *Ecological Imperialism*.

19 Michael Heckenberger and others have identified large urban clusters in Amazonia that suggest advanced urban development and civilizations well beyond the European mythology of empty lands. See Heckenberger et al., "Pre-Columbian Urbanism."

20 Biello, "Mass Deaths in Americas Start New CO_2 Epoch."

21 Holen et al., "A 130,000-Year-Old Archaeological Site."

22 Sloterdijk, *Spheres*, 3:239.

23 Kato, "Nuclear Globalism."

24 *Anthropos* is a Greek term for "human." *Anthropocene* has been coined to capture the emergence of human beings as a natural force at the geological level.

25 Aron, *Peace and War*, 191.

26 Ranke, *The Theory and Practice of History*.

27 Tunander, "Geopolitics of the North."

28 Kjellén also coined a term to describe humans in the biological milieu, which he termed *liebenspolitsche*, later translated as *biopolitics* in his 1924 book *The State as a Lifeform*.

29 Alfred Crosby and Daniel Headrick provide detailed accounts of how Euro-American empires were uniquely indebted to what they call "ecological imperialism." See Crosby, *Ecological Imperialism*; Headrick, *Power over Peoples*.

30 Aron, *Peace and War*.

31 The U.S. military, already indebted to meteorology as a strategic science most famously employed in the planning of D-Day, pursued strategic weather modification projects throughout the second half of the twentieth century. See Fleming, *Fixing the Sky*.

32 Bashford, *Global Population*, 163–64.

33 Bashford, *Global Population*, 180.

34 Bashford, *Global Population*, 356.

35 Bashford, *Global Population*, 363.

36 I originally wrote this chapter before geoengineering was a serious policy proposal. In just a few years the concept has become "reasonable." However, Plato's recounting of

Atlantis demonstrates that humans had imagined gaining control over the condition of life, on a global scale, before the ecological catastrophes that began after 1492.

37 Lasswell, *Politics.*

38 Zylinska, *Minimal Ethics for the Anthropocene*; Connolly, *Facing the Planetary*; Morton, "How I Learned to Stop Worrying"; Tsing, *The Mushroom at the End of the World*; Scranton, *Learning to Die in the Anthropocene.*

39 Rahnema and Bawtree, *The Post-Development Reader.*

40 Rob Nixon's concept of slow violence is vital to understanding the way that speed and privilege come to elevate some forms of violence like terrorism to the scale of emergency whereas recurrent and much more lethal forms of killing like industrial pollution in racially and spatially marginalized communities is treated as a "regulatory" problem. Elizabeth Povinelli has added to this analysis that contemporary violence for the poor and the marginalized is characterized by a kind of "crudiness" and "eventlessness" executed more often through means of affirmative abandonment than direct applications of lethal force. See Nixon, *Slow Violence and the Environmentalism of the Poor*; Povinelli, *Economies of Abandonment.*

41 The second epigraph of this section is from acoustic6strings, "Jacques Cousteau The Nile 2of6.avi," YouTube, 9:15, May 4, 2011, http://www.youtube.com/watch?v=Q47ZVvOelak.

42 Stewart Brand became famous for having founded the *Whole Earth Catalog*, a compendium of future-oriented possibilities that could, by the estimation of him and his compatriots, benefit the progress of the human race. Since then Brand has founded the Long Now Foundation (with musician Brian Eno), which is focused on extending the survival of the human race. See http://longnow.org/; Asafu-Adjaye, Blomqvist, Brand et al. 2016.

43 Mooney, "'And Then We Wept.'"

44 Karimi, "With 1 Male Left Worldwide, Northern White Rhinos under Guard 24 Hours."

45 Sudan died of an infection in March 2018 before this book was published.

46 Apocalypses are definitively what Timothy Morton (2010) calls hyberobjects. The events themselves defy the perceptive and experiential capabilities of humans. One of the shortcomings of the Anthropocene as a concept is the tone of novelty and presentism. Morton rightly points out that climate change and many of the features of the Anthropocene are likely thousands of years old. So the idea that things are "suddenly" weird is specious—none of us alive has ever lived in "normal" times. See Morton, "Hyperobjects and the End of Common Sense."

47 Rockstrom, et al. "Planetary Boundaries."

48 Kolbert, *Sixth Extinction*, 12–16.

49 Kolbert, *Sixth Extinction*, 259–62.

50 Some scientists argue that these scenarios for collapse are reversible but that presumes unprecedented political action on a global scale. See Worm, et al., "Impacts of Biodiversity Loss on Ocean Ecosystem Services."

51 I assume amphibians think, in their way, of our current epoch as a human apocalypse given how much they predate our existence.

52 Headrick argues that "half of the Americas were still Indian country" in the middle of the nineteenth century and that Africa, the Middle East, and Asia were also beyond the capability of continued European development. The breakthrough that allowed the truly global push of European conquest came with the spread of steam-powered shipping as well as advances in tropical medicine and weaponry. See the chapter titled "Steamboat Imperialism" in Headrick, *Power over Peoples.*

53 Nelson Bomber, "'The World' Cruise Ship Departing Port Nelson, New Zealand at Night," YouTube video, 2:56, April 26, 2014, https://www.youtube.com/watch?v=Xh151ygXyrE.

54 Salter, "To Make Move and Let Stop."

55 Wark, *Molecular Red,* 4.

56 The hydrologic cycle is what moves moisture around the planet through evaporation and condensation. Changes in the rate of evaporation or the amount of moisture that the air can hold due to increased temperatures can mean the difference between floods and droughts.

57 On how big lies require half truths, see Connolly, "The Return of the Big Lie."

58 Benton, *When Life Nearly Died.*

59 Harrison, *When Languages Die.*

60 Harrison, *When Languages Die,* 19.

61 Harrison, *When Languages Die,* 249.

62 Harrison. *When Languages Die,* 225.

63 Harrison. *When Languages Die,* viii.

64 Agamben, *Means without End,* 3–6.

65 "Brazil: Uncontacted Tribe Displaced by Amazon Logging," *Argentina Independent,* July 4, 2014.

66 Viveiros de Castro, *From the Enemy's Point of View,* 29.

67 Garcia, *Form and Object,* 221–23.

68 On the importance of diplomats for bridging the cosmological gaps in our current ecological debate, see Latour, *An Inquiry into Modes of Existence,* 17–19.

69 Goldberg-Hiller and Silva, "Sharks and Pigs."

70 Documentary Channel, "Nature's Engineers the Dam Beaver National Geographic Documentary," YouTube, 42:56, June 16, 2013, https://www.youtube.com/watch?v=aHBCQ_EQovo; Manning, "Rewilding the West."

2. WAR AS A FORM OF LIFE

1 Rabinow, *Marking Time,* 13.

2 Zalasiewicz et al., "Scale and Diversity of the Physical Technosphere."

3 The idea that no form of annihilation is sufficient to remove the trace from its aftermath is inspired by Jacques Derrida's materialist reading of the trace as ash in his book *Cinders.* In *Cinders,* Derrida makes reference to the inability of the Holocaust to entirely erase the victims of the Holocaust. There is always a reminder as a kind of limit to political, technological, and geological cataclysms. Derrida develops ash as a kind of paradigm for modernity. In some sense, Derrida is arguing for a kind of

Anthropocene, that is, the impossibility of material history, fully, to forget. However, unlike Crutzen and others whose concept relies on how "big" the human footprint is upon the planet, Derrida develops his concept around the irrepressibly infinitesimal, which is more in line with the ways I am trying to read the remains of war and homogenization. Derrida writes that the cinder "at present, here and now . . . is something material—visible but scarcely readable—that, referring only to itself, no longer makes a trace, unless it traces only by losing the trace it scarcely leaves—that it just barely remains." Derrida, *Cinders*, 21–24.

4 "This is the perfect condition of Slavery, which is nothing else, but the State of War, between a lawful Conquerer and a Captive." Locke Book II, no. 24: 325.

5 Du Bois, *Black Reconstruction*, 670.

6 See Alexander, *The New Jim Crow*; Wacquant, *Punishing the Poor*.

7 Taussig, "Maleficium."

8 Campbell, O'Rourke, and Slater, *Carving Nature at Its Joints*, 2–4.

9 Merleau-Ponty, *Phenomenology of Perception*, 172–73, emphasis added.

10 Those initial flirtations with complexity theory in the mainstream of Political Science have replicated the same impulse to model the world in ways that can be repeatable. The work that has come out of this work is distinctly unable to explain as much as say, researchers like Thomas Homer-Dixon or others who while conservative in terms of their view of the empirical world, are more open-minded in regard to scholarship. Homer-Dixon's initial forays into climate change and security were only possible because he rejected academic standards of selection size and case selection. In order to satisfy those norms one would have to have multiple inhabited planets. See Homer-Dixon, *Environment, Scarcity, and Violence*; Harrison, *Complexity in World Politics*.

11 See Wendt, "The State as Person in International Theory."

12 Moffat, *Complexity Theory and Network Centric Warfare*; Jackson, *Predicting Malicious Behavior*.

13 See Network of Concerned Anthropologists, *The Counter-Counterinsurgency Manual*.

14 Hamilton, *Earthmasters*, 156.

15 Deleuze and Guattari, *What Is Philosophy?*, 118.

16 Deleuze and Guattari, *What Is Philosophy?*, 118.

17 Deleuze and Guattari, *What Is Philosophy?*, 20, 118.

18 Garcia, *Form and Object*, 13.

19 See chapters 3 and 4.

20 Deleuze, "Mediators," 286–87.

21 Williams, *Keywords*, 110.

22 Bateson, *Steps to an Ecology of Mind*, 317.

23 Whitehead, *The Function of Reason*, 6–7.

24 Schrödinger and Penrose, *What Is Life?*, 70–71.

25 Canguilhem, "The Living and Its Milieu."

26 Morton, *The Ecological Thought*, 39. *Force fields* is Connolly's term from *A World of Becoming*, 14.

27 Connolly, *A World of Becoming*, 27.

28 Connolly, *A World of Becoming*, 22.
29 Deleuze and Guattari, *A Thousand Plateaus*, 313.
30 DeLanda, *Intensive Science and Virtual Philosophy*, 10.
31 Morton, *Dark Ecology*, 67.
32 Deleuze and Guattari, *What Is Philosophy?*, 20.
33 Moon, *Sex among Allies*; Nordstrom, *Shadows of War*, 125; Coward, *Urbicide*.
34 See McNeil, *Plagues and Peoples*, 30; Deleuze and Guattari, *A Thousand Plateaus*, 243.
35 Homer-Dixon, *Environment, Scarcity, and Violence*, 177.
36 See Svensmark, "Cosmic Rays and the Biosphere over 4 Billion Years."
37 Dean Rusk insisted that any single warhead launched was the same as launching all of them. In Rusk's words, "the critical question is [not] whether you get a particular missile before it goes off because if they shoot those missiles we are in general nuclear war." Rusk, *At the Edge of the Abyss*, 94.
38 Forrow et al., "Accidental Nuclear War."
39 Haselkorn, "Iraq's Bio-Warfare Option," 19–20.
40 Van Evera, *Causes of War*, 35–37.
41 Carr, *What Is History?*, 74, 117.
42 Massumi, "Such as It Is," 117.
43 Tarde, *Social Laws*, 24.
44 Tanielian et al., *Invisible Wounds*. Steve Coll puts the numbers closer to 500,000. Coll, "The Disgraceful Truth."
45 Bousquet, "All Your Brain Are Belong to Us."
46 Holley, "The Tiny Pill Fueling Syria's War."
47 Abufarha, *The Making of a Human Bomb*, 137–39.
48 "'Body Bombs' Are a Good Sign, DHS Insider Claims."
49 Deleuze and Guattari, *A Thousand Plateaus*, 358.
50 William McNeill relates that boot camp permanently altered the capacity of his body. For the rest of his life, he experienced the possibility of "moving in time" with others. War calibrates the rhythm of the body, taking advantage of the collective connections among people. Like dance, war mobilizes bodies rather than a body. There is an assemblage of parts: arms, legs, spears, tambourines, and so on, all moving in a particular relation. McNeill sees incipient beginnings to collective rhythm throughout the animal kingdom. See McNeill, *Keeping Together in Time*, 1–15, 101–6.
51 E. Manning, *Always More than One*, 30.
52 E. Manning, *Always More than One*, 30.
53 Recent studies in neuroscience demonstrate that the brain rewards the body for violence the same way that it is rewarded for sex. At the level of brain chemistry, the intensities function in the same way. However, philosophically or socially, what remains between the two rewards is the minimal differences that reside at the crossroads of ethics and aesthetics. See Couppis and Kennedy, "The Rewarding Effect of Aggression."
54 Merleau-Ponty, *Phenomenology of Perception*, 174.
55 Bergson, *The Creative Mind*, 164–65.

56 Merleau-Ponty, *Phenomenology of Perception*, 189.
57 Massumi, *What Animals Teach Us about Politics*, 25.
58 Manning, *Always More than One*, 19.
59 The decision to bomb cities was not a forgone conclusion. Some military leaders referred to it as murder as there was no possibility of defense of combat. The use of airpower, like all new technologies, required the development of new sensibilities, a new sensorium in which razing cities to the ground made *sense.* See Sloterdijk, *Terror from the Air.*
60 Nixon, *Slow Violence and the Environmentalism of the Poor.*
61 Sloterdijk, *Spheres*, 3:101–2.

3. FROM EXHAUSTION TO ANNIHILATION

1 Deleuze, *Foucault*, 39.
2 Foucault, *Security, Territory, Population*, 45, 96, 98.
3 What I mean by *evolution* throughout the book is best captured by Eva Jablonka et al.'s concept of evolution in four dimensions. Jablonka and Zeligowski eschew the deadlock between Darwinist and Lamarchian theories in favor of a transversal reading of natural selection, genetics, epigenetics, and the creative processes at the individual and group levels that among humans is called culture. Jablonka et al., *Evolution in Four Dimensions.*
4 Dunbar-Ortiz, *An Indigenous Peoples' History of the United States*; Shapiro, *Violent Cartographies*; Grovogui, *Beyond Eurocentrism and Anarchy*; McNeill, *The Pursuit of Power.*
5 Vargas Machuca, *The Indian Militia and Description of the Indies.*
6 Tarak Barakawi argues that war was a formative part of the development of the tactics of modern European warfare as well as an agent of globalization during the eighteenth and nineteenth centuries through to the contemporary war on terrorism. See Barakawi, *Globalization and War*, 16–25. In particular, the glaring contradiction between the fraternity of the French Revolution and the continuation of slavery and colonialism came to a head in the 1791 Haiti slave revolution. The international cooperation between Jefferson's United States, France, and England was unprecedented but unwavering in its opposition and military and commercial punishment of Haiti and its newly formed constitutional republic. See Grovogui, "Mind, Body, and Gut!," 186.
7 See Arrighi, *The Long Twentieth Century*, 59–64.
8 On the Ottoman Empire, see Abu-Lughod, *Before European Hegemony*; on China, see Greenhalgh and Winckler, *Governing China's Population*; on Soviet imperialism and genocide, see Rummel, *Lethal Politics.*
9 Fuller, *Armament and History*, 85.
10 From a Foucauldian perspective, this tracks theoretical along the lines of the shift from juridical or sovereign power to biopolitical forms of governmentality; however, the timeline is somewhat different, and as I will argue later, the biopolitical transformation of the Ancien Régime to governments may have been driven by transformations in warfare rather than the other way around.

11 DeLanda, *War in the Age of Intelligent Machines*, 71–72.
12 Sloterdijk, *Spheres*, 3:101.
13 Sloterdijk, *Spheres*, 3:93.
14 Sloterdijk, *Spheres*, 3:81.
15 Fuller, *Armament and History*, 93.
16 Delbrück explains that the Teutonic Order of Knights succeeded in its conquest of Prussia because of its reliance on castles. According to Delbrück, "In all such colonial conquests, it is not so much the first subjugation, which often enough can succeed rather easily by surprise and deception, that is the deciding factor, but rather the effective overcoming of the rebellion which is sure to follow as soon as the subjugated people has become familiar with the foreign hegemony. . . . In 1242 . . . the first great rebellion broke out, and it lasted eleven years. . . . It was survived and overcome not with field armies and open battles but because the Prussians were not able to conquer the fortified places of the knights and thus drive them out of the country." Delbrück, *The History of the Art of War*, 3:378–79. Deleuze and Guattari refer to this as the spatiogeographic aspect of the war machine that is its new, innovative means for occupying space in the territorialization of the field of battle. In the case of Teutonic Knights, the distribution of fortifications throughout Prussia was aided by nomadic pilgrims returning from the Crusades. The importance of the sedimentation of the Teutonic warrior class and subsequent striation of the Prussian lands in the rise of the future of the Prussian and then German state speaks to the uneasy but essential role the war machine plays in the constitution of the state apparatus that is the transition from ground to land, contrary to the normally deterritorializing effect the warrior class has on the state apparatus. Deleuze and Guattari posit that the "war function" is exterior to the state apparatus; however, the Prussian case may suggest that mutational possibilities always exist for sedimentation and rigidification of the war function into the state form. Deleuze and Guattari, *A Thousand Plateaus*, 380, 424.
17 Fuller, *Armament and History*, 88–89.
18 Fuller, *Armament and History*, 91.
19 Fuller, *Armament and History*, 92.
20 Fuller, *Armament and History*, 92.
21 Marx's general intellect becomes a central feature of martial capability. See Marx, "Grundrisse: Notebook VII—The Chapter on Capital," https://www.marxists.org/archive/marx/works/1857/grundrisse/ch14.htm.
22 Fernand Braudel makes a similar point about the competitive advantage of states over city-states during the sixteenth century. The difference in wealth between the two different organizational types had much of an effect on the periodic fights between states and city-states and commercial compacts—that is, until the wealth could be converted into artillery and as a result war could not be fought successfully without artillery. Braudel, *The Mediterranean and the Mediterranean World in the Age of Philip*, 2:658–61.
23 Mumford, *Technics and Civilization.*
24 Fuller, *Armament and History*, 93; McNeill notes the same moral transformation of war. See McNeill, *The Pursuit of Power*, 86–87.

25 McNeill, *The Pursuit of Power*, 88–89.
26 Fuller, *Armament and History*, 100; McNeill, *The Pursuit of Power*, 173–74.
27 McNeill, *The Pursuit of Power*, 173–74.
28 Mumford, *Technics and Civilization*, 24; Fuller, *Armament and History*, 107; see also McNeill, *The Pursuit of Power*, 150.
29 Fuller, *Armament and History*, 102–3.
30 Mumford, *Technics and Civilization*, 24.
31 Fuller, *Armament and History*, 107.
32 Fuller, *Armament and History*, 110. For a discussion of the ways the state form and Napoleon in particular captured the *esprit de corps* of revolutionary and martial forces, see Deleuze and Guattari, *A Thousand Plateaus*, 366–86.
33 Delbrück, *History of the Art of War*, 4:396.
34 Delbrück, *History of the Art of War*, 4:397.
35 McNeill, *The Pursuit of Power*, 159.
36 McNeill, *The Pursuit of Power*, 158.
37 Fraser, *Napoleon's Cursed War.*
38 Delbrück, *History of the Art of War*, 4:411.
39 McNeill, *The Pursuit of Power*, 159.
40 DeLanda, *War in the Age of Intelligent Machines*, 72.
41 Fuller, *Armament and History*, 111.
42 Fuller, *Armament and History*, 120.
43 Fuller, *Armament and History*, 117, 120.
44 Fuller, *Armament and History*, 112. See also Aron, *A Century of Total War*, 41.
45 Geggus, *The Haitian Revolution*, 184.
46 Fuller, *Armament and History*, 140.
47 Fuller, *Armament and History*, 140.
48 Blight, *The Fog of War*, 220.
49 Halpern, *Beautiful Data*, 41–43.
50 Grossman, *On Combat*, 124.
51 I do not want to suggest that this process was unique to the Americas. Africa, Oceania, and Asia were also laboratories for new modes of warfare. However, the exceptionalism here is not without some reason. As is argued throughout this book, the centrality of the U.S. as an inheritor of the Eurocene is all too often obscured by attempts to "spread the blame" of the current world order. The process of horizontalizing the cause of the current martial and environmental catastrophe is, to me, all too often an attempt to depoliticize the geopolitical character of the crisis. In the benevolent version, the depoliticization is carried out in the name of a new cosmopolitanism to "save the planet"; in more cynical versions, the "accident" of American power is used to defend renewed arguments for colonial trusteeship and colonialism.
52 On Assyrian wall carvings, see Bahrani, *Rituals of War*. Parker is discussed by Kris Lane, "Preface," in Vargas Machuca, *The Indian Militia and Description of the Indies*, xi.
53 For exploration of Kant's geography, see Harvey, *Cosmopolitanism and the Geographies of Freedom*, 25–26; Humboldt, *Cosmos.*

54 Machiavelli, *The Prince*, chapter 17.
55 Althusser, *Philosophy of the Encounter*, 167–69.
56 Neel Ahuja provides the best account of the nonhuman animal and microbial assemblages at work in settler colonialism. See Ahuja, *Bioinsecurities*.
57 Vargas Machuca, *The Indian Militia and Description of the Indies*, 76–78.
58 Vargas Machuca, *The Indian Militia and Description of the Indies*, 178.
59 Lindqvist, *A History of Bombing*; Sloterdijk, *Terror from the Air*.
60 Vargas Machuca, *The Indian Militia and Description of the Indies*, 133.
61 Vargas Machuca, *The Indian Militia and Description of the Indies*, 147.
62 Vargas Machuca, *The Indian Militia and Description of the Indies*, 157.
63 Vargas Machuca, *The Indian Militia and Description of the Indies*, 252.
64 Vargas Machuca, *The Indian Militia and Description of the Indies*, 254, 257.
65 Crosby, *Ecological Imperialism*, 3–6.
66 Hoffman, *The War Machines*.
67 Lane, "Preface," in Vargas Machuca, *The Indian Militia and Description of the Indies*, xii.
68 W. Brown, *Walled States, Waning Sovereignty*, 80–82.
69 Weigley, *The American Way of War*, 114.
70 Deleuze and Guattari, *A Thousand Plateaus*, 354–57.
71 Foucault, "Preface," in Deleuze and Guattari, *Anti-Oedipus*.
72 On Custer and Hays, see Dunbar-Ortiz, *An Indigenous Peoples' History of the United States*, 129.
73 United States, Department of the Army, *The U.S. Army/Marine Corps Counterinsurgency Field Manual*, xxiii.
74 Shapiro, *Violent Cartographies*, 136–70.
75 Blackhawk, *Violence over the Land*, 9–15.
76 Brown, Parrish, and Speri, "Leaked Documents Reveal Counterterrorism Tactics."
77 Brown, Parrish, and Speri, "Leaked Documents Reveal Counterterrorism Tactics."
78 Wax, "The Uses of Anthropology in the Insurgent Age," 153–60. For a description of COINTELPRO operations against the American Indian Movement, see Michael Apted's film *Incident at Oglala* (2004).
79 Owens, *The Economy of Force*, 164.
80 Adams, *Education for Extinction*, 335–37.
81 Owens, *The Economy of Force*, 165.
82 Weigley, *The American Way of War*, 153.
83 Dunbar-Ortiz, *An Indigenous Peoples' History of the United States*, 166–67.
84 U.S. Immigration and Customs Enforcement, "ICE Shadow Wolves."
85 U.S. Immigration and Customs Enforcement, "ICE Shadow Wolves."
86 Byrd, *The Transit of Empire*, 225–29.
87 I unfortunately discovered Laleh Khalihi's book after this manuscript was finished. I do not engage her book sufficiently. Importantly, Khalihi makes a connection between the Indian Wars and the Philippines as well as the continued legal connection between Indian removal and executive power invoked by John You in the extralegal regimes created after September 11, 2001. I hope analysis is complementary to rather than merely redundant with Khalihi's excellent work on the connections between

practices of external security and the domestic pacification of activists. One point of disagreement with her very powerful genealogy of counterinsurgency is her agreement with Andrew Birtle that the Indian Wars became an "unwritten tradition" based on Birtle's assessment that the curriculum at West Point no longer taught the Indian Wars. The curriculum at West Point is a very small part of the overall pedagogical structure of the U.S. Army, much less the U.S. military more broadly. A review of the digital archive at Ft. Leavenworth has many master's theses and dissertations engaging the Indian Wars as both a case study and a significant part of U.S. military development. Special thanks to an anonymous commenter responding to a version of this section on the blog *The Disorder of Things* suggesting I read Khalili's book. See Khalili, *Time in the Shadows*, esp. 16, 46. See also Birtle, *U.S. Army Counterinsurgency and Contingency Operations Doctrine 1860–1941*.

88 Russell, "Going Native." See also Vizenor, *Manifest Manners*, 8–9.

89 Marx, "Division of Labour and Manufacture."

90 Du Bois, *Black Reconstruction in America*.

91 Foucault, *Society Must Be Defended*.

92 Galli and Sitze, *Political Spaces and Global War*; Tiqqun, *Introduction to Civil War*.

93 Foucault, *Security, Territory, Population*, 10–18.

94 Foucault, *The Birth of Biopolitics*, 6–9, 52–54; see also Foucault, *Security, Territory, Population*, 90.

95 Fuller, *Armament and History*, 171–72.

96 Negri, *Porcelain Workshop*, 55.

97 Hounshell, *From the American System to Mass Production*, 32–35, 43.

98 Ann Laura Stoler's account of the biopolitics lectures was the first to make note of the fact that Foucault has an underappreciated interest in race. However, Foucault's use of the term is something closer to ethnicity or class and draws very little from the colonial or settler archive of history. Instead, Foucault's genealogy of race as a problem for security works from the interior of the congealing nation-state rather than in the projects of extermination taking place in the periphery. It is worth considering how differently Foucault's genealogy would be if the story of the rise of the police in response to food riots during the seventeenth century were told in dialogue with the carceral settlements being built throughout the rest of the planet. *Security, Territory, Population* might have seen a great continuity between martial character of sovereign violence and the biopolitical ordering of security if a more global context had been appreciated by Foucault. See Stoler, *Race and the Education of Desire*, 55–56; Foucault, *Security, Territory, Population*, 20–21. This reading of Foucault is indebted to the late Robin Kilson's class on race and empire taught at the University of Texas at Austin in fall 1999.

99 Negri, *Porcelain Workshop*, 57.

100 Foucault, *The History of Sexuality*, vol. 1, 189–91.

101 See Dillon and Reid, *The Liberal Way of War*; see also Odysseos and Petito, *The International Political Thought of Carl Schmitt*.

102 Sloterdijk, "The Time of the Crime of the Monstrous," 167.

103 Arendt, *The Origins of Totalitarianism*, 223.

4. BOMBS

1 Gallagher, *Kaboom*, 132.
2 Morton, *Realist Magic*, 44.
3 Shaviro, *The Universe of Things*, 52.
4 Morton, *Realist Magic*, 44–45.
5 Higginbotham, "U.S. Military Learns to Fight Deadliest Weapons."
6 Barker, "Improvised Explosive Devices"; Briscoe, Weiss, Whitaker, and Trewhitt, "A System-Level Understanding"; McFate, "Iraq."
7 Jones et al., "The Psychological Effects."
8 Jones et al., "The Psychological Effects," 471.
9 Chivers, *The Gun*.
10 Simondon, *On the Mode of Existence*, 13.
11 Simondon, *On the Mode of Existence*, 59.
12 Simondon, *On the Mode of Existence*, 66–67.
13 Guattari, *Chaosmosis*, 33.
14 Guattari, *Chaosmosis*, 34.
15 Goldman and Eliason, *The Diffusion of Military Technology and Ideas*, 7–9.
16 Deleuze and Guattari, *A Thousand Plateaus*, 437.
17 Deleuze, *Bergsonism*, 117.
18 Uexküll, *A Foray into the Worlds*, 47.
19 Whitehead, *Process and Reality*, 19–20.
20 Caesar, *The Conquest of Gaul*, 192; Delbrück, *Warfare in Antiquity*, 498–99.
21 Goldman and Eliason, *The Diffusion of Military Technology and Ideas*.
22 Rancière, *Politics of Aesthetics*, 29.
23 Hoffmann, *The State of War*.
24 Simondon, *On the Mode of Existence*, 58–59.
25 Connolly, "Evangelical-Capitalist Resonance Machine," 33.
26 Connolly, *A World of Becoming*, 138.
27 Ashby, *Design for a Brain*, 265.
28 DeLanda, *A Thousand Years of Nonlinear History*, 16, 258–59.
29 Guattari, *The Machinic Unconscious*, 152.
30 Deleuze and Guattari, *What Is Philosophy?*, 118.
31 Deleuze and Guattari, *A Thousand Plateaus*, 511.
32 Connolly, *A World of Becoming*, 22.
33 Croll, *The History of Landmines*, 20.
34 Whitehead, *An Enquiry Concerning the Principles*, 65.
35 Morton, *The Ecological Thought*, 39.
36 Bergson, *Creative Evolution*; Bennet, *Vibrant Matter*.
37 Whitehead, *An Enquiry Concerning the Principles*, 196.
38 Whitehead, *An Enquiry Concerning the Principles*, 196.
39 Atkinson, "IED Is Insurgents' Iconic Device."
40 Atkinson, "IED Is Insurgents' Iconic Device."
41 Greenpeace, "E-Waste."
42 C. P. Baldé, V. Forti, V. Gray, R. Kuehr, P. Stegmann, *The Global E-waste Monitor*.

43 Small Arms Survey, "Direct Conflict Deaths."
44 UN-HABITAT, "The Challenge of Slums."
45 Davis, *Planet of Slums*, 31.
46 Davis, *Planet of Slums*, 1–5.
47 Davis, *Planet of Slums*, 67.
48 Virilio, *The Futurism of the Instant*, 32.
49 Davis, *Planet of Slums*, 67–69; see also Pacione, *Urban Geography*, 485.
50 Fuller, *Armament and History*, 123.
51 Der Derian, *Virtuous War*, 209.
52 Clausewitz, *On War*, 67.
53 Walsh et al. "Wikileaks Afghanistan Files: Every IED Attack, with Co-ordinates."
54 Afghan Conflict Monitor, "U.S. Casualty from IEDs Skyrocket."
55 Keyes, "U.S. Won't Join Landmine Ban."
56 DeLanda, *Intensive Science and Virtual Philosophy*, 10.
57 Pickering, *The Cybernetic Brain*, 74; Bennett, *Vibrant Matter*, 31.
58 On actants, see Latour, *Science in Action*, 91.
59 Pickering, *The Cybernetic Brain*, 31.
60 Bennett, *Vibrant Matter*, 20.

5. BLOOD

1 Weheliye, *Habeas Viscus*, 49. Weheliye also drew my attention to a fantastic quote from Deleuze and Guattari's *A Thousand Plateaus*: "A race is defined not by its purity but rather by the impurity conferred upon it by a system of domination. Bastard and mixed blood are the true names of race." Weheliye, *Habeas Viscus*, 49.
2 Department of the Army, *Planning for Health Service Support*, Field Manual 8-55.
3 Starr, *Blood*, 53; Berseus, Hervig, and Seghatchian, "Military Walking Blood Bank and the Civilian Blood Service," 341.
4 Hedley-Whyte and Milamed, "Blood and War."
5 Lem, *Summa Technologiae*, 5.
6 Bennett, *Vibrant Matter*, 2.
7 Bogost, *Alien Phenomenology*, 24.
8 Morton, *Realist Magic*, 44.
9 Hedley-Whyte and Milamed, "Blood and War," 132.
10 Starr, *Blood*, 31–32.
11 Hedley-Whyte and Milamed, "Blood and War," 132.
12 Starr, *Blood*, 90–92; Hess and Thomas, "Blood Use in War and Disaster."
13 Starr, *Blood*, 60.
14 Starr, *Blood*, 65–66.
15 Starr, *Blood*, 71.
16 Wynes, *Charles Richard Drew*.
17 Wynes, *Charles Richard Drew*, 67; "Jim Crow Blood Policy Laid to War Department," *The Afro-American*, May 5, 1945.
18 Starr, *Blood*, 96–97; Wynes, *Charles Richard Drew*.

19 "Jim Crow Blood Policy Laid to War Department," *The Afro-American*, May 5, 1945.
20 Starr, *Blood*, 96–97.
21 Starr, *Blood*, 72.
22 Starr, *Blood*, 72.
23 Hess and Thomas, "Blood Use in War and Disaster," 1623.
24 Hess and Thomas, "Blood Use in War and Disaster," 1623.
25 Wynes, *Charles Richard Drew*.
26 Starr, *Blood*, 108.
27 Foucault, *Society Must Be Defended*, 237.
38 Foucault, *Society Must Be Defended*, 237.
29 For a description of "air hunger" after plasma transfusions, see Starr, *Blood*, 124.
30 I do not want to suggest in any way that the lack of blood racism meant that the French were not racist. Instead, blood provided a bond that militated against forms of civilizational and colonial superiority in ways that American and particularly Nazi racism did not.
31 Starr, *Blood*, 136–37.
32 Starr, *Blood*, xi.
33 Office of the Inspector General, Department of Defense, "Armed Services Blood Program Readiness," Report No. D-2001-059, February 23, 2001; Assistant Secretary of Defense for Health Affairs, "Armed Services Blood Program (ASBP) Operational Procedures," DODI-6480.4, August 5, 1996; Department of the Army, *Planning for Health Service Support*.
34 Assistant Secretary of Defense for Health Affairs, "Armed Services Blood Program."
35 Kauvar et al., "Fresh Whole Blood Transfusion."
36 Hess and Thomas, "Blood Use in War and Disaster," 1626.
37 Kauvar et al., "Fresh Whole Blood Transfusion," 182.
38 Starr, *Blood*, 53; Berseus, Hervig, and Seghatchian, "Military Walking Blood Bank and the Civilian Blood Service," 341.
39 Department of the Army, *Planning for Health Service Support*, 3–4.
40 Mansoor et al., "A National Mapping Assessment," 69, 74.

6. BRAINS

1 Sloterdijk, *Spheres*, 3:81.
2 Malabou, *What Should We Do with Our Brain?*, 58.
3 Kant, *Anthropology from a Pragmatic Point of View*.
4 Kant, *Anthropology from a Pragmatic Point of View*, 5.
5 The account of maturity here bears little resemblance to Kant's description of maturity in *Was Ist Aufklärung?*
6 Kant, *Anthropology from a Pragmatic Point of View*, 15.
7 Kant, *Anthropology from a Pragmatic Point of View*, 34–35.
8 Malabou, *Future of Hegel*.
9 Ashby, *Design for a Brain*, v.

10 Pickering, *The Cybernetic Brain*, 5–7.

11 Ashby, *Design for a Brain*, 382.

12 "The Blue Brain Project—A Swiss Brain Initiative," http://bluebrain.epfl.ch/page-56882-en.html.

13 Ashby, *Design for a Brain*, 233.

14 Ashby, *Design for a Brain*, 236–37.

15 Malafouris, *How Things Shape the Mind*; Noë, *Out of Our Heads*.

16 Lilly, *Programming and Metaprogramming in the Human Biocomputer*, iv.

17 Nicolas Langlitz, *Neuropsychedelia*.

18 Lehrer, "The Forgetting Pill."

19 Dupuy, *On the Origins of Cognitive Science*, 34–36.

20 Von Neumann, *The Computer and the Brain*, 40.

21 The two best criticisms of mechanistic and reductionist images of the human brain have been developed by Jean-Pierre Dupuy and Evan Thompson. See Dupuy, *On the Origins of Cognitive Science*; Thompson, *Waking, Dreaming, Being*.

22 Wolfram, "The Principle of Computational Equivalence."

23 Deutsch, *The Nerves of Government*.

24 Choi, "Brain Researchers Can Detect."

25 Ma, "UW Study Shows Direct Brain Interface."

26 Leys, "Turn to Affect," 434–72; Rose and Abi-Rached, *Neuro*.

27 For an in-depth discussion of Habermas's resistance to neuropolitics, see chapter 9.

28 Žižek, "Philips Mental Jacket," 4.

29 Bogost, *Alien Phenomenology*, 48.

30 It is worth considering that the pill popping was the result of the migraine as alea.

31 Habermas, *The Future of Human Nature*, 25.

32 Deleuze, *Negotiations*, 178–79.

33 Coming to the conclusion of the bankruptcy of ethical theories that attack us with ought was helped a great deal by a conversation with Levi Bryant on his blog *Larval Subjects*. Bryant is always a generous and helpful inspiration and sounding board for ideas. See Bryant, "Ethics and Politics."

34 Pickering, *The Cybernetic Brain*, 8.

35 For an explanation of the importance of the black box approach to cybernetic experimentation, see Ashby, *An Introduction to Cybernetics*, 86–88. For a very provocative critique of black box thinking, see Galloway, "Black Box."

36 For an excellent history and analysis of the formative role of cybernetics in the development of contemporary cognitive science and neuroscience, see Dupuy, *On the Origins of Cognitive Science*.

37 Wiener, *The Human Use of Human Beings*, 27.

38 Ashby, *Design for a Brain*, 1.

39 There has been a resurgence in the interest in cybernetics as it becomes more and more evident that thinkers and scientists from Stuart Kaufman to Gilles Deleuze were informed by cybernetics. The significance of cybernetics for the increasing sophistication of artificial intelligence also contributes to this fascination. See Clarke and Hansen, *Emergence and Embodiment*; Pickering, *The Cybernetic Brain*.

40 For a history of Stafford Beer's role in Salvador Allende's almost cybernetic revolution in Chile, see Medina, *Cybernetic Revolutionaries.*
41 Deutsch, *The Nerves of Government*, xiii.
42 Deutsch, *The Nerves of Government*, viii.
43 Deutsch, *The Nerves of Government*, xiii.
44 Deutsch, *The Nerves of Government*, 129.
45 Deutsch, *The Nerves of Government*, 134.
46 Deutsch, *The Nerves of Government*, 137, 129.
47 Foucault, *Power/Knowledge*, 90–91.
48 Wiener, *The Human Use of Human Beings*, 131.
49 Wiener, *The Human Use of Human Beings*, 161–62.
50 Galloway and Thacker, *The Exploit*, 6–7.
51 Gilbert Simondon, the French engineer who was the inspiration for Gilles Deleuze and Félix Guattari, similarly described this process, naming the individuals "logics of individuation." Simondon is trying to capture the formation of something like an individual but through the complex process of identity/difference produced in the systemic orders in which such an event takes place. Simondon, *La individuación.*
52 Wiener, *The Human Use of Human Beings*, 181.
53 Deutsch, *The Nerves of Government*, ix–x.
54 Delgado, *Physical Control of the Mind*, 281.
55 Delgado, *Physical Control of the Mind*, 287.
56 Delgado, *Physical Control of the Mind*, 3.
57 Delgado, *Physical Control of the Mind*, 4.
58 Delgado, *Physical Control of the Mind*, 6.
59 Delgado, *Physical Control of the Mind*, 7–8.
60 Delgado, *Physical Control of the Mind*, 8.
61 Delgado, *Physical Control of the Mind*, 9.
62 Delgado, *Physical Control of the Mind*, 11.
63 Delgado's vision is creepy but maybe not as insidious as so-called benign imperialists who defend the colonial project as a precondition for global peace. At least in Delgado's case there is wholly absent from the text and argument the racial geographies of superiority and inferiority that underwrite many liberal visions of cosmopolitanism. See Harvey, *Cosmopolitanism and the Geographies of Freedom.*
64 As fellow travelers in the hope for species control, it is important to note the connection between James Watson and Francis Crick and cybernetics. When Watson and Crick announced their research agenda to discover the language of heredity, they said they hoped to do for genetics what Norbert Wiener had done for cybernetics: to understand the information or code that constituted life or, in their words, to show the presence "of cybernetics on the bacterial level." It is then not surprising that in his later life, James Watson has embarrassed himself as an advocate of racial eugenics. The search for the genetic code was animated by the desire to steer it. See Conway and Siegelman, *Dark Hero*, 278; Milmo, "Fury."
65 Malabou, *What Should We Do with Our Brain?*, 82.
66 Malabou, *What Should We Do with Our Brain?*, 82.

67 Malabou, *What Should We Do with Our Brain?*, 79.

68 Simondon instructively distinguishes mechanical versus machinic objects on the basis of the capacity of an object to adapt to or ignore its environment. This helps overcome the presumption that all reductions to process are reductionist. Rather, machinic objects have the capacity for creativity and emergent properties despite being made up of parts. Simondon, *On the Mode of Existence of Technical Objects*.

69 For video footage of the demonstration, see Crosby, "José Delgado, Implants, and Electromagnetic Mind Control," https://www.youtube.com/watch?v=23pXqY3X6c8, accessed December 20, 2018.

70 Delgado, *Physical Control of the Mind*, 214. So far Delgado is certainly right. Further, the research has been driven primarily by defense-related funding sources. The U.S. military in particular has led the way in brain research with application for brain-machine interfaces and brain-behavior modification. As demonstrated by the exhaustive research of medical ethicist Jonathan Moreno, very little attention has been given to the military-brain nexus by the public or the civilian scientist whose research depends on military financing. See Moreno, *Mind Wars*.

71 Moreno, *Mind Wars*.

72 Delgado, *Physical Control of the Mind*, 215.

73 See also Bennett and Connolly, "Contesting Nature/Culture."

74 Malabou, *What Should We Do with Our Brain?*, 7.

75 Delgado, *Physical Control of the Mind*, 215.

76 Kant, *On Education*, 6–7.

77 Kant, *On Education*, 4.

78 Delgado, *Physical Control of the Mind*, 215.

79 Delgado, *Physical Control of the Mind*, 233.

80 Delgado, *Physical Control of the Mind*, 244.

81 Kurzweil, *Singularity Is Near*.

82 Malabou, *What Should We Do with Our Brain?*, 82.

83 Burroughs, *Limits of Control*, 38. Deleuze and Guattari took a long road trip across the United States in the summer of 1975. During that trip, in addition to seeing the Grateful Dead, Deleuze and Guattari met with William Burroughs. The meeting, as well as Burroughs's literature more generally, is reflected throughout their collaborative writing. See Demers, "American Excursion."

84 Demers, "American Excursion."

85 Demers, "American Excursion."

86 Burroughs, *Limits of Control*, 339.

87 Burroughs, *Limits of Control*, 339.

88 Malabou, *Ontology of the Accident*, 14.

89 Burroughs, *Limits of Control*, 40.

90 Connolly, *Neuropolitics*, 56–57.

91 Galloway and Thacker, *The Exploit*, 6.

92 Galloway, *Protocol*, 74–75.

93 The explanation of neural Darwinism that follows is taken from Daniel Lord Smail's provocative challenge to Stephen Jay Gould's neo-Lamarckian theory of cultural

evolution. Smail's attempt is to prevent the "backdoor" Cartesian dualism that often occurs when culture is privileged as something opposed to genetics on the basis that the dissolution of this opposition would result in genetic determinism. See Smail, *Deep History*, 112–15.

94 Wade, "Dark Side."

95 Kwinter, "Notes," 100–101.

96 Burroughs, *Limits of Control*, 38.

97 Connolly calls this the "wild" element that traverses the multiply layered and intercalated systems of the universe. This wild element is the limit to order or that which cannot be ordered, leaving open the possibility of creativity and the new. See Connolly, *Neuropolitics*, 95.

98 Malabou, *Ontology of the Accident*, 39, 52.

99 William Connolly's account in *Neuropolitics* of his father's brain damage shows how we are beholden to our brains for thought. At any moment, the contingency of thought, memory, recognitions, and connections, which we desperately wish could transcend the matter of life, can all unravel. The very real confrontation with the limits of organic life makes neuropolitics more than a trend or a new discourse of the human. Instead, neuropolitics is what we all must face in medias res; see Connolly, *Neuropolitics*, preface. On the concept of fragility as the modern predicament, see "First Interlude: Melancholia," in Connolly, *The Fragility of Things*.

100 Moten and Harney, "Blackness and Governance," 357.

101 Moreno, *Mind Wars*.

102 Orders go awry in both senses of the word: the command and the underlying organization in which orders take place. Despite the rigor and conditioning of boot camp and the bodily discipline of years of military service, steering soldiers between the purposeful devastation of Fallujah in 2004 and the killing spree of U.S. Army staff sergeant Robert Bales in southern Afghanistan is unpredictable at best, despite increasingly sophisticated behavioral and neuroinvasive technics for control. With both Fallujah and Bales, one cannot but wonder how the bodies involved had been altered by go pills for night patrols, oxytocin for morale, antidepressants to fight battle stress, and so on. Furthermore, invasive intracranial intervention into soldiers' brains is no longer speculative. The Defense Advanced Research Projects Agency (DARPA) has begun experiments with mood-regulating brain implants to treat veterans suffering from PTSD. The program, armed with a $12 million budget, hopes to have an approved "cybernetic implant" within five years. If DARPA continues to make progress, it will receive an additional $20 million. Tucker, "The Military Is Building Brain Chips."

103 Malabou, *Ontology of the Accident*, 88.

104 For an exploration of the relationship between cultivation and the arts of the self required for what I mean by freedom, see Connolly, *Neuropolitics*, 106–8.

105 Malabou, *Ontology of the Accident*, 17.

106 Malabou, *Ontology of the Accident*, 30.

107 Malabou, *Ontology of the Accident*, 30.

108 Vargas Machuca, *The Indian Militia and Description of the Indies*, 6.

109 Malabou, *Ontology of the Accident*, 18.

110 Thacker, *In the Dust*, 104.

111 Thacker, *In the Dust*, 104.

112 Malabou, *Ontology of the Accident*, 17.

113 Deleuze, *Pure Immanence.*

114 Cioran, *Short History*, 47.

115 Malabou, *Ontology of the Accident*, 7.

7. THREE IMAGES OF TRANSFORMATION AS HOMOGENIZATION

1 Benjamin Bratton's deconstruction of TED Talk aesthetics captures this point best. According to Bratton, while the topics of TED Talks vary, the form of the talks is nearly identical. Like a magic show, a great TED Talk directs attention with an overly simplistic take on a counterintuitive idea about a big problem only to return to an aesthetic of common sense to support an inspiring "bright idea" that can save us in the end. TEDx Talks, "New Perspectives—What's Wrong with TED Talks? Benjamin Bratton at TEDxSanDiego 2013—Re:Think," YouTube video, 11:50, December 30, 2013, https://www.youtube.com/watch?v=Yo5cKRmJafo.

2 Honig, *Emergency Politics*, 12–14.

3 Bruce Clarke and Mark Hansen note that this strain of neocybernetics adopts a view of emergence that reduces "the chaotically complex to the manageably complex," which they argue is an essential move for justifying the possibility of environmental management. According to Clarke and Hansen, the tame version of emergence is adopted as a way to "reduce the complexity of the environment by processing it through systemic constraints." See Clarke and Hansen, *Emergence and Embodiment*, 11.

4 Brand, *The Essential Whole Earth Catalog.*

5 Meadows, Randers, and Meadows, *The Limits to Growth.*

6 Biello, "Stratospheric Pollution Helps Slow Global Warming."

7 Asafu-Adjaye, Blomqvist, et al., "An Ecomodernist Manifesto," 7.

8 Asafu-Adjaye, Blomqvist, et al., "An Ecomodernist Manifesto," 16.

9 Asafu-Adjaye, Blomqvist, et al., "An Ecomodernist Manifesto," 15.

10 Asafu-Adjaye, Blomqvist, et al., "An Ecomodernist Manifesto," 23.

11 See Latour, *We Have Never Been Modern*; Bennett and Connolly, "Contesting Nature/Culture"; Descola, *Beyond Nature and Culture*; Haraway, *Simians, Cyborgs, and Women.*

12 Asafu-Adjaye, Blomqvist, et al., "An Ecomodernist Manifesto," 31.

13 Purdy, *After Nature*, 436.

14 Purdy, *After Nature*, 35.

15 Purdy, *After Nature*, 481–82.

16 Purdy, *After Nature*, 483–84.

17 Rockstrom, et al., "Planetary Boundaries: Exploring the Safe Operating Space for Humanity."

18 See chapter 1 for the imbrication of expansionist geopolitics and environmental crisis.

19 Wark, "Molecular Red."

20 Purdy, "The New Nature."
21 Du Bois, *Color and Democracy*, 245–50.
22 Purdy, "The New Nature."
23 Kurzweil. *The Age of Spiritual Machines.*
24 Singularity University, "Founders," accessed August 4, 2014, http://singularityu.org/community/founders/.
25 Mumford, *Technics and Civilization.*
26 Dyson. *Darwin among the Machines.*
27 Worstall, "DARPA Bigwig and Intel Fellow."
28 Foster, "Marx and the Rift in the Universal Metabolism of Nature."
29 Foster, "Marx and the Rift in the Universal Metabolism of Nature."
30 Moore, "Anthropocene or Capitaloscene," 11.
31 Moore continues, "As capital progressively internalizes the costs of climate change, massive biodiversity loss, toxification, epidemic disease, and many other biophysical costs, new movements are gaining strength. These are challenging not only capitalism's unequal distribution—pay the 'ecological debt'!—but the very way we think about what is being distributed. The exhaustion of capitalism's valuation of reality is simultaneously internal to capital and giving rise to the new ontological politics outside that value system—and in direct response to its breakdown. We see as never before the lowering of an ontological imagination beyond Cartesian dualism, one that carries forth the possibility of alternative valuations of food, climate, nature, and everything else. They are revealing capitalism's law of value as the value of nothing—or at any rate, of nothing particularly valuable (Patel 2009). And they point toward a world-ecology in which power, wealth, and re/production are forged in conversation with needs of the web of life, and humanity's place within it." Moore, "Anthropocene or Capitaloscene," 11.
32 "The contradictions of capitalism dramatized by biospheric instability reveal modernity's accomplishment as premised on an active and ongoing theft: of our times, of planetary life, of our—and our children's—futures." Moore, "Anthropocene or Capitaloscene," 10.
33 Balakrishnan, "The Abolitionist—II,"100.
34 Balakrishnan, "The Abolitionist—II,"100.
35 Malm, *Fossil Capital*, 786–88; Klein, *This Changes Everything*, 222–24.
36 Balakrishnan, "The Coming Contradiction," 53.
37 Foucault, *Security, Territory, Population*, 32–34.
38 Foucault, *The History of Sexuality*, 1:289.
39 Guattari, *The Three Ecologies*, 43.
40 Marx, *The Eighteenth Brumaire of Louis Bonaparte*, 3.
41 Frey and Osborne, "The Future of Employment."
42 Sassen, *Expulsions*, 6–10.
43 Coulthard, *Red Skin, White Masks*, 208–10.
44 Mbembe, "Necropolitics."
45 Guattari, *The Three Ecologies.*
46 Sloterdijk, *In the World Interior of Capital*, 31.
47 Sloterdijk, *In the World Interior of Capital*, 30.

48 Williams and Srnicek, "#ACCELERATE MANIFESTO."

49 Brander, Du, and Hellmann, "The Effects of Government-Sponsored Venture Capital," 607–9.

50 Wark, "The Vectoralist Class"; Wark, "The Vectoralist Class, Part II."

51 Greenberg, "How a 'Deviant' Philosopher Built Palantir."

52 Shaviro, *No Speed Limit*, 10.

53 Williams and Srnicek, "#ACCELERATE MANIFESTO."

54 Williams and Srnicek, "#ACCELERATE MANIFESTO."

55 Williams and Srnicek, "#ACCELERATE MANIFESTO."

56 The Neanderthal line in Asia.

57 Jameson, *Valences of the Dialectic*, 433–34.

58 Warren, "Some Last Words on Pessimism."

59 "Remarks by Deputy Secretary Work on Third Offset Strategy," U.S. Department of Defense, April 28, 2016, https://www.defense.gov/News/Speeches/Speech-View/Article/753482/remarks-by-deputy-secretary-work-on-third-offset-strategy/; Ahmed, "Pentagon Preparing for Mass Civil Breakdown."

60 Mearsheimer, "Nuclear Weapons and Deterrence in Europe," 19.

61 Kroenig, "Facing Reality."

62 See Shaw, "The Rise of the Predator Empire in the Vietnam War."

63 Bitzinger, "Third Offset Strategy and Chinese A2/AD-Capabilities."

64 Michael Adas has detailed how the faith in technological innovation has driven U.S. war planning in the twentieth century. Vietnam in particular was meant to showcase how precision weapons and nonhuman surveillance could bridge the gaps in human intelligence and popular support. See Adas, *Dominance by Design*.

65 For the best account of how cybernetics and complexity theory came to drive technological approaches to military affairs, see Bousquet, *The Scientific Way of Warfare Order and Chaos*, 121–63.

66 Freedberg, "War without Fear."

67 Wait, "Government Hiring Practices Hamper Cybersecurity Efforts."

68 There is an extraordinary resonance between the temporal shifts between the Cold War and War on Terrorism traced by Brian Massumi and the obsession now with innovation and deployment at the speed of machines. The logic of preemption is not explicit in the Third Offset but the temporality of the crisis—responding to not-yet threats because their potentiality is a threat—is explicit in the offset approach. See Massumi, *Ontopower*.

69 Crampton, "Collect It All."

70 Freedberg, "Centaur Army."

71 Freedberg, "Tiny Drones Win over Army Grunts."

72 Ananthaswamy, "Nerve Probe Controls Cyborg Moth in Flight"; Christensen, "Military Plans Cyborg Sharks."

73 Allen, "The Future of Military Robotics Looks Like a Nature Documentary."

74 Bienaimé, "The US Navy's Combat Dolphins Are Serious Military Assets"; Cudworth and Hobden, "The Posthuman Way of War."

75 Parpinelli and Lopes, "New Inspirations in Swarm Intelligence," 1.

76 It is important to note that "military researchers" is not restricted here to active duty soldiers researching in the military. Instead, by military researchers I mean all of those researchers whose work is funded by the military and therefore can be taken and made top secret such that only the military can use it.

77 Harman, *Quentin Meillassoux*, 187.

78 Harman, *Quentin Meillassoux*, 185.

79 Parikka, *A Geology of Media*, 92.

80 Morton, "Hyperobjects and the End of Common Sense."

81 See chapter 4 for a detailed investigation of the mechanic ecology of war waste.

82 See Gitte du Plessis, "When Pathogens Determine the Territory: Toward a Concept of Non-Human Borders."

83 Vidal, "Iraqi Children Pay High Health Cost."

84 See Ballard, *War Fever*; Haldeman, *The Forever War*.

85 Jefferson, "The Science in Science Fiction."

86 Sloterdijk, *Spheres*, 3:119.

87 Nowell and Davidson, *Stone Tools*, 199.

88 Sloterdijk, *Spheres*, 3:343.

89 Thacker, *After Life*, 38–39.

90 Basalla, *The Evolution of Technology*, 135–64.

91 In Peter Watts's *Echopraxia*, he imagines a class of soldiers and mercenaries that choose to become automatons. In the story soldiers sign contracts to give up consciousness and be operated as cyborg drones in exchange for their families receiving large financial payments. See Watts, *Echopraxia*.

92 Coker, "Targeting in Context."

93 Bargu, *Starve and Immolate*, 272.

94 "ASELSAN Fire Support Automation System—AFSAS," accessed December 1, 2018, http://www.aselsan.com.tr/en-us/capabilities/command-control-communications-computer-and-intelligence-systems/fire-direction-systems/aselsan-fire-support-automation-system-afsas.

95 Chamayou, *A Theory of the Drone*, 127–29.

96 Chamayou, *A Theory of the Drone*, 46–48.

97 Review the section titled "Aphorisms for a New Realism" in the beginning of this book. See Cavell, *The Claim of Reason*, 434.

98 After some mild public concern, the DOD issued an order clarifying that weapons were not currently, autonomously killing targets. At first, press coverage of the order suggested that the DOD had "banned" this ability. However, closer investigation of the order suggests that the DOD merely claims regulatory power over autonomous killing. This means that the DOD will make the decision and is capable of deploying weapons that can kill without human oversight. In fact, what the order claimed to prevent is the unpreventable. The order claimed that the DOD would not allow machines to activate themselves for autonomous killing. See Gubrud, "DOD Directive on Autonomy in Weapon Systems."

99 Philosopher David Roden makes the point that actually posthuman posthumans will be so only insofar as we and they are no longer legible to one another. That is, true

speciation is marked materially and philosophically by the inability to make sense to one another. Roden, *Posthuman Life*, 76–78.

100 Harrison, *When Languages Die*, 55.

101 Viveiros de Castro, *The Relative Native*, 85.

102 Singleton, "Maximum Jailbreak."

103 Giedion, *Mechanization Takes Command*, 41–43.

104 A NASA-funded research project concluded that the confluence of intensive social inequality and natural limits to economic growth represents a "perfect storm" for conflict and the collapse of industrial civilization. See Motesharrei, Rivas, and Kalnay, "Human and Nature Dynamics (HANDY)."

105 Glen Coulthard provides an excellent corrective to Marx by extending the practices of primitive accumulation to the present and insisting that place-based and particularistic forms of political resistance are innovative rather than regressive. See Coulthard, *Red Skin, White Masks*.

106 "Dyson Sphere," Wikipedia, accessed December 1, 2018, https://en.wikipedia.org/wiki/Dyson_sphere.

8. APOCALYPSE AS A THEORY OF CHANGE

1 I am dubious of those who hope indigenous ways of life can be mirrored or scaled up to save the world from itself. The recent fascination with indigenous forms of life as policy-relevant research material seems to me obscene. To make indigenous politics into global politics seems to me the last-ditch effort to fully instrumentalize the sacred in hopes that without sacrifice or transformation of the existing distribution of power, we could somehow "discover" new modes of efficiency and sustainability in the surviving communities that were meant by the current geopolitical order to be exterminated. No thanks.

2 DeLanda, "Nonorganic Life," in *Zone*, vol. 6, *Incorporations*, 139.

3 In an act of defiance during the peak of George W. Bush's warmongering leading up to the release of the National Intelligence Report that revealed Iran was not pursuing nuclear weapons, Fallon publicly announced, "This constant drumbeat of conflict is not helpful and not useful. I expect that there will be no war, and that is what we ought to be working for. We ought to try to do our utmost to create different conditions." Barnett, "The Man between War and Peace."

4 Deleuze, *Cinema II*, 187.

5 Deleuze, *Difference and Repetition*, 53.

6 Connolly, *Neuropolitics*.

7 Bataille, *The Cradle of Humanity*, 175–76.

8 See West-Eberhard, *Developmental Plasticity and Evolution*; Jablonka et al., *Evolution in Four Dimensions*.

9 Gould, *The Structure of Evolutionary Theory*, 745–83.

10 Kant, "Idea for a Universal History."

11 Gould, *The Structure of Evolutionary Theory*, 487.

12 Gould, *The Structure of Evolutionary Theory*, 485.

13 David Christian and his predecessor William McNeill have devoted their careers to the "endless waltz of chaos and complexity" as it stretches from the big bang to the urbanized contemporary moment. The "Big History" movement has had a substantial impact on the historians' attention to ecology, geologic change, as well as the formative importance of "prehistory" in contemporary human life. Unfortunately, this attention to "deep time" is less common among political thinkers and remains limited in its impact. See McNeill, "Foreword."

14 Connolly and Grove, "Extinction Events and the Human Sciences."

15 Whitehead, *The Function of Reason*, 3.

16 Whitehead, *Nature and Life*, 33.

17 Connolly, "Freedom, Teleodynamism, Creativity," 64–65.

18 Development, Concepts and Doctrine Centre, "Global Strategic Trends," accessed August 4, 2014, https://www.gov.uk/government/publications/global-strategic-trends-out-to-2045.

19 Department of Defense, "Quadrennial Defense Review Report," February 2010, accessed July 31, 2014, http://www.defense.gov/qdr/images/QDR_as_of_12Feb10_1000.pdf.

20 Ahmed, "Pentagon Preparing for Mass Civil Breakdown." The time frame for catastrophic sea-level rise contracts continuously. The most recent models suggest that it is a problem we will face in our lifetime rather than a problem for the next generation. DeConto and Pollard, "Contribution of Antarctica to Past and Future Sea-Level Rise."

21 Dupuy, *The Mark of the Sacred.*

22 Benton, *When Life Nearly Died*, 8.

23 Deleuze, *Cinema II.*

24 Connolly, *The Fragility of Things*, 45.

25 Bataille, *Accursed Share*, 96.

26 Connolly, *The Fragility of Things*, 172, 167.

27 I think Catherine Malabou makes a similar turn in her work on destructive plasticity, which is a necessary corrective to the tendency of seeing all forms of becoming and transformation as somehow positive. See Grove, "Something Darkly This Way Comes."

28 Whitehead, *Science and Philosophy*, 118–19.

29 Garcia, *Form and Object*, 1.

30 Dole, "The Natural History of Oxygen."

31 Cairns-Smith, *Seven Clues to the Origin of Life*, 35–37.

32 Lane, *Oxygen*, 16–20.

33 Christian, *Maps of Time*, 111.

34 Kopp, Kirschvink, Hilburn, and Nash, "The Paleoproterozoic Snowball Earth."

35 "One and Only Earth."

36 Special issue of Geonature.

37 Gibling, qtd. in Fischetti, "Thanks to Plants."

38 Christian, *Maps of Time*, 121–27. Bilateralism seems a strange "advance" but imagine how different human life would be without a front and a back. We would not possess faces or the capacity to distinguish between engaging one another or turning our

backs on each other—such a simple thing for which social existence as we know it would be unthinkable. The creation of the nucleus is a critical juncture for which "mutualism" is a better explanation for creative advance than competition and selection. See Christian, *Maps of Time*, 112–16.

39 See Benton, *When Life Nearly Died*, 9–10.

40 Pääbo, *Neanderthal Man*. There is an exception to this story, which is the isolated island of Flores in Indonesia. The human inhabitants of the island of Flores may have shared their lifeworld with *Homo floresiensis* until as little as ten thousand years ago. Unlike the forgotten Neanderthals of Europe, the Flores inhabitants have extant oral histories of *Homo floresiensis*.

41 "Cro-Magnons somehow used their far superior technology, and their language skills or brains, to infect, kill, or displace the Neanderthals, leaving behind little or no evidence of hybridization between Neanderthals and Cro-Magnons." Diamond, *Guns, Germs, and Steel*, 39.

42 For an excellent critique of the naturalization of the state of nature as a state of war, see Sahlins, *The Western Illusion of Human Nature*.

43 Ferguson, "What Was Politics to the Denisovan?"

44 My more recent African and Amerindian genetic roots put me in the 82 percentile for Neanderthal DNA. According to Svante Pääbo, director of the Department of Genetics at the Max Planck Institute for Evolutionary Anthropology, there are "genetic models to date the admixture between Neanderthals and modern humans to sometime between 40,000 and 90,000 years ago." Pääbo, *Neanderthal Man*, 251.

45 Huerta-Sánchez et al., "Altitude Adaptation in Tibetans."

46 Weise, "30,000 Years Ago."

47 Stannard, *American Holocaust*, 11.

48 Lem, *Summa Technologiae*.

49 Bennett, "Earthling, Now and Forever?"

9. FREAKS, OR THE INCIPIENCE OF OTHER FORMS OF LIFE

1 Cavell, *The Claim of Reason*, 368–69.

2 Cavell, *The Claim of Reason*, 354.

3 Bataille, *Visions of Excess*, 55.

4 See Miller, "Freaks (1932)," and for the story about Fitzgerald, see Jefferson, "On Writers and Writing."

5 Chandler Swain, "Alternate Opening Crawl for 'Freaks,'" YouTube video, 2:26, January 5, 2014, https://youtu.be/ZPhBLF_Hy6w?t=2m6s.

6 James Vincent, "Twitter Taught Microsoft's AI Chatbot to Be a Racist Asshole in Less Than a Day," *The Verge*, March 24, 2016, https://www.theverge.com/2016/3/24/11297050/tay-microsoft-chatbot-racist.

7 Connolly, *A World of Becoming*, 152.

8 See William Connolly's chapter on Herculine Barbine, "Voices from the Whirlwind."

9 Bunge, "Do Computers Think?"; Herrick, "Mechanism and Organism"; Kantor, "Man and Machine in Science"; Kapp, "Living and Lifeless Machines"; Miles, "On the

Difference Between Man and Machines"; Northrop, Rignano's Hypothesis of a Vital Energy and the Prerequisites of a Sound Theory of Life"; Roberts, "Are We Machines? And What of It?"

10 Putnam, "Robots: Machines or Artificially Created Life?," 690.

11 Albritton, "Mere Robots and Others"; Clack, "Can a Machine Be Conscious?"; Gauld, "Could a Machine Perceive?"; Gunderson, "Robots, Consciousness and Programmed Behaviour"; Lucas, "Human and Machine Logic"; Puccetti, "On Thinking Machines and Feeling Machines"; Rorty, "Functionalism, Machines and Incorrigibility."

12 Thompson, "Can a Machine Be Conscious?," 41.

13 Putnam, "Robots," 691–92.

14 Putnam, "Robots," 691–92.

15 Agamben, *Homo Sacer*, 160–64; Foucault, *Society Must Be Defended*, 254–58; Foucault, *Security, Territory and Populations*, 3, 22–23.

16 Beginning in the late 1950s, MIT scientists in coordination with the U.S. government and Quaker Oats fed institutionalized children classified as "morons" radioactive oatmeal so that they could study the effects of radiation. The term was developed in the early twentieth century by Henry Goddard as part of the vocabulary for the eugenics movement. Those who have not yet died of cancer were awarded US$60,000 apiece. However, the survivors are still petitioning the state of Massachusetts to have the label "moron" removed from their permanent record. Many of the survivors continue to face discrimination because of this label. See Mehren, "Seeking Freedom from Label at Last."

17 NARA, Federal Register, Environmental Protection Agency, September 12, 2005, accessed January 27, 2006, http://www.epa.gov/fedrgstr/EPA-GENERAL/2005/September/Day-12/g18010.htm. See also Public Information and Records Integrity Branch (PIRIB), Office of Pesticide Programs, U.S. Environmental Protection Agency, Mail, Code: 7502C, 1200 Pennsylvania Ave., NW, Washington, DC, 20460-0001, Attention: Docket ID Number OPP-2003-0132.

18 I came to discover after writing this chapter that Sylvia Wynter has a term corresponding to somatic fundamentalism I also quite like: bio-scholasticism. See McKittrick, *Sylvia Wynter*, 20.

19 Canguilhem, *The Normal and the Pathological*, 87.

20 Mori, *The Buddha in the Robot*, 33–35.

21 "Scientists Study Robot-Human Interactions."

22 Mori, "The Uncanny Valley," 33.

23 Connolly, *Neuropolitics*, 86.

24 Connolly, *Neuropolitics*, 196.

25 Connolly, *Neuropolitics*, 60–61.

26 Howarth, "Ethos, Agonism, and Populism," 188.

27 Connolly, *Neuropolitics*, 61.

28 Connolly, *Neuropolitics*, 60–61; Howarth, "Ethos, Agonism, and Populism," 183, 184.

29 Connolly, *Neuropolitics*, 77.

30 Howarth, "Ethos, Agonism, and Populism," 187.

31 Howarth, "Ethos, Agonism, and Populism," 189.

32 Habermas, *The Future of Human Nature*, 35.
33 Habermas, *The Future of Human Nature*, 34, 64.
34 Habermas, *The Future of Human Nature*, 94.
35 Habermas, *The Future of Human Nature*, 39.
36 Habermas, *The Future of Human Nature*, 22.
37 DeLanda, *A Thousand Years of Nonlinear History*; Dennett, *Darwin's Dangerous Idea*.
38 "Sperm Made from Human Bone Marrow."
39 Connolly, *Pluralism*, 121, 126.
40 Connolly, *Neuropolitics*, 54.
41 Connolly, *Neuropolitics*, 162.
42 Connolly, *Neuropolitics*, 162.
43 Connolly, *World of Becoming*, 9.
44 Deleuze, *Negotiations*, 170–71.
45 Haraway, *Staying with the Trouble*, 145–46.
46 Recent research in evolutionary biology is attempting to revise or even scrap the concept of species after learning that many mutations and evolutionary shifts involved sex between two different species, what is termed "species jumping," which invalidates the very definition of species (i.e., two members of a population that can produce fertile offspring). See Owen, "Interspecies Sex."
47 Deleuze and Guattari, *A Thousand Plateaus*, 323.
48 Deleuze and Guattari, *A Thousand Plateaus*, 322.
49 Deleuze and Guattari, *A Thousand Plateaus*, 311.
50 Deleuze and Guattari, *A Thousand Plateaus*, 318.
51 Connolly, *Neuropolitics*, 162.
52 The refrain speaks coldly of the cutting edges of machines when they alter the arrangement of an assemblage. See Deleuze and Guattari, *A Thousand Plateaus*, 332–33.
53 Deleuze and Guattari, *A Thousand Plateaus*, 313.
54 Goodwin, *How the Leopard Changed Its Spots*, 206.
55 Goodwin, *How the Leopard Changed Its Spots*, 232.
56 Habermas, *The Future of Human Nature*, 94.
57 Habermas, *The Future of Human Nature*, 44.
58 Goodwin, *How the Leopard Changed Its Spots*, 2.
59 Cairns-Smith, *Genetics Takeover*, 79–80.
60 Cairns-Smith, *Genetics Takeover*, 81.
61 Cairns-Smith, *Genetics Takeover*, 4–5.
62 Connolly and Bennett, "Contesting Nature/Culture."
63 Bennett, "The Force of Things," 354–55.
64 Bennett and Connolly, "Contesting Nature/Culture," 160–61.
65 Connolly, *Capitalism and Christianity*, 73.
66 Deleuze and Guattari, *A Thousand Plateaus*, 329.
67 Foucault, *The Order of Things*, 387.
68 Connolly, *Neuropolitics*, 106.
69 Connolly, *Neuropolitics*, 104.

70 Pagan Kennedy, "Distrust That Particular Flavor," *New York Times*, January 13, 2012. https://www.nytimes.com/2012/01/15/books/review/distrust-that-particular-flavor-by-william-gibson-book-review.html?mtrref=www.google.com&gwh=2CDA400C034FAB2815DFDFC61E385D1E&gwt=pay.

71 Metz, "The Sadness and Beauty of Watching Google's AI Play Go."

CONCLUSION: RATIO FERITAS

1 Whitehead, *The Function of Reason*, 26.

2 Whitehead, *Nature and Life*, 31.

3 Whitehead, *The Function of Reason*, 23.

4 Connolly, "Freedom, Teleodynamism, Creativity," 63.

5 Connolly, *The Fragility of Things*, 75.

6 Whitehead, *The Function of Reason*, 26.

7 Whitehead, *The Function of Reason*, 89; Connolly, *A World of Becoming*, 167.

8 Connolly, "Immanence, Abundance, Democracy," 244.

9 Whitehead, *The Function of Reason*, 12.

10 Connolly, *The Fragility of Things*, 77.

11 Whitehead, *The Function of Reason*, 19.

12 Dark Mountain Project, "Uncivilisation." The deadening effect of anxiety cannot be overstated. In a compelling essay by Plan C, anxiety is nominated as the "affect" of our time. Rather than boredom or simple alienation, we live with a continuous and stultifying anxiousness. We are stuck between what has happened and what is about to happen and so overcome that we cannot attend to either the past or the future. See Plan C, "We Are All Very Anxious."

13 Smith, "It's the End of the World."

14 Smith, "It's the End of the World."

15 For a description of just how thoroughly indifferent the most powerful nation-states around the planet are to the current ecological catastrophe, see Hamilton, *Requiem for a Species*.

16 Benjamin Bratton asks a similar question about the efficacy of action, arguing that to intervene in our current predicament requires of the "artist/designer" to "speculate upon irreducibly complex interdependencies . . . [that] are really impossible to know (and yet nothing deserves more attention)." Bratton, "Some Trace Effects of the Post-Anthropocene."

17 Kingsnorth, "The Lie of the Land."

18 McKenzie Wark has laid a shot across the bow, demanding of the humanities and social sciences that they reconsider their concepts and conceptions of the world in the age of the Anthropocene before then turning the concepts against the "idea" of the Anthropocene. Wark, "Critical Theory after the Anthropocene."

19 Bennett, "Afterword."

20 Colebrooke, *Death of the PostHuman*, 1:114–15.

21 Dark Mountain Project, "Uncivilisation."

22 See Brassier, *Nihil Unbound*, 49.

23 Brahic, "Meet the Electric Life Forms."

24 I was particularly inspired by a line in Roy Scranton's editorial "Learning How to Die in the Anthropocene": "If we want to learn to live in the Anthropocene, we must first learn how to die."

25 Connolly, "Immanence, Abundance, Democracy," 244.

26 Viveiros de Castro, *The Relative Native*, 172.

27 Keller, *Apocalypse Now and Then.*

28 In a critique of Hegelian theories of history, Dipesh Chakrabarty is inspired by climate change to challenge his colleagues to consider a history "that escapes our capacity to experience the world." Chakrabarty, "The Climate of History," 222.

THE END

1 "Salt Pit," Wikipedia, accessed December 1, 2018, https://en.wikipedia.org/wiki/Salt_Pit.

2 Konikow, "Long-Term Groundwater Depletion in the United States."

3 Michele Guarnieri, "HiBot Amphibious Snake Robot," YouTube video, 2:41, December 18, 2009, https://www.youtube.com/watch?v=_5PplUmtEvA.

4 Babcock, "Surveillance Blimp Takes Flight over Baltimore."

5 Physicians for Social Responsibility, "Body Count."

6 Myers, "U.S. Wields Defter Weapon against Iraq."

7 Alnasrawi, "Iraq."

8 Human Rights Watch, "U.S. Using Cluster Munitions in Iraq"; "US Used White Phosphorus in Iraq."

9 Yacoub, "Iraq."

10 Rogers, "US Military Suicides in Charts."

11 Jamail, "Dogs Eating Bodies in Streets of Fallujah."

12 Senate Select Committee on Intelligence, "Committee Study of the Central Intelligence Agency's Detention and Interrogation Program."

13 Weiner, "Air Force Seeks Bush's Approval for Space Weapons Programs."

14 Carter, "Autonomy in Weapon Systems," Department of Defense Directive, No. 3000.09, November 21, 2012, updated May 8, 2017, https://www.esd.whs.mil/Portals/54/Documents/DD/issuances/dodd/300009p.pdf.

BIBLIOGRAPHY

Abu-Lughod, Janet L. *Before European Hegemony: The World System A.D. 1250–1350*. New York: Oxford University Press, 1991.

Adams, David Wallace. *Education for Extinction: American Indians and the Boarding School Experience, 1875–1928*. Lawrence: University Press of Kansas, 1995.

Adas, Michael. *Dominance by Design: Technological Imperatives and America's Civilizing Mission*. Cambridge, MA: Belknap Press of Harvard University Press, 2006.

Afghan Conflict Monitor. "Casualties from IEDs Skyrocket from 2009 to 2010." http://www.afghanconflictmonitor.org/2011/01/us-casualties-from-ieds-skyrocket-from-2009-to-2010.html.

Agamben, Giorgio. *Means without End: Notes on Politics*. Theory out of Bounds, vol. 20. Minneapolis: University of Minnesota Press, 2000.

Agamben, Giorgio. *Sovereign Power and Bare Life*. Homo Sacer 1. Stanford, CA: Stanford University Press, 1998.

Agamben, Giorgio. *The Use of Bodies*. Homo Sacer IV, 2. Translated by Adam Kotsko. Stanford, CA: Stanford University Press, 2015.

Ahmed, Nafeez. "Pentagon Preparing for Mass Civil Breakdown." *The Guardian*, June 12, 2014.

Ahuja, Neel. *Bioinsecurities: Disease Interventions, Empire, and the Government of Species*. Durham, NC: Duke University Press, 2016.

Albritton, R. "Mere Robots and Others: Comments." *The Journal of Philosophy* 61, no. 21 (1964): 691–94.

Alexander, Michelle. *The New Jim Crow: Mass Incarceration in the Age of Colorblindness*. Rev. ed. New York: New Press, 2012.

Allen, Gregory C. "The Future of Military Robotics Looks Like a Nature Documentary." *War on the Rocks*, July 21, 2017. https://warontherocks.com/2017/07/the-future-of-military-robotics-looks-like-a-nature-documentary/.

Alnasrawi, Abbas. "Iraq: Economic Sanctions and Consequences, 1990–2000." *Third World Quarterly* 22, no. 2 (2001): 205–18. doi:10.1080/01436590120037036.

Althusser, Louis. *Philosophy of the Encounter: Later Writings, 1978–87*. Edited by François Matheron and Olivier Corpet. Translated by G. M. Goshgarian. New York: Verso, 2006.

Ananthaswamy, Anil. "Nerve Probe Controls Cyborg Moth in Flight." *New Scientist*, February 8, 2012. https://www.newscientist.com/article/dn21431-nerve-probe-controls-cyborg-moth-in-flight/.

Apted, Michael, dir. *Incident at Oglala.* Santa Monica, CA: Artisan Home Entertainment, 2004. DVD.

Arendt, Hannah. *The Origins of Totalitarianism.* New ed. New York: Harcourt Brace Jovanovich, 1973.

Aron, Raymond. *The Century of Total War.* Westport, CT: Greenwood Press, 1981.

Aron, Raymond. *Peace and War: A Theory of International Relations.* New Brunswick, NJ: Transaction, 2003.

Arrighi, Giovanni. *The Long Twentieth Century: Money, Power, and the Origins of Our Times.* London: Verso, 2010.

Asafu-Adjaye, John, Linus Blomqvist, Stewart Brand, et al. "An Ecomodernist Manifesto." www.ecomodernism.org. 2015.

Ashby, W. Ross. *Design for a Brain: The Origin of Adaptive Behaviour.* London: Chapman and Hall, 1960.

Ashby, W. Ross. *An Introduction to Cybernetics.* New York: Wiley and Sons, 1965.

Atkinson, Rick. "IED Is Insurgents' Iconic Device." *Star-Telegram* (Fort Worth, TX), September 30, 2007.

Babcock, Stephen. "Surveillance Blimp Takes Flight over Baltimore." *Technically Baltimore,* January 6, 2015. http://technical.ly/baltimore/2015/01/06/surveillance-blimp-begins-float-baltimore/.

Bahrani, Zainab. *Rituals of War: The Body and Violence in Mesopotamia.* New York: Zone Books, 2008.

Balakrishnan, Gopal. "The Abolitionist—II." *New Left Review* 91 (January–February 2015): 69–100.

Balakrishnan, Gopal. "The Coming Contradiction." *New Left Review* 66 (November–December 2010): 31–53.

Baldé, C. P., V. Forti, V. Gray, R. Kuehr, P. Stegmann. *The Global E-waste Monitor.* Bonn: United Nations University (UNU), International Telecommunication Union (ITU), and International Solid Waste Association (ISWA), 2017.

Ballard, J. G., "Interview with Vale and Juno." *Re/Search* 8–9 (1984): 8–15.

Ballard, J. G. *War Fever.* New York: Farrar, Straus and Giroux, 2001.

Barakawi, Tarak. *Globalization and War.* Lanham, MD: Rowman and Littlefield, 2006.

Bargu, Banu. *Starve and Immolate: The Politics of Human Weapons.* New York: Columbia University Press, 2014.

Barker, Alec D. "Improvised Explosive Devices in Southern Afghanistan and Western Pakistan, 2002–2009." *Studies in Conflict and Terrorism* 34, no. 88 (2011): 600–620.

Barnett, Thomas P. M. "The Man between War and Peace." *Esquire,* March 11, 2008.

Basalla, George. *The Evolution of Technology.* New York: Cambridge University Press, 1988.

Bashford, Alison. *Global Population: History, Geopolitics, and Life on Earth.* New York: Columbia University Press, 2014.

Bataille, Georges. *The Accursed Share: An Essay on General Economy.* New York: Zone Books, 1988.

Bataille, Georges. *The Cradle of Humanity: Prehistoric Art and Culture.* Edited by Stuart Kendall. Translated by Stuart Kendall and Michelle Kendall. New York : Zone, 2009.

Bataille, Georges. *The Unfinished System of Nonknowledge*. Edited by Stuart Kendall. Translated by Stuart Kendall and Michelle Kendall. Minneapolis: University of Minnesota Press, 2001.

Bataille, Georges. *Visions of Excess: Selected Writings, 1927–1939*. Edited by Allan Stoekl. Minneapolis: University of Minnesota Press, 1985.

Bateson, Gregory. *Steps to an Ecology of Mind*. San Francisco: Chandler, 1972.

Beer, Stafford. *Brain of the Firm: A Development in Management Cybernetics*. New York: Herder and Herder, 1972.

Beer, Stafford. *Decision and Control: The Meaning of Operational Research and Management Cybernetics*. New York: John Wiley, 1966.

Bennett, Jane. "Afterword: Earthling, Now and Forever?" In *Making the Geologic Now: Responses to Material Conditions of Contemporary Life*, edited by Elizabeth Ellsworth and Jamie Kruse, 244–46. New York: Punctum Books, 2012.

Bennett, Jane. "The Force of Things: Steps Toward an Ecology of Matter." *PoliticalTheory* 32, no. 3 (2004): 347–72.

Bennett, Jane. *Vibrant Matter: A Political Ecology of Things*. Princeton, NJ: Princeton University Press, 2009.

Bennett, Jane, and William E. Connolly. "Contesting Nature/Culture: The Creative Character of Thinking." *Journal of Nietzsche Studies*, no. 24 (2002): 148–63. http://www.jstor.org/stable/20717795.

Benton, M. J. *When Life Nearly Died: The Greatest Mass Extinction of All Time*. New York: Thames and Hudson, 2003.

Bergson, Henri. *Creative Evolution*. New York: Dover, 2010.

Bergson, Henri. *The Creative Mind: An Introduction to Metaphysics*. New York: Dover, 2013.

Berlant, Lauren Gail. *Cruel Optimism*. Durham, NC: Duke University Press, 2011.

Berseus, Olle, Tor Hervig, and Jerard Seghatchian. "Military Walking Blood Bank and the Civilian Blood Service." *Transfusion and Apheresis Science* 46 (2012): 341–42.

Besteman, Catherine Lowe, Network of Concerned Anthropologists, eds. *The Counter-Counterinsurgency Manual: Notes on Demilitarizing American Society*. Chicago: Prickly Paradigm Press, 2009.

Biello, David. "Mass Deaths in Americas Start New CO_2 Epoch." *Scientific American*, March 11, 2015. http://www.scientificamerican.com/article/mass-deaths-in-americas-start-new-co2-epoch/.

Biello, David. "Stratospheric Pollution Helps Slow Global Warming." *Scientific American*, July 22, 2011. http://www.scientificamerican.com/article/stratospheric-pollution-helps-slow-global-warming/.

Bienaimé, Pierre. "The US Navy's Combat Dolphins Are Serious Military Assets." *Business Insider*, March 12, 2015. http://www.businessinsider.com/the-us-navys-combat-dolphins-are-serious-military-assets-2015-3.

Birtle, Andrew J. *U.S. Army Counterinsurgency and Contingency Operations Doctrine, 1860–1941*. St. John's Press, [1998] 2016.

Bitzinger, Richard A. "Third Offset Strategy and Chinese A2/AD-Capabilities." Center for a New American Security, May 2016. https://s3.amazonaws.com/files.cnas.org/documents/CNAS_Third-Offset-Strategy-and-Chinese-A2-AD-Capabilities_FINAL.pdf.

Blackhawk, Ned. *Violence over the Land: Indians and Empires in the Early American West.* Cambridge, MA: Harvard University Press, 2008.
Blight, James G. *The Fog of War: Lessons from the Life of Robert S. McNamara.* New York: Rowman and Littlefield, 2005.
"'Body Bombs' Are a Good Sign, DHS Insider Claims." *Wired,* July 2, 2011. https://www.wired.com/2011/07/body-bombs-are-a-good-sign-dhs-insider-claims/.
Bogost, Ian. *Alien Phenomenology, or What It's Like to Be a Thing.* Minneapolis: University of Minnesota Press, 2012.
Bousquet, Antoine. "All Your Brain Are Belong to Us: Neuroscience Goes to War." *The Disorder of Things* (blog), February 29, 2012. https://thedisorderofthings.com/2012/02/19/all-your-brain-are-belong-to-us/.
Bousquet, Antoine J. *The Scientific Way of Warfare Order and Chaos on the Battlefields of Modernity.* New York: Columbia University Press, 2009.
Bradbury, Ray. *Fahrenheit 451.* New York: Simon and Schuster, 2013.
Brahic, Catherine. "Meet the Electric Life Forms That Live on Pure Energy." *New Scientist,* July 16, 2014. http://www.newscientist.com/article/dn25894-meet-the-electric-life-forms-that-live-on-pure-energy.html#.U_fsJEgRbxs.
Brand, Stewart, ed. *The Essential Whole Earth Catalog: Access to Tools and Ideas.* Garden City, NY: Doubleday, 1986.
Brander, James A., Qianqian Du, and Thomas F. Hellmann. "The Effects of Government-Sponsored Venture Capital: International Evidence." *Review of Finance* 19, no. 2 (2015): 571–618. doi:10.1093/rof/rfu009.
Brassier, Ray. *Nihil Unbound: Enlightenment and Extinction.* Basingstoke, UK: Palgrave Macmillan, 2007.
Bratton, Benjamin H. *Dispute Plan to Prevent Future Luxury Constitution.* Berlin: Sternberg Press, 2015.
Bratton, Benjamin. "Some Trace Effects of the Post-Anthropocene: On Accelerationist Geopolitical Aesthetics." *E-Flux,* no. 46 (June 2013). http://www.e-flux.com/journal/some-trace-effects-of-the-post-anthropocene-on-accelerationist-geopolitical-aesthetics/.
Braudel, Fernand. *A History of Civilizations.* Translated by Richard Mayne. New York: Penguin Books, 1995.
Braudel, Fernand. *The Mediterranean and the Mediterranean World in the Age of Philip II.* Vol. 2. Berkeley: University of California Press, 1996.
Briscoe, Erica, Lora Weiss, Elizabeth Whitaker, and Ethan Trewhitt. "A System-Level Understanding of Insurgent Involvement in Improvised Explosive Devices Activities." *Systems Research and Behavioral Science* 28, no. 4 (2011): 391–400.
Brown, Alleen, Will Parrish, and Alice Speri. "Leaked Documents Reveal Counterterrorism Tactics Used at Standing Rock to 'Defeat Pipeline Insurgencies.'" *The Intercept,* May 27, 2017. https://theintercept.com/2017/05/27/leaked-documents-reveal-security-firms-counterterrorism-tactics-at-standing-rock-to-defeat-pipeline-insurgencies/.
Brown, Wendy. *Walled States, Waning Sovereignty.* New York: Zone Books, 2014.
Bryant, Levi. "Ethics and Politics: What Are You Asking?" *Larval Subjects* (blog), May 29, 2012. http://larvalsubjects.wordpress.com/2012/05/29/ethics-and-politics-what-are-you-asking/.

Bunge, M. "Do Computers Think?" *The British Journal for the Philosophy of Science* 7, no. 26 (1956): 139–48.

Burroughs, William S. "The Limits of Control." *Semiotexte* 3, no. 2 (1978): 38.

Byrd, Jodi A. *The Transit of Empire: Indigenous Critiques of Colonialism*. Minneapolis: University of Minnesota Press, 2011.

Caesar, Julius. *The Conquest of Gaul*. New York: Penguin Classics, 1982.

Cairns-Smith, A. G. *Genetic Takeover and the Mineral Origins of Life*. Cambridge: Cambridge University Press. 1982.

Cairns-Smith, A. G. *Seven Clues to the Origin of Life: A Scientific Detective Story*. New York: Cambridge University Press, 1990.

Campbell, Joseph Keim, Michael O'Rourke, and Matthew H. Slater, eds. *Carving Nature at Its Joints: Natural Kinds in Metaphysics and Science*. Cambridge, MA: MIT Press, 2011.

Canguilhem, Georges. "The Living and Its Milieu." Translated by John Savage. *Grey Room* 3 (March 2001): 7–31. doi:10.1162/152638101300138521.

Canguilhem, Georges. *The Normal and the Pathological*. New York: Zone Books, 1989.

Carr, E. H. *What Is History?* New York: Vintage, 1967.

Carter, Ashton B., John D. Steinbruner, Charles A. Zraket, Brookings Institution, and John F. Kennedy School of Government, eds. *Managing Nuclear Operations*. Washington, DC: Brookings Institution, 1987.

Cavell, Stanley. *The Claim of Reason: Wittgenstein, Skepticism, Morality, and Tragedy*. New York: Oxford University Press, 1999.

Cavell, Stanley. *In Quest of the Ordinary: Lines of Skepticism and Romanticism*. Chicago: University of Chicago Press, 1994.

Chakrabarty, Dipesh. "The Climate of History: Four Theses." *Critical Inquiry* 35, no. 2 (2009): 197–222.

Chamayou, Grégoire. *A Theory of the Drone*. Translated by Janet Lloyd. New York: New Press, 2015.

Chivers, C. J. *The Gun*. New York: Simon and Schuster, 2011.

Choi, Charles Q. "Brain Researchers Can Detect Who We Are Thinking About." *Scientific American*, March 14, 2013. https://www.scientificamerican.com/article/brain-researchers-can-detect-who-we-are-thinking-about/.

Chow, Rey. *The Age of the World Target: Self-Referentiality in War, Theory, and Comparative Work*. Durham, NC: Duke University Press, 2006.

Christensen, Bill. "Military Plans Cyborg Sharks." *Live Science*, March 7, 2006. https://www.livescience.com/603-military-plans-cyborg-sharks.html.

Christian, David. *Maps of Time: An Introduction to Big History*. Berkeley: University of California Press, 2011.

Cioran, E. M. *A Short History of Decay*. New York: Penguin, 2018.

Clack, R. J. "Can a Machine Be Conscious? Discussion of Dennis Thompson." *The British Journal for the Philosophy of Science* 17, no. 3 (1966): 232–34.

Clarke, Bruce, and Mark B. N. Hansen, eds. *Emergence and Embodiment: New Essays on Second-Order Systems Theory*. Durham, NC: Duke University Press, 2009.

Clausewitz, Carl von. *On War*. Edited and translated by Michael Howard and Peter Paret. Princeton, NJ: Princeton University Press, 1989.

Coates, Ta-Nehisi. *Between the World and Me.* New York: Spiegel and Grau, 2015.
Coker, Christopher. “Targeting in Context.” In *Targeting: The Challenges of Modern Warfare,* edited by Paul A. L. Ducheine, Michael N. Schmitt, and Frans P. B. Osinga, 9–25. The Hague: T. M. C. Asser Press, 2016.
Colebrook, Claire. *Death of the PostHuman: Essays on Extinction,* Vol. 1. London: Open Humanities Press, 2014.
Coll, Steve. “We Buried the Disgraceful Truth.” *New York Review of Books,* June 23, 2016. https://www.nybooks.com/articles/2016/06/23/we-buried-disgraceful-truth/.
Connolly, William E. *Capitalism and Christianity, American Style.* Durham, NC: Duke University Press, 2008.
Connolly, William E. *The Ethos of Pluralization.* Minneapolis: University of Minnesota Press, 1995.
Connolly, William E. “Evangelical-Capitalist Resonance Machine.” *Political Theory* 33, no. 6 (2005): 869–86.
Connolly, William E. *Facing the Planetary: Entangled Humanism and the Politics of Swarming.* Durham, NC: Duke University Press, 2017.
Connolly, William E. *The Fragility of Things: Self-Organizing Processes, Neoliberal Fantasies, and Democratic Activism.* Durham, NC: Duke University Press, 2013.
Connolly, William E. “Freedom, Teleodynamism, Creativity.” *Foucault Studies,* no. 17 (April 2014): 60–75.
Connolly, William E. *Identity/Difference: Democratic Negotiations of Political Paradox.* Expanded ed. Minneapolis: University of Minnesota Press, 2002.
Connolly, William E. “Immanence, Abundance, Democracy.” In *Radical Democracy: Politics between Abundance and Lack,* edited by Lars Tonder and Lasse Thomassen, 239–55. Manchester, UK: Manchester University Press, 2005.
Connolly, William E. *Neuropolitics: Thinking, Culture, Speed.* Minneapolis: University of Minnesota Press, 2002.
Connolly, William E. *Pluralism.* Durham, NC: Duke University Press, 2005.
Connolly, William E. “The Return of the Big Lie.” *The Contemporary Condition* (blog), December 3, 2011. http://contemporarycondition.blogspot.com/2011/12/return-of-big-lie.html.
Connolly, William E. “Voices from the Whirlwind.” In *In the Nature of Things: Language, Politics, and the Environment,* edited by Jane Bennett and William Chaloupka, 197–225. Minneapolis: University of Minnesota Press, 1993.
Connolly, William E. *A World of Becoming.* Durham, NC: Duke University Press, 2010.
Connolly, William E., and Jairus Victor Grove. “Extinction Events and the Human Sciences.” *The Contemporary Condition* (blog), July 3, 2014. http://contemporarycondition.blogspot.com/2014/07/extinction-events-and-human-sciences.html.
Conway, Flo, and Jim Siegelman. *Dark Hero of the Information Age: In Search of Norbert Wiener, the Father of Cybernetics.* New York: Basic Books, 2006.
Coulthard, Glen Sean. *Red Skin, White Masks: Rejecting the Colonial Politics of Recognition.* Minneapolis: University of Minnesota Press, 2014.
Couppis, Maria H., and Craig H. Kennedy. “The Rewarding Effect of Aggression Is Reduced by Nucleus Accumbens Dopamine Receptor Antagonism in Mice.” *Psychopharmacology* 197, no. 3 (April 2008): 449–56. https://doi.org/10.1007/s00213-007-1054-y.

Coward, Martin. *Urbicide.* New York: Routledge 2007.
Crampton, Jeremy W. "Collect It All: National Security, Big Data and Governance." *GeoJournal* 80, no. 4 (2015): 519–31. doi:10.1007/s10708-014-9598-y.
Croll, Mike. *The History of Landmines.* New York: Lee Cooper, 1998.
Crosby, Alfred W. *Ecological Imperialism: The Biological Expansion of Europe, 900–1900.* 2nd ed. New York: Cambridge University Press, 2004.
Crosby, Kevin. "José Delgado, Implants, and Electromagnetic Mind Control." YouTube video, 1:07, January 2, 2013, https://www.youtube.com/watch?v=23pXqY3X6c8.
Crutzen, Paul J. "Estimates of Possible Variations in Total Ozone Due to Natural Causes and Human Activities." *Ambio* 3, no. 6 (1974): 201–10.
Crutzen, Paul J., and John W. Birks. "The Atmosphere after a Nuclear War: Twilight at Noon." *Ambio* 11, no. 2–3 (1982): 114–25.
Cudworth, Erika, and Steve Hobden. "The Posthuman Way of War." *Security Dialogue* 46, no. 6 (2015): 513–29. doi:10.1177/0967010615596499.
Dark Mountain Project. "Uncivilisation: The Dark Mountain Manifesto." 2009. https://dark-mountain.net/about/manifesto/.
Davis, Mike. *Planet of Slums.* New York: Verso, 2006.
DeConto, Robert M., and David Pollard. "Contribution of Antarctica to Past and Future Sea-Level Rise." *Nature* 531, no. 7596 (2016): 591–97. doi:10.1038/nature17145.
DeLanda, Manuel. *Intensive Science and Virtual Philosophy.* New York: Continuum, 2005.
DeLanda, Manuel. *A Thousand Years of Nonlinear History.* Cambridge: Zone Books, 2000.
DeLanda, Manuel. *War in the Age of Intelligent Machines.* New York: Zone Books, 2003.
de Latil, Pierre. *La pensee artificielle.* Paris: Gallimard, [1953] 1956.
Delbrück, Hans. *The History of the Art of War.* 4 vols. Translated by Walter J. Renfroe Jr. Lincoln: University of Nebraska Press, 1975.
Delbrück, Hans. *Warfare in Antiquity.* Lincoln: Bison Books, 1990.
Deleuze, Gilles. *Bergsonism.* Cambridge: Zone Books, 1990.
Deleuze, Gilles. *Difference and Repetition.* New York: Columbia University Press, 1994.
Deleuze, Gilles. *Foucault.* Translated by Seán Hand. Minneapolis: University of Minnesota, Press, 1988.
Deleuze, Gilles. "Mediators." In *Zone 6: Incorporations*, 280–295. New York: Zone Books, 1992.
Deleuze, Gilles. "Postscript on Control Societies." In Gilles Deleuze, *Negotiations, 1972–1990*, 177–82. New York: Columbia University Press, 1995.
Deleuze, Gilles. *Pure Immanence: Essays on a Life.* 2nd ed. New York: Zone, 2005.
Deleuze, Gilles, and Félix Guattari. *Anti-Oedipus: Capitalism and Schizophrenia.* Minneapolis: University of Minnesota Press, 1983.
Deleuze, Gilles, and Félix Guattari. *A Thousand Plateaus: Capitalism and Schizophrenia.* Minneapolis: University of Minnesota Press, 1987.
Deleuze, Gilles, and Félix Guattari. *What Is Philosophy?* New York: Columbia University Press, 2003.

Deleuze, Gilles, and Martin Joughin. *Negotiations: 1972–1990.* New York: Columbia University Press, 1997.
Delgado, José Manuel Rodríguez. *Physical Control of the Mind: Toward a Psychocivilized Society.* New York: Harper and Row, 1971.
DeLillo, Don. *Point Omega: A Novel.* New York: Scribner, 2010.
DeLillo, Don. *Zero K: A Novel,* New York: Scribner, 2017.
Demers, Jason. "An American Excursion: Deleuze and Guattari from New York to Chicago." *Theory and Event* 14, no. 1 (2011).
Dennett, D. *Darwin's Dangerous Idea: Evolution and the Meaning of Life.* New York: Simon and Schuster, 1986.
Department of the Army. *Planning for Health Service Support.* Field Manual 8-55. Washington, DC: Department of the Army, 1994. https://www.globalsecurity.org/military/library/policy/army/fm/8-55/index.html.
Der Derian, James. *Virtuous War: Mapping the Military-Industrial-Media-Entertainment-Network.* New York: Routledge, 2009.
Derrida, Jacques. *Cinders.* Translated by Ned Lukacher. Minneapolis: University of Minnesota Press, 2014.
Derrida, Jacques, Catherine Porter, and Philip Lewis. "No Apocalypse, Not Now (Full Speed Ahead, Seven Missiles, Seven Missives)." *Diacritics* 14, no. 2 (1984): 20. https://doi.org/10.2307/464756.
Descola, Philippe, and Marshall David Sahlins. *Beyond Nature and Culture.* Translated by Janet Lloyd. Chicago: University of Chicago Press, 2014.
Deutsch, Karl W. *The Nerves of Government: Models of Political Communication and Control.* New York: Free Press, 1963.
Deutsch, Karl. "The Study of Nation-Building." Foreword to *Nation Building in Comparative Contexts,* edited by Karl W. Deutsch and William J. Foltz, 1–3. New Brunswick, NJ: AldineTransaction, [1966] 2010.
Diamond, Jared M. *Guns, Germs, and Steel: The Fates of Human Societies.* 20th Anniversary edition. New York: W. W. Norton, 2017.
Dick, Philip K. *The Three Stigmata of Palmer Eldritch.* Boston: Mariner Books, 2011.
Dillon, Michael, and Julian Reid. *The Liberal Way of War: Killing to Make Life Live.* New York: Routledge, 2009.
Dole, M. "The Natural History of Oxygen." *Journal of General Physiology* 49, no. 1 (1965): 5–27. doi:10.1085/jgp.49.1.5.
Du Bois, W. E. B. *Black Reconstruction in America, 1860–1880.* New York: Free Press, 1998.
Du Bois, W. E. B. *Color and Democracy: Colonies and Peace.* Millwood, NY: Kraus-Thomson Organization, 1975.
Dunbar-Ortiz, Roxanne. *An Indigenous Peoples' History of the United States.* Boston: Beacon, 2014.
Dupuy, Jean Pierre. *The Mark of the Sacred.* Translated by M. B. DeBevoise. Stanford, CA: Stanford University Press, 2013.
Dupuy, Jean-Pierre. *On the Origins of Cognitive Science: The Mechanization of the Mind.* Cambridge, MA: MIT Press, 2009.

Dyson, George. *Darwin among the Machines: The Evolution of Global Intelligence.* New York: Basic Books, 2012.

Ferguson, K. "What Was Politics to the Denisovan?" *Political Theory* 42, no. 2 (2014): 167–87. doi:10.1177/0090591713506714.

Fischetti, Mark. "Thanks to Plants, We Will Never Find a Planet Like Earth." *Scientific American*, February 1, 2012. https://www.scientificamerican.com/article/plants-created-earth-landscapel/.

Flank, Lenny, Jr. *At the Edge of the Abyss: A Declassified Documentary History of the Cuban Missile Crisis.* St. Petersburg, FL: Red and Black, 2010.

Fleming, James Rodger. *Fixing the Sky: The Checkered History of Weather and Climate Control.* New York: Columbia University Press, 2010.

Forrow, Lachlan, et al. "Accidental Nuclear War: A Post–Cold War Assessment." *New England Journal of Medicine* (1998): 1326–32.

Foster, John Bellamy. "Marx and the Rift in the Universal Metabolism of Nature." *Monthly Review*, December 1, 2013. https://monthlyreview.org/2013/12/01/marx-rift-universal-metabolism-nature/.

Foucault, Michel. *The Birth of Biopolitics: Lectures at the Collège de France, 1978–79.* Edited by Michel Senellart. Translated by Graham Burchell. New York: Palgrave Macmillan, 2008.

Foucault, Michel. *The History of Sexuality.* Vol. 1, *An Introduction.* New York: Vintage, 1990.

Foucault, Michel. *The Order of Things: An Archaeology of the Human Sciences.* New York: Vintage Books, 1994.

Foucault, Michel. *Security, Territory, Population: Lectures at the Collège de France 1977–78.* Edited by Michel Senellart. Translated by Graham Burchell. New York: Palgrave Macmillan, 2009.

Foucault, Michel. *Society Must Be Defended: Lectures at the Collège de France, 1975–76.* Edited by Mauro Bertani and Alessandro Fontana. Translated by David Macey. New York: Picador, 2003.

Fraser, Ronald. *Napoleon's Cursed War: Spanish Popular Resistance in the Peninsular War, 1808–1814.* New York: Verso, 2008.

Freedberg, Sydney J., Jr. "Centaur Army: Bob Work, Robotics, and the Third Offset Strategy." *Breaking Defense*, November 9, 2015. https://breakingdefense.com/2015/11/centaur-army-bob-work-robotics-the-third-offset-strategy/.

Freedberg, Sydney J., Jr. "Tiny Drones Win over Army Grunts; Big Bots? Not So Much." *Breaking Defense*, November 9, 2015. https://breakingdefense.com/2016/08/palmtop-drones-win-over-army-infantry-big-bots-not-so-much/.

Freedberg, Sydney J., Jr. "War without Fear: DepSecDef Work on How AI Changes Conflict." *Breaking Defense*, May 31, 2017. http://breakingdefense.com/2017/05/killer-robots-arent-the-problem-its-unpredictable-ai/.

Frey, Carl Benedikt, and Michael A. Osborne. "The Future of Employment: How Susceptible Are Jobs to Computerisation?" Working paper, September 17, 2013. http://www.oxfordmartin.ox.ac.uk/downloads/academic/future-of-employment.pdf.

Fuller, J. F. C. *Armament and History.* New York: De Capo, 1998.

Fuller, R. Buckminster. *Operating Manual for Spaceship Earth.* New York: Pocket Books, 1973.
Fuller, Richard Buckminster, and Jaime Snyder. *Utopia or Oblivion: The Prospects for Humanity.* New ed. Baden: L. Müller, 2008.
Gabrys, Jennifer. *Program Earth: Environmental Sensing Technology and the Making of a Computational Planet.* Minneapolis: University of Minnesota Press, 2016.
Gallagher, Matt. *Kaboom: Embracing the Suck in a Savage Little War.* Cambridge, MA: Da Capo, 2010.
Galli, Carlo, and Adam Sitze. *Political Spaces and Global War.* Translated by Elisabeth Fay. Minneapolis: University of Minnesota Press, 2010.
Galloway, Alexander. "Black Box, Black Bloc." In *Communization and Its Discontents: Contestation, Critique, and Contemporary Struggles.*, ed. Benjamin Noys, 237–52. Wivenhoe, UK: Minor Compositions, 2012.
Galloway, Alexander R. *Protocol: How Control Exists after Decentralization.* Cambridge, MA: MIT Press, 2004.
Galloway, Alexander R., and Eugene Thacker. *The Exploit: A Theory of Networks.* Minneapolis: University of Minnesota Press, 2007.
Garcia, Tristan. *Form and Object: A Treatise on Things.* Edinburgh: Edinburgh University Press, 2014.
Gauld, A. "Could a Machine Perceive?" *The British Journal for the Philosophy of Science* 17, no. 1 (1966): 44–58.
Geggus, David Patrick. *The Haitian Revolution: A Documentary History.* Indianapolis: Hackett, 2014.
Giedion, S. *Mechanization Takes Command: A Contribution to Anonymous History.* Minneapolis: University of Minnesota Press, 2013.
Goldberg-Hiller, J., and N. K. Silva. "Sharks and Pigs: Animating Hawaiian Sovereignty against the Anthropological Machine." *South Atlantic Quarterly* 110, no. 2 (April 1, 2011): 429–46. https://doi.org/10.1215/00382876-1162525.
Goldman, Emily, and Leslie Eliason. *The Diffusion of Military Technology and Ideas.* Palo Alto, CA: Stanford University Press, 2003.
Goodwin, B. *How the Leopard Changed Its Spots: The Evolution of Complexity.* Princeton, NJ: Princeton University Press, 1994.
Gould, Stephen Jay. *The Structure of Evolutionary Theory.* Cambridge, MA: Belknap Press of Harvard University Press, 2002.
Greenberg, Andy. "How a 'Deviant' Philosopher Built Palantir, a CIA-Funded Data-Mining Juggernaut." *Forbes*, September 2, 2013. https://www.forbes.com/sites/andygreenberg/2013/08/14/agent-of-intelligence-how-a-deviant-philosopher-built-palantir-a-cia-funded-data-mining-juggernaut/#7ca5bd1d7785.
Greenhalgh, Susan, and Edwin A. Winckler. *Governing China's Population: From Leninist to Neoliberal Biopolitics.* Stanford, CA: Stanford University Press, 2005.
Greenpeace. "Where Does the E-Waste Go?" http://www.greenpeace.org/usa/en/campaigns/toxics/hi-tech-highly-toxic/e-waste/.
Greenwald, Glenn. *No Place to Hide: Edward Snowden, the NSA, and the U.S. Surveillance State.* New York: Picador, 2015.

Grossman, Dave, and Loren W. Christensen. *On Combat: The Psychology and Physiology of Deadly Conflict in War and in Peace*. 3rd ed., [Nachdr.], 4. Nachdruck. Millstadt, IL: Warrior Science, 2008.

Grove, Jairus. "Something Darkly This Way Comes." In *Plastic Materialities: Politics, Legality, and Metamorphosis in the Work of Catherine Malabou*, edited by Brenna Bhandar and Jonathan Goldberg-Hiller, 233–63. Durham, NC: Duke University Press, 2015.

Grove, Jairus Victor. *Target Practice: Automated Killing and the Deformation of American Sovereignty*. Lanham, MD: Rowman and Littlefield, forthcoming.

Grove, Nicole Sunday. "The Cartographic Ambiguities of HarassMap: Crowdmapping Security and Sexual Violence in Egypt." Edited by Anthony Amicelle, Claudia Aradau, and Julien Jeandesboz. *Security Dialogue* 46, no. 4 (August 2015): 345–64. https://doi.org/10.1177/0967010615583039.

Grovogui, Siba. "Mind, Body, and Gut!" In *Decolonizing International Relations Theory*, edited by Branwen Gruffydd Jones. Lanham, MD: Rowman and Littlefield, 2006.

Grovogui, Siba N'Zatioula. *Beyond Eurocentrism and Anarchy: Memories of International Order and Institutions*. New York: Palgrave Macmillan, 2006.

Guattari, Félix. *Chaosmosis: An Ethico-Aesthetic Paradigm*. Bloomington: Indiana University Press, 1995.

Guattari, Félix. *The Machinic Unconscious: Essays in Schizoanalysis*. Los Angeles: Semiotext(e), 2011.

Guattari, Félix. *Schizoanalytic Cartographies*. Translated by Andrew Goffey. New York: Bloomsbury, 2013.

Guattari, Félix. *The Three Ecologies*. London: Bloomsbury Academic, 2014.

Gubrud, Mark. "DOD Directive on Autonomy in Weapon Systems." ICRAC, November 27, 2012. https://www.icrac.net/dod-directive-on-autonomy-in-weapon-systems/.

Gunderson, K. "Robots, Consciousness and Programmed Behaviour." *The British Journal for the Philosophy of Science* 19 (1968): 109–22.

Habermas, Jürgen. *The Future of Human Nature*. Cambridge: Polity Press, 2006.

Halberstam, Judith [Jack]. *The Queer Art of Failure*. Durham, NC: Duke University Press, 2011.

Haldeman, Joe W. *The Forever War*. New York: Thomas Dunne Books, 2009.

Halpern, Orit. *Beautiful Data: A History of Vision and Reason since 1945*. Durham, NC: Duke University Press, 2014.

Hamilton, Clive. *Earthmasters: The Dawn of the Age of Climate Engineering*. New Haven, CT: Yale University Press, 2013.

Hamilton, Clive. *Requiem for a Species: Why We Resist the Truth about Climate Change*. Crows Nest, NSW: Allen and Unwin, 2010.

Haraway, Donna Jeanne. *Simians, Cyborgs, and Women: The Reinvention of Nature*. London: Free Association Books, 1998.

Haraway, Donna J. *Staying with the Trouble: Making Kin in the Chthulucene*. Durham, NC: Duke University Press, 2016.

Harman, Graham. *Quentin Meillassoux: Philosophy in the Making*. Edinburgh: Edinburgh University Press, 2011.

Harney, Stefano, and Fred Moten. *The Undercommons: Fugitive Planning and Black Study*. New York: Minor Compositions, 2013. http://www.minorcompositions.info/wp-content/uploads/2013/04/undercommons-web.pdf.

Harrison, K. David. *When Languages Die: The Extinction of the World's Languages and the Erosion of Human Knowledge*. Oxford: Oxford University Press, 2008.

Harrison, Neil E., ed. *Complexity in World Politics: Concepts and Methods of a New Paradigm*. Albany: State University of New York Press, 2006.

Harvey, David. *Cosmopolitanism and the Geographies of Freedom*. New York: Columbia University Press, 2009.

Haselkorn, Avigdor. "Iraq's Bio-Warfare Option: Last Resort, Preemption, or a Blackmail Weapon?" *Biosecurity and Bioterrorism: Biodefense Strategy, Practice, and Science* 1, no. 1 (2003): 19–26.

Headrick, Daniel R. *Power over Peoples: Technology, Environments, and Western Imperialism, 1400 to the Present*. Princeton, NJ: Princeton University Press, 2012.

Heckenberger, Michael J., J. Christian Russell, Carlos Fausto, Joshua R. Toney, Morgan J. Schmidt, Edithe Pereira, Bruna Franchetto, and Afukaka Kuikuro. "Pre-Columbian Urbanism, Anthropogenic Landscapes, and the Future of the Amazon." *Science* 321, no. 5893 (2008): 1214–17. doi:10.1126/science.1159769.

Hedley-Whyte, John, and Debra R. Milamed. "Blood and War." *Ulster Medical Journal* 70 (2010): 125–34.

Heidegger, Martin. *The Question Concerning Technology and Other Essays*. Translated by William Lovitt. New York: Harper and Row, 1996.

Hernández, Anabel, Iain Bruce, and Lorna Scott Fox. *Narcoland: The Mexican Drug Lords and Their Godfathers*. Rev. ed. New York: Verso, 2013.

Herrick, C. J. "Mechanism and Organism." *The Journal of Philosophy* 26, no. 22 (1929): 589–97.

Hess, J. R., and M. J. G. Thomas. "Blood Use in War and Disaster: Lessons from the Past Century." *Transfusion* 43, no. 11 (2003): 1622–33.

Higginbotham, Adam. "U.S. Military Learns to Fight Deadliest Weapons." *Wired*, July 28, 2010. http://www.wired.com/2010/07/ff_roadside_bombs/all/1.

Hoffman, Danny. *The War Machines: Young Men and Violence in Sierra Leone and Liberia*. Durham, NC: Duke University Press, 2011.

Hoffmann, Stanley. "An American Social Science: International Relations." *Daedalus* 106, no. 3 (1977): 41–60. www.jstor.org/stable/20024493.

Hoffmann, Stanley. *The State of War: Essays on the Theory and Practice of International Relations*. New York: Praeger, 1965.

Holen, Steven R., et al. "A 130,000-Year-Old Archaeological Site in Southern California, USA." *Nature* 544, no. 7651 (April 26, 2017): 479–83. doi:10.1038/nature22065.

Holley, Peter. "The Tiny Pill Fueling Syria's War and Turning Fighters into Superhuman Soldiers." *Washington Post*, November 19, 2015.

Homer-Dixon, Thomas. *Environmental Scarcity and Global Security*. New York: Foreign Policy Association, 1993.

Homer-Dixon, Thomas. *Environment, Scarcity, and Violence*. Princeton, NJ: Princeton University Press, 2001.

Honig, Bonnie. *Emergency Politics: Paradox, Law, Democracy.* Princeton, NJ: Princeton University Press, 2009.

Hounshell, David A. *From the American System to Mass Production, 1800–1932: The Development of Manufacturing Technology in the United States.* Baltimore: Johns Hopkins University Press, 1984.

Howarth, D. "Ethos, Agonism and Populism: William Connolly and the Case for Radical Democracy." *British Journal of Politics and International Relations* 10, no. 2 (2008): 171–93.

Huerta-Sánchez, Emilia, et al. "Altitude Adaptation in Tibetans Caused by Introgression of Denisovan-like DNA." *Nature,* no. 512 (July 2, 2014): 194–97. doi:10.1038/nature13408.

Human Rights Watch. "U.S. Using Cluster Munitions in Iraq." April 1, 2003. https://www.hrw.org/news/2003/04/01/us-using-cluster-munitions-iraq.

Humboldt, Alexander von. *Cosmos: A Sketch of a Physical Description of the Universe.* Baltimore: Johns Hopkins University Press, 1997.

Huntington, Samuel P. *The Clash of Civilizations and the Remaking of World Order.* New York: Simon and Schuster, 2011.

Jablonka, Eva, Marion J. Lamb, and Anna Zeligowski. *Evolution in Four Dimensions: Genetic, Epigenetic, Behavioral, and Symbolic Variation in the History of Life.* Cambridge, MA: MIT Press, 2006.

Jackson, Gary M. *Predicting Malicious Behavior: Tools and Techniques for Ensuring Global Security.* Indianapolis: John Wiley, 2012.

Jamail, Dahr. "Dogs Eating Bodies in Streets of Fallujah." *Antiwar,* November 16, 2004. http://www.antiwar.com/jamail/?articleid=3984.

Jameson, Fredric. *Valences of the Dialectic.* Brooklyn, NY: Verso, 2010.

Jefferson, Margo. "On Writers and Writing; The Sideshow Must Go On." *The New York Times,* March 17, 2002. https://www.nytimes.com/2002/03/17/books/on-writers-and-writing-the-sideshow-must-go-on.html?mtrref=www.google.com&gwh=8ABA2F6B126BE6559DA4EE55A5514ADB&gwt=pay.

Jones, Norman, Gursimran Thandi, Nicola T. Fear, Simon Wessely, and Neil Greenberg. "The Psychological Effects of Improvised Explosive Devices (IEDs) on UK Military Personnel in Afghanistan." *Occupational and Environmental Medicine* 71, no. 7 (2014): 466–71.

Kant, Immanuel. "Idea for a Universal History from a Cosmopolitan Point of View." In *On History,* 11–26. Translated by Lewis White Beck. Indianapolis: Bobbs-Merrill, 1963.

Kant, Immanuel. *On Education.* Newburyport, MA: Dover Publications, 2012.

Kant, Immanuel. "Was Ist Aufklärung?" *Berlinische Monatsschrift,* December 1784.

Kant, Immanuel, Katharina Holger, and Eduard Gerreshelm. *Vorlesungen über Physische Geographie. Gesammelte Schriften.* Berlin: G. Reimer, 1902.

Kant, Immanuel, and Robert B. Louden. *Anthropology from a Pragmatic Point of View.* Cambridge: Cambridge University Press, 2006.

Kantor, J. R. "Man and Machine in Science." *The Journal of Philosophy* 32, no. 25 (1935): 673–84.

Kapp, R. O. "Living and Lifeless Machines." *The British Journal for the Philosophy of Science,* 5, no. 18 (1954): 91–103.

Karimi, Faith. "With 1 Male Left Worldwide, Northern White Rhinos under Guard 24 Hours." CNN, April 16, 2015. http://www.cnn.com/2015/04/16/africa/kenya-northern-white-rhino/.

Kato, Masahide. "Nuclear Globalism: Traversing Rockets, Satellites, and Nuclear War via the Strategic Gaze." *Alternatives* 18, no. 3 (1993): 339–60.

Kauvar, David, John Holcomb, Gary Norris, and John Hess. "Fresh Whole Blood Transfusion: A Controversial Military Practice." *Journal of Trauma: Injury, Infection, and Critical Care* 61 (2006): 181–84.

Keller, Catherine. *Apocalypse Now and Then: A Feminist Guide to the End of the World.* Minneapolis: Fortress Press, 2005.

Kelly, John D., ed. *Anthropology and Global Counterinsurgency.* Chicago: University of Chicago Press, 2010.

Keyes, Charley. "U.S. Won't Join Landmine Ban, Administration Decides." CNN, November 24, 2009. http://edition.cnn.com/2009/POLITICS/11/24/us.landmines/index.html.

Khalili, Laleh. *Time in the Shadows: Confinement in Counterinsurgencies.* Stanford, CA: Stanford University Press, 2013.

Kingsnorth, Paul. "The Lie of the Land: Does Environmentalism Have a Future in the Age of Trump?" *The Guardian*, March 18, 2017.

Kjellén, Rudolf, and J. Sandmeier. *Der Staat Als Lebensform*, 4. aufl. Berlin-Grunewald: K. Vowinckel, 1924.

Klein, Naomi. *This Changes Everything: Capitalism vs. the Climate.* Toronto: Alfred A. Knopf Canada, 2014.

Klingan, Katrin, Ashkan Sepahvand, Christoph Rosol, and Bernd M. Scherer, eds. *Textures of the Anthropocene: Grain Vapor Ray.* Cambridge, MA: MIT Press, 2015.

Kolbert, Elizabeth. *The Sixth Extinction: An Unnatural History.* New York: Picador, 2015.

Konikow, Leonard F. "Long-Term Groundwater Depletion in the United States." *Groundwater* 53, no. 1 (2015): 2–9. https://www.documentcloud.org/documents/1674356-konikow-2015-groundwater.html.

Kopp, R. E., J. L. Kirschvink, I. A. Hilburn, and C. Z. Nash. "The Paleoproterozoic Snowball Earth: A Climate Disaster Triggered by the Evolution of Oxygenic Photosynthesis." *Proceedings of the National Academy of Sciences* 102, no. 32 (2005): 11131–36. doi:10.1073/pnas.0504878102.

Kroenig, Matthew. "Facing Reality: Getting NATO Ready for a New Cold War." *Survival* 57, no. 1 (2015): 49–70. doi:10.1080/00396338.2015.1008295.

Kuletz, Valerie. *The Tainted Desert: Environmental Ruin in the American West.* New York: Routledge, 1998.

Kurzweil, Ray. *The Age of Spiritual Machines: When Computers Exceed Human Intelligence.* New York: Penguin, 2000.

Kurzweil, Ray. *The Singularity Is Near: When Humans Transcend Biology.* New York: Viking, 2005.

Lane, Nick. *Oxygen: The Molecule That Made the World.* New York: Oxford University Press, 2002.

Langerglitz, Nicolas. *Neuropsychedelia: The Revival of Hallucinogen Research since the Decade of the Brain.* Berkeley: University of California Press, 2012.

Lasswell, Harold D. *Politics: Who Gets What, When, How.* New York: Whittlesey House, 1936.

Latour, Bruno. *An Inquiry into Modes of Existence: An Anthropology of the Moderns.* Cambridge, MA: Harvard University Press, 2013.

Latour, Bruno. *Science in Action: How to Follow Scientists and Engineers through Society.* Cambridge, MA: Harvard University Press, 1988.

Latour, Bruno. *We Have Never Been Modern.* Cambridge, MA: Harvard University Press, 1993.

Law, John. 2004. *After Method: Mess in Social Science Research.* New York: Routledge, 2004.

Lehrer, Jonah. "The Forgetting Pill Erases Painful Memories Forever." *Wired*, February 17, 2012. http://www.wired.com/2012/02/ff_forgettingpill/.

Lem, Stanisław. *Summa Technologiae.* Minneapolis: University of Minnesota Press, 2013.

Lenin, Vladimir Il'ich. *Imperialism, the Highest Stage of Capitalism.* Eastford, CT: Martino Fine Books, 2011.

Lévinas, Emmanuel. *Totality and Infinity: An Essay on Exteriority.* Pittsburgh: Duquesne University Press, 2011.

Leys, Ruth. "The Turn to Affect: A Critique." *Critical Inquiry* 37, no. 3 (2011): 434–72. doi:10.1086/659353.

Lilly, John Cunningham. *Programming and Metaprogramming in the Human Biocomputer: Theory and Experiments.* Portland, OR: Coincidence Control Publishing, 2014.

Lindqvist, Sven. *A History of Bombing.* New York: New Press, 2011.

Litfin, Karen. *Ozone Discourses: Science and Politics in Global Environmental Cooperation.* New York: Columbia University Press, 1994.

Lucas, J. R. "Human and Machine Logic: A Rejoinder." *The British Journal for the Philosophy of Science* 19, no. 2 (1968): 155–56.

Lyotard, Jean-François. *The Postmodern Condition: A Report on Knowledge.* Minneapolis: University of Minnesota Press, 1984.

Ma, Michelle. "UW Study Shows Direct Brain Interface between Humans." UW News, November 5, 2014. http://www.washington.edu/news/2014/11/05/uw-study-shows-direct-brain-interface-between-humans/.

Machiavelli, Niccolò. *The Prince.* 2nd ed. Chicago: University of Chicago Press, 1998.

Malabou, Catherine. *The Future of Hegel: Plasticity, Temporality, and Dialectic.* New York: Routledge, 2005.

Malabou, Catherine. *Ontology of the Accident: An Essay on Destructive Plasticity.* Translated by Carolyn Shread. Cambridge: Polity Press, 2012.

Malabou, Catherine, with introduction by Marc Jeannerod. Translated by Sebastian Rand. *What Should We Do with Our Brain?* New York: Fordham University Press, 2008.

Malafouris, Lambros. *How Things Shape the Mind: A Theory of Material Engagement.* Cambridge, MA: MIT Press, 2013.

Malm, Andreas. *Fossil Capital: The Rise of Steam Power and the Roots of Global Warming.* New York: Verso, 2016.

Manning, Erin. *Always More than One: Individuation's Dance*. Durham, NC: Duke University Press, 2013.
Manning, Richard. *Rewilding the West: Restoration in a Prairie Landscape*. Berkeley: University of California Press, 2009.
Mansoor, Farooq, Ahmad Masoud Rahmani, M. Aziz Kakar, Pashtoon Hashimy, Parwiz Abrahimi, Paul T. Scott, Sheila A. Peel, Francisco J. Rentas, and Catherine S. Todd. "A National Mapping Assessment of Blood Collection and Transfusion Service Facilities in Afghanistan." *Transfusion* 53, no. 1 (2012): 69–75.
Marx, Karl. "Division of Labour and Manufacture." Chapter 14 of *Capital*, vol. 1. Accessed June 7, 2017. https://www.marxists.org/archive/marx/works/1867-c1/ch14.htm.
Marx, Karl. *The Eighteenth Brumaire of Louis Bonaparte*. New York: International, 1990.
Massumi, Brian. *Ontopower: War, Powers, and the State of Perception*. Durham, NC: Duke University Press, 2015.
Massumi, Brian. "Such as It Is: A Short Essay in Extreme Realism." *Body and Society* 22, no. 1 (2016): 115–27. doi:10.1177/1357034X15612896.
Massumi, Brian. *What Animals Teach Us about Politics*. Durham, NC: Duke University Press, 2014.
Maxwell, James Clerk, and P. M. Harman. *The Scientific Letters and Papers of James Clerk Maxwell*. Cambridge: Cambridge University Press, 1990.
Mbembe, Achille. "Necropolitics." Translated by Libby Meintjes. *Public Culture* 15, no. 1 (2003): 11–40.
McFate, Montgomery. "Iraq: The Social Context of IEDs." *Military Review*, May–June 2005.
McKittrick, Katherine, ed. *Sylvia Wynter: On Being Human as Praxis*. Durham, NC: Duke University Press, 2015.
McNeill, William H. "Foreword." In *Maps of Time: An Introduction to Big History*, by David Christian, 1–16, Berkeley: University of California Press, 2011.
McNeill, William H. *Keeping Together in Time: Dance and Drill in Human History*. Cambridge, MA: Harvard University Press, 1995.
McNeill, William H. *Plagues and Peoples*. New York: Anchor, 1977.
McNeill, William H. *The Pursuit of Power: Technology, Armed Force, and Society since A.D. 1000*. Chicago: University of Chicago Press, 1982.
Meadows, Donella H., Jørgen Randers, and Dennis L. Meadows. *The Limits to Growth: The 30-Year Update*. White River Junction, VT: Chelsea Green, 2004.
Mearsheimer, John J. "Benign Hegemony." *International Studies Review* 18, no. 1 (2016): 147–49. doi:10.1093/isr/viv021.
Mearsheimer, John J. "Nuclear Weapons and Deterrence in Europe." *International Security* 9, no. 3 (1984): 19–46. doi:10.2307/2538586.
Mehren, Elizabeth. "Seeking Freedom from Label at Last." *The Nation*, August 23, 2004.
Merleau-Ponty, Maurice. *Phenomenology of Perception*. New York: Routledge, 2002.
Metz, Cade. "The Sadness and Beauty of Watching Google's AI Play Go." *Wired*, March 11, 2016. https://www.wired.com/2016/03/sadness-beauty-watching-googles-ai-play-go/.
Miles, T. R. "On the Difference between Men and Machines." *The British Journal for the Philosophy of Science* 7, no. 28 (1957): 277–92.

Miller, Frank. "Freaks (1932)." Turner Classic Movies. Accessed May 25, 2016. http://www.tcm.com/tcmdb/title/163/Freaks/articles.html.

Milmo, Cahal. "Fury at DNA Pioneer's Theory: Africans Are Less Intelligent than Westerners." *Independent*, October 17, 2007. https://www.independent.co.uk/news/science/fury-at-dna-pioneers-theory-africans-are-less-intelligent-than-westerners-394898.html.

Moffat, James. *Complexity Theory and Network Centric Warfare.* Washington, DC: CCRP Publication Series, 2003.

Moon, Katherine H. S. *Sex among Allies: Military Prostitution in U.S.-Korea Relations.* New York: Columbia University Press, 1997.

Mooney, Chris. "'And Then We Wept': Scientists Say 93 Percent of the Great Barrier Reef Now Bleached." *Washington Post*, April 20, 2016.

Moore, Jason W. "Introduction: Anthropocene or Capitalocene? Nature, History, and the Crisis of Capitalism." In *Anthropocene or Capitalocene? Nature, History, and the Crisis of Capitalism*, edited by Jason W. Moore, 1–14. Oakland, CA: PM Press, 2016.

Moreno, Jonathan D. *Mind Wars: Brain Science and the Military in the Twenty-First Century.* New York: Bellevue Literary Press, 2012.

Mori, Masahiro. *The Buddha in the Robot: A Robot Engineer's Thoughts on Science and Religion.* Tokyo: Kosei Publishing, 1989.

Morton, Timothy. *Dark Ecology: For a Logic of Future Coexistence.* New York: Columbia University Press, 2016.

Morton, Timothy. *The Ecological Thought.* Cambridge, MA: Harvard University Press, 2010.

Morton, Timothy. "How I Learned to Stop Worrying and Love the Term *Anthropocene*." *Cambridge Journal of Postcolonial Literary Inquiry* 1, no. 2 (2014): 257–64. doi:10.1017/pli.2014.15.

Morton, Timothy. "Hyperobjects and the End of Common Sense." *The Contemporary Condition*, March 18, 2010. http://contemporarycondition.blogspot.com/2010/03/hyperobjects-and-end-of-common-sense.html.

Morton, Timothy. *Realist Magic: Objects, Ontology, Causality.* Ann Arbor, MI: Open Humanities Press, 2013.

Moten, Fred, and Stefano Harney. "Blackness and Governance." In *Beyond Biopolitics: Essays on the Governance of Life and Death*, edited by Patricia Ticineto Clough and Craig Willse, 351–62. Durham, NC: Duke University Press, 2011.

Motesharrei, Safa, Jorge Rivas, and Eugenia Kalnay. "Human and Nature Dynamics (HANDY): Modeling Inequality and Use of Resources in the Collapse or Sustainability of Societies." *Ecological Economics* 101 (May 2014): 90–102. doi:10.1016/j.ecolecon.2014.02.014.

Muecke, Stephen. "Wolfe Creek Meteorite Crater—Indigenous Science Queers Western Science." *Ctrl-Z: New Media Philosophy*. Accessed December 12, 2018. http://www.ctrl-z.net.au/articles/issue-7/muecke-wolfe-creek-meteorite-crater/.

Mumford, Lewis. *Technics and Civilization.* Chicago: University of Chicago Press, 2010.

Myers, Steven Lee. "U.S. Wields Defter Weapon against Iraq: Concrete Bomb." *New York Times*, October 7, 1999.

Negri, Antonio, and Noura Wedell. *The Porcelain Workshop: For a New Grammar of Politics.* Los Angeles: Semiotext(e), 2008.

Network of Concerned Anthropologists. *The Counter-counterinsurgency Manual: Or, Notes on Demilitarizing American Society.* Edited by Catherine Besteman. Chicago: Prickly Paradigm Press, 2009.
Nietzsche, Friedrich Wilhelm. *Human, All Too Human.* Translated by R. J. Hollingdale. New York: Cambridge University Press, 1996.
Nixon, Rob. *Slow Violence and the Environmentalism of the Poor.* Cambridge, MA: Harvard University Press, 2013.
Noë, Alva. *Out of Our Heads: Why You Are Not Your Brain, and Other Lessons from the Biology of Consciousness.* New York: Hill and Wang, 2010.
Nordstrom, Carolyn. *Shadows of War: Violence, Power, and International Profiteering in the Twenty-First Century.* Berkeley: University of California Press, 2004.
Nowell, April, and Iain Davidson. *Stone Tools: And the Evolution of Human Cognition.* Boulder: University Press of Colorado, 2011.
Odysseos, Louiza, and Fabio Petito. *The International Political Thought of Carl Schmitt: Terror, Liberal War and the Crisis of Global Order.* New York: Routledge, 2007.
"One and Only Earth." *Nature Geoscience*, no. 5 (2012): 81. doi:10.1038/ngeo1400.
Owen, James. "Interspecies Sex: Evolution's Hidden Secret?" *National Geographic News*, March 14, 2007.
Owens, Patricia. *Economy of Force: Counterinsurgency and the Historical Rise of the Social.* Cambridge: Cambridge University Press, 2016.
Pääbo, Svante. *Neanderthal Man: In Search of Lost Genomes.* New York: Basic Books, 2014.
Pacione, Michael. *Urban Geography: A Global Perspective.* London: Routledge, 2009.
Parikka, Jussi. *A Geology of Media.* Minneapolis: University of Minnesota Press, 2015.
Parpinelli, R. S., and H. S. Lopes. "New Inspirations in Swarm Intelligence: A Survey." *International Journal of Bio-Inspired Computation* 3, no. 1 (2011): 1–16. doi:10.1504/IJBIC.2011.038700.
Physicians for Social Responsibility. *Body Count: Casualty Figures after 10 Years of the "War on Terror" Iraq, Afghanistan, Pakistan*, 2015. https://www.psr.org/wp-content/uploads/2018/05/body-count.pdf.
Pickering, Andrew. *The Cybernetic Brain: Sketches of Another Future.* Chicago: University of Chicago Press, 2010.
Pickering, Andrew. "Cyborg History and the World War II Regime." *Perspectives on Science* 3, no. 1 (1995): 32–36.
Plan C. "We Are All Very Anxious." *We Are Plan C* (blog), April 4, 2014. https://www.weareplanc.org/blog/we-are-all-very-anxious/.
Plessis, Gitte du. "When Pathogens Determine the Territory: Toward a Concept of Non-Human Borders." *European Journal of International Relations* 24, no. 2 (June 2018): 391–413. doi:10.1177/1354066117710998.
Povinelli, Elizabeth A. *Economies of Abandonment: Social Belonging and Endurance in Late Liberalism.* Durham, NC: Duke University Press, 2011.
Puccetti, R. "On Thinking Machines and Feeling Machines." *The British Journal for the Philosophy of Science* 18, no. 1 (1967): 39–51.
Purdy, Jedediah. *After Nature: A Politics for the Anthropocene.* Cambridge, MA: Harvard University Press, 2015.

Purdy, Jedediah. "The New Nature." *Boston Review*, January 11, 2016.

Putnam, H. "Robots: Machines or Artificially Created Life?" *The Journal of Philosophy*, 61, no. 21 (1964): 668–91.

Pynchon, Thomas. *Gravity's Rainbow*. New York: Viking Press, 1973.

Rabinow, Paul. *Marking Time: On the Anthropology of the Contemporary*. Princeton, NJ: Princeton University Press, 2008.

Rahnema, Majid, and Victoria Bawtree, eds. *The Post-Development Reader*. London: Zed Books, 1997.

Rancière, Jacques. *Disagreement: Politics and Philosophy*. Minneapolis: University of Minnesota Press, 1994.

Rancière, Jacques. *The Politics of Aesthetics: The Distribution of the Sensible*. London: Continuum, 2006.

Ranke, Leopold von. *The Theory and Practice of History*. Edited by Georg G. Iggers. New translations by Wilma A. Iggers. New York: Routledge, 2011.

Ripple, William J., Christopher Wolf, Thomas M. Newsome, Mauro Galetti, Mohammed Alamgir, Eileen Crist, Mahmoud I. Mahmoud, and William F. Laurance. "World Scientists' Warning to Humanity: A Second Notice." *BioScience* 67, no. 12 (2017): 1026–28. https://doi.org/10.1093/biosci/bix125.

Roberts, W. H. "Are We Machines? And What of It?" *The Journal of Philosophy* 28, no. 13 (1931): 347–56.

Rockstrom, Steffen, et al. "Planetary Boundaries: Exploring the Safe Operating Space for Humanity." *Ecology and Society* 14, no. 2: 32. http://www.ecologyandsociety.org/vol14/iss2/art32/.

Roden, David. *Posthuman Life: Philosophy at the Edge of the Human*. New York: Routledge, 2015.

Rogers, Simon. "US Military Suicides in Charts: How They Overtook Combat Deaths." *The Guardian*, February 1, 2013.

Rorty, R. "Functionalism, Machines and Incorrigibility." *The Journal of Philosophy* 69, no. 8: (1972): 203–20.

Rose, Nikolas S., and Joelle M. Abi-Rached. *Neuro: The New Brain Sciences and the Management of the Mind*. Princeton, N.J: Princeton University Press, 2013.

Rucker, Rudy. "A Transrealist Manifesto." *Bulletin of the Science Fiction Writers of America* 83 (winter 1983). Reprinted on author's website. http://www.rudyrucker.com/pdf/transrealistmanifesto.pdf.

Rummel, R. J. *Lethal Politics: Soviet Genocide and Mass Murder Since 1917*, 2017. http://search.ebscohost.com/login.aspx?direct=true&scope=site&db=nlebk&db=nlabk&AN=1550741.

Russell, Steve. "Going Native: Why bin Laden Wanted Major Jim Gant Dead." *Indian Country Today*, July 13, 2014. https://newsmaven.io/indiancountrytoday/archive/going-native-why-bin-laden-wanted-major-jim-gant-dead.

Sahlins, Marshall David. *The Western Illusion of Human Nature*. Chicago: Prickly Paradigm Press, 2008.

Salter, Mark B. "To Make Move and Let Stop: Mobility and the Assemblage of Circulation." *Mobilities* (January 10, 2013): 7–19.

Sassen, Saskia. *Expulsions: Brutality and Complexity in the Global Economy.* Cambridge, MA: Belknap Press of Harvard University Press, 2014.
Scahill, Jeremy. *Dirty Wars: The World Is a Battlefield.* New York: Nation Books, 2014.
Schelling, Thomas C. *Arms and Influence.* New Haven, CT: Yale University Press, 2008.
Schlosser, Eric. *Command and Control: Nuclear Weapons, the Damascus Accident, and the Illusion of Safety.* New York: Penguin Books, 2014.
"The Science in Science Fiction." *Talk of the Nation,* NPR, November 30, 1999. http://www.npr.org/templates/story/story.php?storyId=1067220.
"Scientists Study Robot-Human Interactions." *Space Daily,* August 30, 2006. http://www.spacedaily.com/reports/Scientists_Study_Robot_Human_Interactions_999.html.
Scranton, Roy. "Learning How to Die in the Anthropocene." *New York Times* (blog), November 10, 2013. http://opinionator.blogs.nytimes.com/2013/11/10/learning-how-to-die-in-the-anthropocene/.
Scranton, Roy. *Learning to Die in the Anthropocene: Reflections on the End of a Civilization.* San Francisco: City Lights Books, 2015.
Senate Select Committee on Intelligence. "Committee Study of the Central Intelligence Agency's Detention and Interrogation Program." December 13, 2012. https://web.archive.org/web/20141209165504/http://www.intelligence.senate.gov/study2014/sscistudy1.pdf.
Schrödinger, Erwin, and Roger Penrose. *What Is Life? South Asian Edition: With Mind and Matter and Autobiographical Sketches.* Cambridge: Cambridge University Press, 2013.
Shapiro, Michael J. *The Time of the City: Politics, Philosophy and Genre.* New York: Routledge, 2010.
Shapiro, Michael J. *Violent Cartographies: Mapping Cultures of War.* Minneapolis: University of Minnesota Press, 1997.
Shaviro, Steven. *No Speed Limit: Three Essays on Accelerationism.* Minneapolis: University of Minnesota Press, 2015.
Shaviro, Steven. *The Universe of Things: On Speculative Realism.* Minneapolis: University of Minnesota Press, 2014.
Shaw, Ian G. R. "The Rise of the Predator Empire in the Vietnam War." In *Predator Empire: Drone Warfare and Full Spectrum Dominance.* Minneapolis: University of Minnesota Press, 2016.
Simondon, Gilbert. *La individuación.* Buenos Aires: Cactus, 2015.
Simondon, Gilbert. *La individuación a la luz de las nociones de forma y de información: incluye tres articulos inéditos: Las consecuencias de la noción de individuación; Allagmática; Forma, información y potenciales,* 2015.
Simondon, Gilbert. *On the Mode of Existence of Technical Objects.* Translated by Ninian Mellamphy. London: University of Western Ontario, 1980.
Simondon, Gilbert. *On the Mode of Existence of Technical Objects.* Translated by Cécile Malaspina and John Rogove. Minneapolis: University of Minnesota Press, 2017.
Singleton, Benedict. "Maximum Jailbreak." *E-Flux,* no. 46 (June 2013). http://www.e-flux.com/journal/maximum-jailbreak/.
Sloterdijk, Peter. "European Earth-users as the Primary Medium." In *Not Saved: Essays after Heidegger.*

Sloterdijk, Peter. *In the World Interior of Capital: Towards a Philosophical Theory of Globalization.* Cambridge: Polity Press, 2013.
Sloterdijk, Peter. *Not Saved: Essays after Heidegger.* Translated by Ian Alexander Moore and Christopher Turner. Malden, MA: Polity, 2017.
Sloterdijk, Peter. *Spheres.* Vol. 3, *Foams: Plural Spherology.* Los Angeles: Semiotext(e), 2016.
Sloterdijk, Peter. *Terror from the Air.* Translated by Amy Patton and Steve Corcoran. Los Angeles: Semiotext(e), 2009.
Smail, Daniel Lord. *On Deep History and the Brain.* Berkeley: University of California Press, 2008.
Small Arms Survey. "Armed Violence." http://www.smallarmssurvey.org/armed-violence.html.
Small Arms Survey. "Direct Conflict Deaths." http://www.smallarmssurvey.org/index.php?id=296.
Small Arms Survey. "Weapons and Markets." http://www.smallarmssurvey.org/weapons-and-markets.html.
Smith, Daniel. "It's the End of the World as We Know It . . . and He Feels Fine." *New York Times*, April 17, 2014.
"Sperm Made from Human Bone Marrow." BBC news, April 13, 2007. http://news.bbc.co.uk/go/pr/fr/-/2/hi/health/6547675.stm.
Stannard, David E. *American Holocaust: The Conquest of the New World.* New York: Oxford University Press, 1993.
Starr, Douglas. *Blood: An Epic History of Medicine and Commerce.* New York: Knopf, 1999.
Steinbruner, John D. *The Cybernetic Theory of Decision: New Dimensions of Political Analysis; with a New Preface by the Author.* Princeton, NJ: Princeton University Press, 2002.
Stoler, Ann Laura. *Race and the Education of Desire: Foucault's History of Sexuality and the Colonial Order of Things.* Durham, NC: Duke University Press, 1995.
Svensmark, Henrik. "Cosmic Rays and the Biosphere over 4 Billion Years." *Astronomische Nachriten* 327, no. 9 (2006): 871–75.
Tanielian, Terri, et al. *Invisible Wounds: Mental Health and Cognitive Care Needs of America's Returning Veterans.* Santa Monica, CA: RAND, 2008.
Tarde, G. *Social Laws: An Outline of Sociology.* Plano, TX: Read Books Ltd., 2013.
Taussig, Michael. "Maleficium: State Fetishism." In *The Nervous System.* New York: Routledge, 1991.
Thacker, Eugene. *After Life.* Chicago: University of Chicago Press, 2010.
Thacker, Eugene. *Cosmic Pessimism.* Minneapolis: University of Minnesota Press, 2015.
Thacker, Eugene. *In the Dust of This Planet.* Alresford, UK: Zero Books, 2011.
Thompson, D. "Can a Machine Be Conscious?" *The British Journal for the Philosophy of Science* 16, no. 61 (1965): 33–43.
Thompson, Evan. *Waking, Dreaming, Being: Self and Consciousness in Neuroscience, Meditation, and Philosophy.* New York: Columbia University Press, 2015.
Tiqqun. *Introduction to Civil War.* Translated by Alexander R. Galloway and Jason E. Smith. Los Angeles: Semiotext(e), 2010.

Tsing, Anna Lowenhaupt. *The Mushroom at the End of the World: On the Possibility of Life in Capitalist Ruins*. Princeton, NJ: Princeton University Press, 2015.
Tucker, Patrick. "The Military Is Building Brain Chips to Treat PTSD." *Defense One*, May 28, 2014. https://www.defenseone.com/technology/2014/05/D1-Tucker-military-building-brain-chips-treat-ptsd/85360/.
Tunander, Ola. "Geopolitics of the North: *Geopolitik* of the Weak: A Post-Cold War Return to Rudolf Kjellén." *Cooperation and Conflict* 43, no. 2 (2008): 164–84. doi:10.1177/0010836708089081.
Uexküll, Jacob von. *A Foray into the Worlds of Animals and Humans with a Theory of Meaning*. Minneapolis: University of Minnesota Press, 2010.
UN-HABITAT. "The Challenge of Slums." 2003. http://www.unhabitat.org/pmss/listItemDetails.aspx?publicationID=1156.
Union of Concerned Scientists. "World Scientists' Warning to Humanity." November 1992. https://www.ucsusa.org/about/1992-world-scientists.html#.Wm038FPwbMJ.
United States, Department of the Army. *The U.S. Army/Marine Corps Counterinsurgency Field Manual*. U.S. Army Field Manual No. 3-24. Marine Corps Warfighting Publication No. 3-33.5. Chicago: University of Chicago Press, 2007.
US Immigration and Customs Enforcement. "ICE Shadow Wolves." June 1, 2007. Accessed June 22, 2017. https://www.ice.gov/factsheets/shadow-wolves.
"US Used White Phosphorus in Iraq." *BBC News*, November 16, 2005. http://news.bbc.co.uk/2/hi/middle_east/4440664.stm.
Valencia, Sayak. *Capitalismo gore*. Mexico City: Paidós, 2016.
VanderMeer, Jeff. *Acceptance*. New York: Farrar, Straus and Giroux, 2014.
Van Evera, Stephen. *Causes of War: Power and the Roots of Conflict*. Ithaca, NY: Cornell University Press, 2001.
Vargas Machuca, Bernardo de. *The Indian Militia and Description of the Indies*. Edited by Kris Lane. Translated by Timothy F. Johnson. Durham, NC: Duke University Press, 2008.
Vidal, John. "Iraqi Children Pay High Health Cost of War-Induced Air Pollution, Study Finds." *The Guardian*, August 22, 2016.
Virilio, Paul. *The Futurism of the Instant: Stop-Eject*. Cambridge: Polity, 2011.
Viveiros de Castro, Eduardo Batalha. *From the Enemy's Point of View: Humanity and Divinity in an Amazonian Society*. Chicago: University of Chicago Press, 1992.
Viveiros de Castro, Eduardo. *The Relative Native: Essays on Indigenous Conceptual Worlds*. Chicago: HAU Books, 2015.
Vizenor, Gerald Robert. *Manifest Manners: Narratives on Postindian Survivance*. Lincoln: University of Nebraska Press, 2010.
von Neumann, John. *The Computer and the Brain*. New Haven, CT: Yale University Press, 1958.
Wacquant, Loïc J. D. *Punishing the Poor: The Neoliberal Government of Social Insecurity*. Durham, NC: Duke University Press, 2009.
Wade, Nicholas. "The Dark Side of Oxytocin, the Hormone of Love: Ethnocentrism." *New York Times*, Science, January 10, 2011. Accessed November 7, 2013. www.nytimes.com/2011/01/11/science/11hormone.html.

Wait, Patience. "Government Hiring Practices Hamper Cybersecurity Efforts." *InformationWeek*, May 20, 2014. http://www.informationweek.com/government/cybersecurity/government-hiring-practices-hamper-cybersecurity-efforts/d/d-id/1252939.

Walsh, Declan, Simon Rogers, and Paul Scruton. "Wikileaks Afghanistan Files: Every IED Attack, with Co-ordinates." *The Guardian*, July 26, 2010.

Walter, W. Grey. *The Living Brain*. New York: W. W. Norton, 1953.

Wark, McKenzie. "Critical Theory after the Anthropocene." *Public Seminar*, August 9, 2014. http://www.publicseminar.org/2014/08/critical-theory-after-the-anthropocene/#.U-b-jkgRbxt.

Wark, McKenzie. "Molecular Red: Theory for the Anthropocene (On Alexander Bogdanov and Kim Stanley Robinson)." *E-Flux*, no. 63 (March 2015). http://www.e-flux.com/journal/63/60889/molecular-red-theory-for-the-anthropocene-on-alexander-bogdanov-and-kim-stanley-robinson/.

Wark, McKenzie. *Molecular Red: Theory for the Anthropocene*. London: Verso, 2016.

Wark, McKenzie. "The Vectoralist Class." "Supercommunity." Special issue, *E-Flux*, August 29, 2015. http://supercommunity.e-flux.com/texts/the-vectoralist-class.

Wark, McKenzie. "The Vectoralist Class, Part II." *E-Flux*, no. 70 (February 2016). https://www.e-flux.com/journal/70/60567/the-vectoralist-class-part-ii/.

Warren, Rosie. "Some Last Words on Pessimism." *Salvage*, January 4, 2016. http://salvage.zone/in-print/some-last-words-on-pessimism/.

Watts, Peter. *Echopraxia*. New York: Tor Books, 2015.

Wax, Dustin M. "The Uses of Anthropology in the Insurgent Age." In *Anthropology and Global Counterinsurgency*, edited by John D. Kelly, 153–68. Chicago: The University of Chicago Press, 2010.

Weheliye, Alexander G. *Habeas Viscus: Racializing Assemblages, Biopolitics, and Black Feminist Theories of the Human*. Durham, NC: Duke University Press, 2014.

Weigley, Russell F. *The American Way of War: A History of United States Military Strategy and Policy*. Bloomington: Indiana University Press, 1977.

Weiner, Tim. "Air Force Seeks Bush's Approval for Space Weapons Programs." *New York Times*, May 18, 2005.

Weise, Elizabeth. "30,000 Years Ago, as Few as 1,000 Humans in Asia, Europe." *USA Today*, July 15, 2011.

Weizman, Eyal. "Lethal Theory." *Log*, no. 7 (2006): 53–77. www.jstor.org/stable/41765087.

Wendt, Alexander. "The State as Person in International Theory." *Review of International Studies* 30, no. 2 (2004): 289–316.

West-Eberhard, Mary Jane. *Developmental Plasticity and Evolution*. New York: Oxford University Press, 2003.

Whitehead, Alfred North. *An Enquiry Concerning the Principles of Natural Knowledge*. Cambridge: Cambridge University Press, 1919.

Whitehead, Alfred North. *Essays in Science and Philosophy*. New York: Philosophical Library, 1947.

Whitehead, Alfred North. *The Function of Reason*. Boston: Beacon Press, 1971.

Whitehead, Alfred North. *Nature and Life*. Cambridge: Cambridge University Press, 2011.

Whitehead, Alfred North. *Process and Reality: An Essay in Cosmology.* Edited by David Ray Griffin and Donald W. Sherburne. Gifford Lectures, 1927–28. New York: Free Press, 1985.

Whitehead, Alfred North. *Science and the Modern World.* New York: Free Press, 1925.

Whitman, Walt. *Leaves of Grass.* Mineola, NY: Dover, 2007.

Williams, Alex, and Nick Srnicek. "#ACCELERATE MANIFESTO for an Accelerationist Politics." *Critical Legal Thinking* (blog), May 14, 2013. Accessed February 11, 2019. http://criticallegalthinking.com/2013/05/14/accelerate-manifesto-for-an-accelerationist-politics.

Williams, Mark, Jan Zalasiewicz, P. K. Haff, Christian Schwägerl, Anthony D. Barnosky, and Erle C. Ellis. "The Anthropocene Biosphere." *The Anthropocene Review* 2, no. 3 (December 2015): 196–219.

Williams, Raymond. *Keywords: A Vocabulary of Culture and Society.* Oxford: Oxford University Press, 1985.

Wittgenstein, Ludwig. *Culture and Value.* Translated by Peter Winch. Chicago: University of Chicago Press, 1980.

Wittgenstein, Ludwig, and G. E. M. Anscombe. *Philosophical Investigations: The German Text, with a Revised English Translation.* 3rd ed. Malden, MA: Blackwell, 2003.

Wolfram, Stephen. "The Principle of Computational Equivalence." In *A New Kind of Science*, 715–848. Champaign, IL: Wolfram Media, 2002.

Worm, B., et al. "Impacts of Biodiversity Loss on Ocean Ecosystem Services." *Science* 314, no. 5800 (November 3, 2006): 787–90. https://doi.org/10.1126/science.1132294.

Worstall, Tim. "Darpa Bigwig and Intel Fellow: Moore's Law Is Ending Soon." *Forbes*, August 29, 2013. http://www.forbes.com/sites/timworstall/2013/08/29/darpa-chief-and-intel-fellow-moores-law-is-ending-soon/.

Wynes, Charles. *Charles Richard Drew: The Man and the Myth.* Urbana: University of Illinois Press, 1988.

Yacoub, Sameer N. "Iraq: Overall Death Toll Down, Uptick in 'Execution-Style' Killings." *Christian Science Monitor*, December 1, 2013. http://www.csmonitor.com/World/Latest-News-Wires/2013/1201/Iraq-Overall-death-toll-down-uptick-in-execution-style-killings.

Zalasiewicz, Jan, et al. "Scale and Diversity of the Physical Technosphere: A Geological Perspective." *Anthropocene Review* 4, no. 1 (2017): 9–22. doi:10.1177/2053019616677743.

Ziff, Paul. "The Feelings of Robots." *Analysis* 19, no. 3 (January 1, 1959): 64–68.

Žižek, Slavoj. "Bring Me My Philips Mental Jacket." *London Review of Books* 25, no. 10 (2003): 3–5. https://www.lrb.co.uk/v25/n10/slavoj-zizek/bring-me-my-philips-mental-jacket.

Zylinska, Joanna. *Minimal Ethics for the Anthropocene.* Ann Arbor, MI: Open Humanities Press, 2014. http://www.oapen.org/download?type=document&docid=502334.

INDEX

CPSIA information can be obtained
at www.ICGtesting.com
Printed in the USA
JSHW051423271222
35305JS00004B/16

9 781478 004844